Clean Energy and Fuel (Hydrogen) Storage

Clean Energy and Fuel (Hydrogen) Storage

Special Issue Editors

Elias K. Stefanakos
Sesha S. Srinivasan

MDPI • Basel • Beijing • Wuhan • Barcelona • Belgrade

Special Issue Editors

Elias K. Stefanakos
University of South Florida
USA

Sesha S. Srinivasan
Florida Polytechnic University
USA

Editorial Office
MDPI
St. Alban-Anlage 66
4052 Basel, Switzerland

This is a reprint of articles from the Special Issue published online in the open access journal *Applied Sciences* (ISSN 2076-3417) from 2017 to 2019 (available at: https://www.mdpi.com/journal/applsci/special_issues/fuel_hydrogen_storage)

For citation purposes, cite each article independently as indicated on the article page online and as indicated below:

LastName, A.A.; LastName, B.B.; LastName, C.C. Article Title. *Journal Name* **Year**, *Article Number*, Page Range.

ISBN 978-3-03921-630-7 (Pbk)
ISBN 978-3-03921-631-4 (PDF)

Contents

About the Special Issue Editors

Elias K. Stefanakos is Professor of Electrical Engineering and Director of the Clean Energy Research Center (CERC) at the University of South Florida (USF) in Tampa, Florida. For 13 years, until August 2003, he was Chairman of the Department of Electrical Engineering at USF. He is Editor-in-Chief of the *Journal of Power and Energy Engineering* (*JPEE*) and Associate Editor (PV) of the *Journal of Solar Energy*. He has a number of patents and has published over 200 research papers in refereed journals and international conferences in the areas of smart materials, solar energy production and storage, hydrogen storage, to name but a few. He has received contracts and grants from agencies such as the National Science Foundation (NSF), US Department of Energy (USDOE, ARPAe), NASA, and others. CERC is an interdisciplinary center whose mission is to develop new systems for clean energy and energy storage, with an emphasis on technology development and technology transfer.

Sesha S. Srinivasan is currently Assistant Professor of Physics at Florida Polytechnic University, Florida, USA. Before moving to FPU in 2014, he was a Tenure Track Assistant Professor of Physics at Tuskegee University, Alabama, USA. Dr. Srinivasan has more than a decade of research experience in the interdisciplinary areas of solid state and condensed matter physics, inorganic chemistry, and chemical and materials science engineering. His Ph.D. studies focused on the development of various rare-earth transition metals and intermetallic alloys, composites, nanoparticles, and complex hydrides for reversible hydrogen storage applications. Dr. Srinivasan and his Ph.D. advisor, Professor O.N. Srivastava (BHU, Varanasi), have successfully converted a 4-stroke, 100 cc Honda motorcycle to run on hydrogen gas delivered from the onboard metal hydride canister. After completing his Ph.D., Dr. Srinivasan joined the research team of Professor Craig Jensen as Post-Doctoral Fellow in the Department of (Inorganic) Chemistry, University of Hawaii, Honolulu, Hawaii, USA. Together with his post-doctoral advisor, he has collaborated extensively with scientists around the world on the storage of hydrogen in light weight complex hydrides, an endeavor which was funded by the US Department of Energy (DOE) and WE-NET, Japan. After two years at University of Hawaii, he has joined the Clean Energy Research Center (CERC) at University of South Florida as a Research Scientist under the leaderships of Professor Elias Stefanakos and Professor Yogi Goswami. He has established a state-of-the-art research laboratory at the CERC and supervised several graduate and undergraduate students in completing their master's and Ph.D. dissertations. He has also served as Associate Director of Florida Energy Systems Consortium (FESC) at USF and coordinated a number of research projects on clean energy and the environment, which was funded by the State Energy Office Florida ($9M grant). In his current and previous positions at TU and FPU, Dr. Srinivasan was awarded numerous research grants, worth $1M, from federal (DOE, NSF, ONR), state (FESC, FHI, FPU), and private (BP-Oil Spill, QuantumSphere Inc.) funding sources. He was recently awarded two US patents on nanomaterial development and methodologies in hydrogen storage. He has published eight book chapters and review articles, more than 80 journal publications, and countless more peer-reviewed conference proceedings. Dr. Srinivasan has served as a reviewer in the panel review committee of the National Science Foundation (NSF), SMART, and NDSEG panels of ASEE, an ad hoc merit review committee of the US Department of Energy, and panelist for Qatar National Research Fund (QNRF). Dr. Srinivasan currently serves as Associate Editor of deGruyter Open Book publications in Physics, Materials Science, and Astronomy and Guest Editor for numerous Special Issues in the journals *Nanomaterials*, *Sustainability*, and *Applied Sciences*.

Editorial

Clean Energy and Fuel Storage

Sesha S. Srinivasan [1,*,†] **and Elias K. Stefanakos** [2,*,†]

1   Department of Natural Sciences, Florida Polytechnic University, Lakeland, FL 33810, USA
2   Clean Energy Research Center, College of Engineering, University of South Florida, Tampa, FL 33620, USA
*   Correspondence: ssrinivasan@floridapoly.edu (S.S.S.); estefana@usf.edu (E.K.S.)
†   Guest Editors of the Special Issue 'Clean Energy and Fuel Storage'.

Received: 1 August 2019; Accepted: 5 August 2019; Published: 9 August 2019

Abstract: Clean energy and fuel storage is often required for both stationary and automotive applications. Some of the clean energy and fuel storage technologies currently under extensive research and development are hydrogen storage, direct electric storage, mechanical energy storage, solar-thermal energy storage, electrochemical (batteries and supercapacitors), and thermochemical storage. The gravimetric and volumetric storage capacity, energy storage density, power output, operating temperature and pressure, cycle life, recyclability, and cost of clean energy or fuel storage are some of the factors that govern efficient energy and fuel storage technologies for potential deployment in energy harvesting (solar and wind farms) stations and on-board vehicular transportation. This Special Issue thus serves the need to promote exploratory research and development on clean energy and fuel storage technologies while addressing their challenges to a practical and sustainable infrastructure.

Keywords: thermal energy storage; electrochemical energy storage; hydrogen energy storage; salt cavern energy storage

---

The major focus of this Special Issue is to explore and innovate various clean energy and fuel storage systems that can be deployed successfully for the enhancement of clean energy portfolios for both stationary and automotive applications. It also seeks to address the underlying factors affecting the sequestration and capture of carbon by developing thermochemical energy storage or alternative fuel storage approaches, such as the atomic or molecular storage of hydrogen via solid-state materials. The submitted research papers were initially screened for their originality and novelty and then reviewed by three peer-reviewers, who are experts in the field, before their acceptance for publication in this Special Issue. One review paper and 15 journal articles were published in this Special Issue of *Applied Sciences*, a journal of MDPI publications. The contents of the published papers are categorized according to their nature and application type as (i) thermal energy storage (TES), (ii) electrochemical energy storage (EES), (iii) hydrogen energy storage (HES), and (iv) salt cavern energy storage (SCES) options. The following excerpts are based on the editorial briefing about the novelty and innovation of these 16 peer-reviewed published papers in *Applied Sciences*.

Thermal Energy Storage: The storage of heat energy in materials or other forms for chemical reactions is currently employed for concentrated solar power plants. During the daytime, the heat obtained from the abundant solar irradiation is focused on to dish/trough-type or central receiver-type antennas, where the heat is carried over by a fluid. The available heat is then circulated over the heat storage materials, for example phase change materials or waste slag. The heat is absorbed and hence stored in the material due to endothermic reactions, leading to the release of a gas or a change in the materials' phase structure. In the night hours, there is no constant heat available from the sun. Therefore, the material will release the heat due to external variations such as gaseous absorption or phase change due to cooling; this is referred to as an exothermic process. The reversibility of such TES is then augmented for Concentrated Solar Power (CSP) and other potential applications. In this Special Issue, at least three research articles address the design and optimization of TES systems involving various materials. Michael Kruger et al. [1] demonstrated a type of new slag as an inventory material derived from an Electric Arc Furnace (EAF). Since the physical characteristics of these slag materials are similar to those of other refractory materials and ceramics, their performance with a unique design optimization for TES is evaluated for the Julich Power Tower Company in Germany. Based on the different optimization designs, the vertical TES with axial flow direction was found to perform the best in terms of aptitude value and lower risk value, as evaluated by Quality Function Deployment (QFD) and Failure Mode and Effect Analysis (FMEA), respectively. Since the conclusive decision and techno economic optimum was not made, further work is needed to understand the usability of such low-cost waste slag material for high-end CSP TES applications. Bilin Shao et al. [2] successfully developed and employed an enthalpy-based multiple relaxation time (MRT) lattice Boltzmann method (LBM) to calculate the transient heat characteristics of phase change material (PCM) roofs of green buildings in both hot summer and cold winter areas of China. The authors analyzed the life cycle energy savings of the PCM roof under an intermittent energy utilization condition by comparing it with the performance of a roof filled with sensible insulating materials (SIM). Paraffin was chosen as the PCM, whereas perlite was considered for the SIM since both of these materials have similar thermophysical properties and hence it is fair to compare them for roofing performance and energy savings purposes. Overall, the PCM roof was found to exhibit a better thermal performance than the SIM for energy consumption when the appropriate melting point and layer thickness of the PCM were selected. According to the authors, since the current paraffin PCM is expensive, new low-cost, high-chemical stability PCMs are yet to be explored for viable green building applications. Marie Duqusene et al. [3] reported a seasonal thermal energy storage material based on bio-glass such as Xylitol. The major drawback of utilizing Xylitol as a viable TES is due to its difficulty in the nucleation triggering (energy discharge triggering) process and hence its subsequent low crystallization rates (discharge power delivery). The authors devised a method to address the abovementioned obstacles by developing air-lift reactors that have sequential processes—namely, air bubble generation, the transportation of nucleation sites, and subsequent crystallization. The nucleation and crystal growth of Xylitol was successfully monitored at various times such as the induction period, the recalescence period, and the end period via thermal analysis measurements. The authors concluded that the new air-lift reactor and its underlying techniques are very promising to discharge the TES system based on pure Xylitol at the required power when needed. Further optimization of the bubbling conditions and air injection ratios will enhance the capabilities of Xylitol as a compatible phase change material for TES applications.

Electrochemical Energy Storage: An electrochemical energy storage device or a rechargeable battery is comprised of two electrodes (a cathode and an anode) and an electrolyte (liquid or solid). The secondary batteries mentioned above primarily convert the chemical energy contained in the active materials deposited on the electrodes into electrical energy by electrochemical reduction–oxidation (redox) reactions. There are many types of batteries based on the nature of the elements or compounds of the electrodes, such as nickel–cadmium (Ni–Cd), nickel–metal hydride (Ni–MH), lithium-ion (Li-ion), lithium–polymer (Li–pol), silver–zinc (Ag–Zn), nickel–hydrogen (Ni–H$_2$), sodium–sulfur (Na–S), zinc–air (Zn–O$_2$), etc. The energy density and power density are the two governing factors for the selection of the appropriate battery type for the desired applications. Additionally, the life cycle testing and charge/discharge rates, as well as the associated memory effects, must be considered for day-to-day applications in automotive operations, mobile phones, laptops, etc. In this Special Issue, one review paper on Li-ion batteries and three research articles on various EES systems were included, and are briefly described below. Dervis Emre Demirocak et al. [4] extensively reviewed the nanocomposite materials used in rechargeable Li-ion batteries. In this paper, the authors compared Li-ion batteries with other battery technologies such as lead–acid, Ni–Cd, and Ni–MH batteries in terms of chemistry, specific energy density, cycle life, cell voltage, self-discharge characteristics, safety, and toxicity. Based on this comparison, it is very clear that Li-ion batteries surpass the energy densities of the other samples and are less toxic when compared to Ni–MH batteries. Moreover, Li-ion batteries are currently in demand and will certainly meet or exceed the 2022 US Department of Energy's target in terms of cost estimate thanks to engineering improvements and material advancements, as per the authors' review. Various low-cost, high-performance composite nanomaterials will replace the precious, high-cost, and toxic materials, for example in cathodic and anodic electrodes, binders, and separators within and throughout Li-ion batteries. The innovation of such composite nanomaterials will undoubtedly enhance the specific capacity, rate capability, and cycle life of Li-ion batteries while ensuring their overall safety. Additionally, this review captured the ideal electrochemical impedance spectra (EIS) of standard indium vanadium oxide anode electrodes with different voltages during charge/discharge cycles and the resistance-capacitance equivalent circuit. An advanced equivalent circuit model (ECM) was developed to determine the accurate state-of-charge (SOC) by Xin Lai and co-workers [5]. Some of the unique contributions of the abovementioned paper include: (i) it addresses the New European Driving Cycle (NEDC) and the Dynamic Stress Test (DST) and hence obtains a more suitable battery model for the entire SOC area, (ii) it identifies the subarea parametric classification method using the particle swarm optimization (PSO) algorithm to improve the global accuracy, and (iii) it extends the SOC estimator using a Kalman filter to improve the ECM's accuracy and robustness. As per the single particle model, the authors derived the SOC at the particle surface, $SOC_{surf} = SOC_{avg} + \Delta SOC$, where SOC_{avg} is the average SOC, meaning the average concentration of Li^+ in the electrode particle, and ΔSOC is the difference between SOC_{avg} and SOC_{surf}. An extended Kalman filter (EKF) model proposed by the authors showed greater accuracy than the ECM, specifically in low SOC areas. This is because the EKF model considers the noise characteristics of the current and voltage sensors, and effectively overcomes the problem of random errors. An optimal sizing of the electrochemical (battery) energy storage system for large-scale wind farms, considering auxiliary services compensation options such as scheduling plans, is outlined by Xin Jiang et al. [6]. This optimization model not only smoothens the fluctuations of wind output based on the wind peaking demand, but also compensates the wind output to make up the wind forecast error. The multiple steps of the abovementioned optimization model are given here. Firstly, the uncertainty regarding hourly fluctuations due to wind output are analyzed and then the associated auxiliary service cost of EES mitigation is quantified. Secondly, the equivalent life loss is introduced by considering the impact of the battery storage systems' irregular charge/discharge characteristics on the cycle life. It is shown by the authors that the cycle life exponentially decreases with the increase of the charge/discharge depth in terms of percentage. An optimum of 9000 cycles was demonstrated by this model for the charge/discharge depth of 20%. The results also showed that the EES system can be integrated with large-scale wind farms with investment costs of less than $ 360/kWh. Moreover, cycle life of more than 2800 times can be achieved by taking into consideration of the auxiliary service compensation. In their excellent research paper, Pengfei Lu et al. [7] reported on the structure and capacitance of electrical double layers (EDL) at the graphene–ionic liquid interface via molecular dynamics simulations. Using this model, the distribution and migration of ions on the non-rough and rough electrode surfaces with different charge densities were compared and analyzed. The capacitance vs. voltage characteristic curves of EDL were obtained and corroborated with the electrode surface morphology on the capacitance of the EDL. Based on the detailed simulation results, it is clear that alternate distributions of anions and cations in several consecutive layers in the EDL could plausibly obtain the decaying amplitude characteristics along the direction perpendicular to the charged/uncharged electrode surface. In this alternate configuration, while charging occurs, the anions—for example, $[BF_4]$—have a greater migration of ions when compared to the layers of cations—for example, $[EMIM]^+$—due to the smaller size and steric effects of the anions. Finally, whether positively or negatively charging, the rough surface electrode with effective contact area between ions and electrodes possesses larger capacitance values when compared to the non-rough surface electrodes. An electrochemical method involving the anodization of TiO_2 foils was used to prepare titanium oxide nano arrays (TiO_2-NTAs) for the application of dye-sensitized solar cells (DSSC) by Ho-Sub Kim et al. [8]. The preparation of photoanodes for the DSSC was carried out with TiO_2-NTAs coated on the Fluorinated Tin Oxide (FTO) glass substrate using the sonication of precursor mixtures followed by the doctor blade process. The coating of silver nanoparticles (Ag NPs) or carbon materials (CMs) on the TiO_2-NTAs was achieved using wet-chemistry and chemical vapor deposition (CVD) techniques, respectively. This individual coating of Ag NPs or CMs on the TiO_2-NTAs led to an increase in DSSC efficiency by 15%. However, it was reported by the authors that Ag NPs and CMs were simultaneously coated on the TiO_2-NTAs, the DSSC efficiency doubled (30%) due to the coexistence and additive effects of two mechanistic phenomena—plasmonic interactions (due to the Ag NPs coating) and π–π conjugation (due to the CMs coating). Overall, the he design characteristics and material's performance of the EES was optimized so that it can serve as an adequate storage unit for practical applications in solar, wind, and other potential sources such as DSSC.

Hydrogen Energy Storage: In order to fulfill the current demand of primary or secondary energy sources from non-fossil fuel options, hydrogen stands as the first candidate due its many advantages. Hydrogen as a lean burning fuel, has a higher energy content than carbon-based fuels, and, as the by-product of hydrogen is water vapor, it has no carbon footprint. Hydrogen is the lightest gas and can be generated by various primary sources such as solar, wind, etc., and utilized as a fuel in both residential and industrial sectors. However, its intermediate storage faces challenges due to its gravimetric and volumetric densities. There are at least three types of hydrogen storage—gaseous storage or underground H_2 storage, liquid hydrogen storage, and storage in metal/complex hydride or sorbent systems. Based on these classifications of hydrogen storage, in addition to hydrogen storage densities, other vital parameters such as thermodynamics, kinetics, cycle life, refilling time, toxicity, and safety must be considered for its full deployment as an alternative fuel. In this Special Issue, four papers focus on hydrogen storage for potential stationary or automotive fuel cell applications. Christina Hemme et al. [9] reported the potential risk factors associated with underground hydrogen storage in depleted gas fields via hydrogeochemical modeling (HGCM). This one-dimensional diffusive mass-transport model is based on the equilibrium reactions among gas–water–rock interactions and kinetic reactions for sulfate reduction and methanogenesis. The main risk of underground H_2 storage in depleted gas fields predicted by HGCM is the conversion of hydrogen to methane, $CH_{4(gas)}$, and $H_2S_{(gas)}$ due to microbial activity, gas–water–rock interactions in the reservoir, and the loss of aqueous hydrogen by diffusion through the cap rock brine. According to the results obtained from this one-dimensional modeling, the authors recommended that depleted gas fields for underground H_2 storage possess low residual $CO_{2(gas)}$ concentrations. Additionally, the mineralogical compositions of the reservoir rocks should contain low amounts of sulfate- and carbonate-bearing minerals, but high amounts of iron-bearing minerals. Metal hydrides employed as hydrogen storage systems, often demand the fulfillment of two important criteria such as thermodynamics and the kinetics. Though suitable alloying or mechanical milling can optimize the thermodynamics, the kinetics have intrinsic limitations, which require novel methods such as catalytic doping and nanoparticle formation, etc. Shahine Shafiee et al. [10] employed a novel magnetic field and studied its effects on the thermal reaction kinetics of a metal hydride storage bed. Since the candidate metal hydrides such as lanthanum pentanickellide ($LaNi_5$) and magnesium hydride (MgH_2) are paramagnetic in nature, the effect of an external magnetic field on the heat conduction and reaction kinetics of such metal hydrides were investigated by the authors. The rapid heat transfer in a metal hydride bed was observed when these hydrides were subjected to magnetic fields of 3–5 kGauss or less than half a Tesla. Regarding hydrogen absorption, the capacity of metal hydrides with enhanced kinetics was obtained for a high-intensity magnetic field (0.5 T), however, poor kinetics was obtained for the samples subjected with no field (0 T) or a low-intensity field (0.3 T). Overall, the deployment of a magnetic field in a metal hydride storage test bed was found to enhance the reaction kinetics to due to the greater heat transfer and solid–gas reactions. Since the abovementioned metal hydrides have limitations in terms of hydrogen storage capacity ($LaNi_5$ <1.5 wt.% H_2) and a faster sorption rate at moderate to high temperatures (MgH_2 ~300 °C), a new class of complex hydrides with better temperature and hydrogen storage capacities were demonstrated by Sesha Srinivasan et al. [11]. The new complex hydrides, involving lithium amide and pre-processed magnesium hydride with different stoichiometric rations ($xLiNH_2$-nanoMgH$_2$; x = 1, 2), were synthesized using a mechanochemical process under an inert or hydrogen ambient atmosphere. Another modification of the nano-catalyst additives on the base complex hydrides achieved faster desorption kinetics when compared to pristine (no catalyst additive) complex hydrides. Based on a detailed thermal gravimetric analysis (TGA) and thermal programed desorption (TPD), kinetics enhancement occurred in the following order of catalyst type: TiF_3 > nanoNi > nanoTi > nanoCo > nanoFe. Furthermore, the highest reduction of the on-set decomposition temperature was obtained in the order of nanoCo > TiF_3 > nanoTi > nanoFe > nanoNi. Additionally, the absorption kinetics of 2 wt.% nanoNi on the base complex hydride ($2LiNH_2$-nanoMgH$_2$) was rapid (within the first 60 min of hydrogen absorption) at 200 °C when compared to the base material with no catalyst additive. The structural, microstructural, and chemical characterization supported the hydrogen sorption behavior of the complex hydrides. The abovementioned metal and complex hydrides were investigated for the purpose of using the stored hydrogen for either internal combustion (IC) engines as a lean burning fuel with available atmospheric oxygen or to be electrochemically combined with O_2 in a fuel cell to provide electrical output for stationary or automotive applications. One such combustion characteristic of hydrogen/air fuel for combustion via a four-point lean direct injection (LDI) system was simulated by Jianzhong Li et al. [12]. In this research paper, the authors designed and demonstrated a swirl-venturi 2 by 2 array four-point LDI combustor that can mix a hydrogen and air combustible mixture with different equivalence ratios. This simulation via the Reynolds-averaged Navier–Stokes code for steady-state computations enabled the authors to measure various parameters such as the axial velocity, swirl number, velocity angle, effective area, total pressure drop coefficient, total temperature, mass fraction of hydroxide (OH), and emission of the pollutant nitrous oxide (NO). As the equivalence ratios of hydrogen to air increased, the total temperatures and temperature rises of the co-swirling four-point LDI combustors also increased at approximately the same rate. When the equivalence ratio decreased from 1.0 to 0.6, the emission index of NO (EI$_{NO}$) reduced as well. Overall, this section of hydrogen energy storage, all of the described research papers addresses the deployment of hydrogen as an alternative fuel that can be readily available from solid-state hydride storage beds or hydride canisters.

Salt Cavern Energy Storage: In this Special Issue, problems related to the efficient storage of gases other than hydrogen via rock salt, carbonates, synthesis gas, etc. are considered as well. In addition to material characterization and property measurement, design aspects such as the cantilever fluid flow type or slurry bubble column type reactors are also important. Hongwu Yin et al. [13] determined and successfully demonstrated the optimum permeability and porosity of synthetic and mudstone rock salts for the assessment of natural gas storage facilities. Their study of a storage cavern and mudstone interlayer permeability was based on Jintan, in Jinansu Province of China. Gas slippage or the Klinkenberg effect was widely used by the authors to understand the gas flow in the low-permeability porous medium of the mudstone rock salt and/or layered rock salt. The permeability increased greatly when the porosity exceeded the value of 10%; hence, the permeability depends on the internal pores and microcracks as well as their interconnectivity. In a related work to the previous article on natural gas storage, Feifei Fang et al. [14] reported on carbonate gas reservoirs and the effects of water invasion in such reservoirs. An experimental system of water invasion in gas reservoirs with edge and bottom aquifers was established to simulate the process of water invasion. The effects of gas reservoir properties and production parameters such as water invasion energy, aquifer, production rate, permeability, and fracture were investigated systematically. The authors found that because of the physical simulation of water invasion, the elastic expansion of water–gas bubbles has greater influence. Additionally, the size of the aquifer and its production rate affect the water invasion in gas reservoirs with dissolved pores. The gas flow optimization for Fischer–Tropsch (F–T) synthesis was demonstrated by Siavash Seyednejadian et al. [15] via comprehensive computer modeling of a lab-scale slurry bubble column reactor (SBCR). According to this modeling, a set of partial differential equations in terms of mass transfer and chemical reactions was coupled successfully to predict the behavior of all F–T components in both gas and liquid phases over the SBCR bed. The authors found that the temperature distribution of the slurry reactor remains constant under a base load and varying load conditions The experimental findings agreed with the computer modeling in this study. Lastly, the theoretical work of Xinbo Ge et al. [16] established the dynamics of a hanging vertical cantilever that is subjected concurrently to internal and external axial flows. Additionally, theoretical predictions were obtained for a long leaching-tubing-like system with parameters related to salt cavern energy storage (SCES).

To summarize, in this Special Issue on "Clean Energy and Fuel Storage" features 16 peer-reviewed articles that can be divided in to four major types of energy storage: TES, EES, HES, and SCES. The discovery of novel materials and innovative design aspects are briefly discussed in this editorial section. The readers are encouraged to refer the full articles for detailed information.

Author Contributions: The authors of this editorial served as Guest Editors for this Special Issue on "Clean Energy and Fuel Storage". Additionally, the authors S.S.S. and E.K.S. contributed to the review article on Li-ion batteries [4] and one research article on complex hydrides for hydrogen storage [11]. Both the authors contributed equally to this editorial and proofread for publication.

Acknowledgments: S.S.S. would like to acknowledge Florida Polytechnic University, Florida Industrial and Phosphate Research Institute, and the Hinkley Center for Solid and Hazardous Waste Management for funding support. E.K.S. would like to acknowledge the University of South Florida for funding support.

Conflicts of Interest: The authors declare no conflict of interest.

References

1. Krüger, M.; Haunstetter, J.; Knödler, P.; Zunft, S. Slag as an Inventory Material for Heat Storage in a Concentrated Solar Tower Power Plant: Design Studies and Systematic Comparative Assessment. *Appl. Sci.* **2019**, *9*, 1833. [CrossRef]
2. Shao, B.; Du, X.; Ren, Q. Numerical Investigation of Energy Saving Characteristic in Building Roof Coupled with PCM Using Lattice Boltzmann Method with Economic Analysis. *Appl. Sci.* **2018**, *8*, 1739. [CrossRef]
3. Duquesne, M.; Palomo Del Barrio, E.; Godin, A. Nucleation Triggering of Highly Undercooled Xylitol Using an Air Lift Reactor for Seasonal Thermal Energy Storage. *Appl. Sci.* **2019**, *9*, 267. [CrossRef]
4. Demirocak, D.; Srinivasan, S.; Stefanakos, E. A Review on Nanocomposite Materials for Rechargeable Li-ion Batteries. *Appl. Sci.* **2017**, *7*, 731. [CrossRef]
5. Lai, X.; Qin, C.; Gao, W.; Zheng, Y.; Yi, W. A State of Charge Estimator Based Extended Kalman Filter Using an Electrochemistry-Based Equivalent Circuit Model for Lithium-Ion Batteries. *Appl. Sci.* **2018**, *8*, 1592. [CrossRef]
6. Jiang, X.; Nan, G.; Liu, H.; Guo, Z.; Zeng, Q.; Jin, Y. Optimization of Battery Energy Storage System Capacity for Wind Farm with Considering Auxiliary Services Compensation. *Appl. Sci.* **2018**, *8*, 1957. [CrossRef]
7. Lu, P.; Dai, Q.; Wu, L.; Liu, X. Structure and Capacitance of Electrical Double Layers at the Graphene–Ionic Liquid Interface. *Appl. Sci.* **2017**, *7*, 939. [CrossRef]
8. Kim, H.; Chun, M.; Suh, J.; Jun, B.; Rho, W. Dual Functionalized Freestanding TiO_2 Nanotube Arrays Coated with Ag Nanoparticles and Carbon Materials for Dye-Sensitized Solar Cells. *Appl. Sci.* **2017**, *7*, 576. [CrossRef]
9. Hemme, C.; Van Berk, W. Hydrogeochemical Modeling to Identify Potential Risks of Underground Hydrogen Storage in Depleted Gas Fields. *Appl. Sci.* **2018**, *8*, 2282. [CrossRef]
10. Shafiee, S.; McCay, M.; Kuravi, S. The Effect of Magnetic Field on Thermal-Reaction Kinetics of a Paramagnetic Metal Hydride Storage Bed. *Appl. Sci.* **2017**, *7*, 1006. [CrossRef]
11. Srinivasan, S.; Demirocak, D.; Goswami, Y.; Stefanakos, E. Investigation of Catalytic Effects and Compositional Variations in Desorption Characteristics of $LiNH_2$-nanoMgH$_2$. *Appl. Sci.* **2017**, *7*, 701. [CrossRef]
12. Li, J.; Yuan, L.; Mongia, H. Simulation Investigation on Combustion Characteristics in a Four-Point Lean Direct Injection Combustor with Hydrogen/Air. *Appl. Sci.* **2017**, *7*, 619. [CrossRef]
13. Yin, H.; Ma, H.; Chen, X.; Shi, X.; Yang, C.; Dusseault, M.; Zhang, Y. Synthetic Rock Analogue for Permeability Studies of Rock Salt with Mudstone. *Appl. Sci.* **2017**, *7*, 946. [CrossRef]
14. Fang, F.; Shen, W.; Gao, S.; Liu, H.; Wang, Q.; Li, Y. Experimental Study on the Physical Simulation of Water Invasion in Carbonate Gas Reservoirs. *Appl. Sci.* **2017**, *7*, 697. [CrossRef]

15. Seyednejadian, S.; Rauch, R.; Bensaid, S.; Hofbauer, H.; Weber, G.; Saracco, G. Power to Fuels: Dynamic Modeling of a Slurry Bubble Column Reactor in Lab-Scale for Fischer Tropsch Synthesis under Variable Load of Synthesis Gas. *Appl. Sci.* **2018**, *8*, 514. [CrossRef]
16. Ge, X.; Li, Y.; Chen, X.; Shi, X.; Ma, H.; Yin, H.; Zhang, N.; Yang, C. Dynamics of a Partially Confined, Vertical Upward-Fluid-Conveying, Slender Cantilever Pipe with Reverse External Flow. *Appl. Sci.* **2019**, *9*, 1425. [CrossRef]

Article

Slag as an Inventory Material for Heat Storage in a Concentrated Solar Tower Power Plant: Design Studies and Systematic Comparative Assessment

Michael Krüger *, Jürgen Haunstetter, Philipp Knödler and Stefan Zunft

DLR (German Aerospace Centre), Institute of Technical Thermodynamics, Pfaffenwaldring 38-40, 70569 Stuttgart, Germany; Juergen.Haunstetter@dlr.de (J.H.); Philipp.Knoedler@dlr.de (P.K.); Stefan.Zunft@dlr.de (S.Z.)
* Correspondence: Michael.Krueger@dlr.de; Tel.: +49-(0)711-6862-417

Received: 25 February 2019; Accepted: 25 April 2019; Published: 3 May 2019

Featured Application: Concentrated solar power (CSP) plant with open volumetric receiver.

Abstract: By using metallurgical slag from an electric arc furnace that is otherwise not recycled but deposited as an inventory material in thermal energy storage for concentrated solar power plants, it is possible to make a significant step forward in two transformation processes: energy and raw materials. As this type of slag has not been considered as an inventory material for this purpose, it is important to clarify fundamental questions about this low-cost material and its storage design. In this paper, design studies of slag-based thermal energy storage are carried out. Different slag-specific design concepts are developed, calculated and evaluated by a method based on established management tools. Finally, concepts for further investigations are defined. The highest aptitude value and the lowest risk value are achieved by the vertical storage design with axial flow direction. Therefore, it is taken as the lead concept and will be considered in complete detail in further research. Also, a closer look, but not as detailed as the lead concept, is taken at the horizontal storage with axial flow and the vertical storage with radial flow direction.

Keywords: thermal energy storage (TES); slag; regenerator; concentrated solar power (CSP); quality function deployment (QFD); failure mode and effect analysis (FMEA)

---

1. Introduction

Concentrated solar power (CSP) plants, in conjunction with photovoltaic systems, can contribute to a safe, clean and cost-effective electric power supply in the Earth's equatorial sun-belt [1]. The use of heat storage allows CSP plants to generate dispatchable electricity, and thus make an important contribution to the global energy transition [2].

To protect the environment and conserve primary raw materials, a raw material transition is also required [3–5]. This succeeds only with the efficient use of primary raw materials and the most complete possible use of secondary raw materials. This is addressed by German and European politicians through the German Resource Efficiency Programme II [6] and the EU Initiative for a Resource Efficient Europe [7].

By using metallurgical slag from an electric arc furnace (EAF), that is otherwise not recycled but deposited as an inventory material in thermal energy storage (TES) for CSP plants, it is possible to make a significant step forward in two transformation processes: energy and raw materials.

2. State of the Art of TES in CSP Plants

In CSP plants, molten salt technologies have been extensively deployed in CSP applications for the storage of thermal energy prior to steam or electricity generation. An example among many is the Solana power plant with 280 MWe and a storage system able to supply full power for six hours. [2]

The technology of regenerator-type storage is less developed, but has the potential for higher efficiency and lower costs. In particular, the application of regenerators in CSP power plants with an open volumetric receiver seems to be a promising approach, as shown in Figure 1.

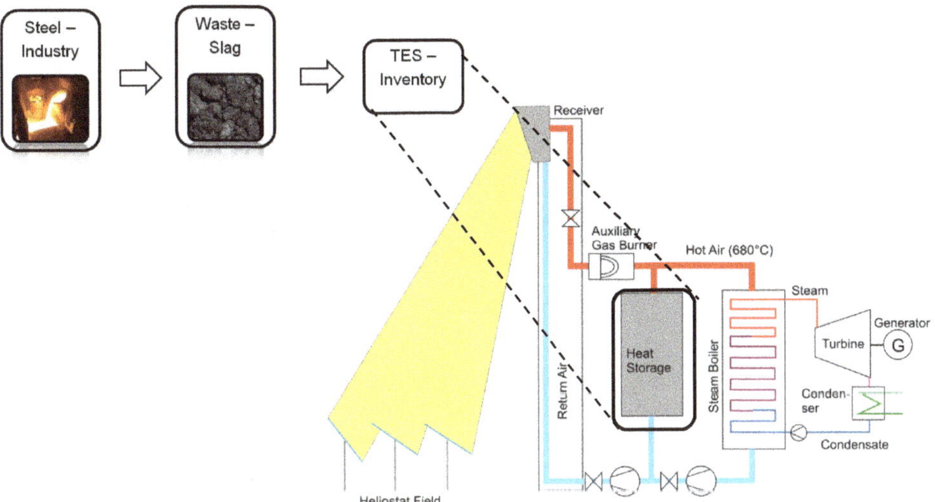

Figure 1. Flowsheet of slag reuse as a thermal energy storage (TES) inventory for a concentrated solar power (CSP) plant with an open volumetric receiver.

The Jülich solar power tower in Germany, seen in Figure 2, is currently the only plant in the world to be based on this technology. This experimental central receiver plant was inaugurated in 2009 to facilitate the further development of this technology. The erection and operation of the plant have been accompanied by a research program, whose objective is to cover the remaining uncertainties around design and operation and to further promote the development of the technology towards a commercial deployment. As a part of this work program, heat storage operation and design are also addressed [8].

As mentioned before, the plant uses an open volumetric receiver technology developed at DLR (German Aerospace Centre). In its primary cycle, air at atmospheric pressure is heated up to temperatures of about 700 °C. This solar heat then powers a steam generator, producing steam at 100 bars and 500 °C and driving a 1.5 MWel turbine-generator set. In parallel to the steam generator and receiver, thermal energy storage is integrated into the power cycle, as seen in Figure 1. It is implemented as an air-cooled regenerator storage system, an installation that is still unique in this application. The state of the art is set with the recently completed HOTSPOT project [9].

Generally, with regenerator-type storage, temperatures of up to 1000 °C and even more can be realized by using solid storage material such as commercially available bricks or beds of smaller particles made of oxide ceramic material [10]. Recent work [11–14] is investigating low-cost alternative inventory materials such as low-temperature-fired clay bricks and magmatic natural stones.

Figure 2. Jülich solar power tower (Germany).

An alternative to this is slag from an electric arc furnace (EAF). Since this type of slag has not been considered as an inventory material for this purpose, it is important to clarify fundamental questions about the material and TES design. These questions are addressed by the REslag project. The main objective of the project is to make an effective valorization of the steel slag and reuse it as a feedstock for four innovative applications, one of which is thermal energy storage systems in CSP applications [15,16].

The degradation effects of the fluid by direct contact with the slag, which are reported, for example, by Grosu et al. [17] in another type of slag, are, in contrast to molten salt, not of importance in the air or can be equalized by correspondingly high air exchange rates. The issue of safety in the context of molten salt is also to be assessed much more critically than with air. The safety regulations for slag are also manageable with this system, as it is a closed circuit and the slag used is classified by the manufacturer as a non-hazardous substance.

In this paper, design studies of slag-based thermal energy storage are carried out. Different design concepts are developed, calculated and evaluated. Finally, the lead concepts for further investigation in the project are defined.

3. Design of Slag-Based TES

3.1. CSP Plant Target Specifications

Since the Jülich solar power tower (see Figure 2) is the only facility of its kind, and is only a demonstration plant, no targets for large systems can be defined. Under these circumstances, literature research was independently performed. However, as no consistent record of a specialized plant exists, the collected data can only serve as guide values. Accordingly, the target figures were determined on the basis of existing knowledge, as shown in Table 1.

Table 1. Concentrated solar power (CSP) plant specifications.

Description	Characterization	Comments
Rated net power output of CSP plant	150 MW$_{el}$	
TES capacity	6.5 h (2.21 GWh)	Considering the solar multiple factor and charging duration.
Temperature at TES inlet while charging	700 °C	
Temperature at TES inlet while discharging	120 °C	
Max. temperature drop at TES outlet while discharging	60 °C	
Max. pressure loss through the TES while discharging	100 mbar	
Discharging mass flow	780 kg/s	At design point (12 p.m., 21 March)
Max. charging mass flow through TES	1080 kg/s	At design point (12 p.m., 21 March)
Mean charging mass flow through TES	706 kg/s	On design day (21 March) Assumption: sinusoidal course of the sun
Charging duration	8 h	Considering the solar multiple factor and sunshine hours
Hours of sunshine on design day (21 March)	12.2 h	Location: Huelva (Spain)
Solar multiple	2	At design point (12 p.m., 21 March)

3.2. Considered TES Designs

Various possible basic TES designs are tested for suitability and compared to determine the most suitable option for slag-based TES. Figure 3 shows an overview of the options considered. They differ in their positioning and flow direction, namely, axial, radial and meandering.

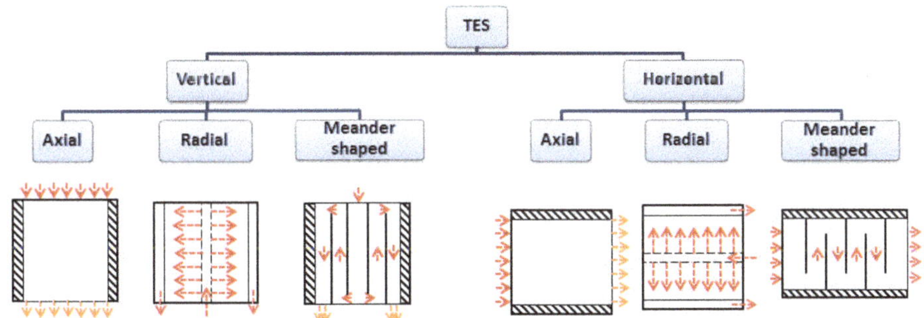

Figure 3. Matrix of variants for TES.

The comparison of the vertical and the horizontal variants reveals that the axial and radial concepts are rotated only by 90 degrees, while the meander-shaped variant is fundamentally different. In the vertical variant, it goes radially from the inside to the outside, thus also increasing the flow cross-sectional area. In the horizontal variant, the size of the cross-sectional area always remains the same; there is only a change in the direction of flow from one chamber to the next.

While the meander-shaped variant is a fundamental innovation, the axial variant has previously been considered and represents the state of the art in vertical position. The radial flow is also very innovative but has been studied in [11,12]. The special feature here is the constantly changing cross-sectional area along the flow path. In order to get a better understanding of each concept and the reason why it is considered, the main advantages and disadvantages are listed in Table 2.

Table 2. Advantages and disadvantages of considered thermal energy storage (TES) designs.

TES Option	Advantage	Disadvantage
Vertical TES	• Easy to fill • Insulation can be placed easily • Space saving • Most previous knowledge	• Bottom piping can be complex
Horizontal TES	• Reduction in height • Distributor is easier to mount • Rectangle-shaped instead of cylindrical, easier to support storage walls	• More complicated to fill • Needs more ground space than vertical
Axial flow	• Basic concept • Most previous knowledge	• Lower degree of uniformity compared with radial flow
Radial flow	• Reduced distributor space • Better degree of uniformity compared with axial flow	• Overflow losses
Meander-shaped flow	• Higher storage degree of utilization • Reduced storage height	• Increased filling effort • Higher inventory costs (additional installations)

3.3. Pre-Design of Storage

By using simplified models based on the Λ-Π-method [18] at a specific site taking into consideration the course of the sun, the geometrical properties for each thermal energy storage design from Figure 3 under different parameter variations are calculated. Here, just the results for the vertical TES are presented as an example.

Dimensioning of TES

The given CSP plant specifications (Table 1) and the slag properties listed in Table 3 are included in the calculation of the heat storage dimensions. The results are shown as fields in Figure 4.

Table 3. Electric arc furnace (EAF) slag characteristics [19].

Description	Characterization	Comments
Density	3430 kg/m^3	
Thermal conductivity	1.43 W/(m K)	at 500 °C
Specific heat capacity	0.933 kJ/(kg K)	at 500 °C

Depending on the slag particle diameter used, the maximum pressure loss is 100 mbar with a storage diameter of approx. 38 to 44 m and a storage height of 8 to 12 m. The large diameters are structurally more challenging, so the heat storage is divided into three smaller modules connected in parallel.

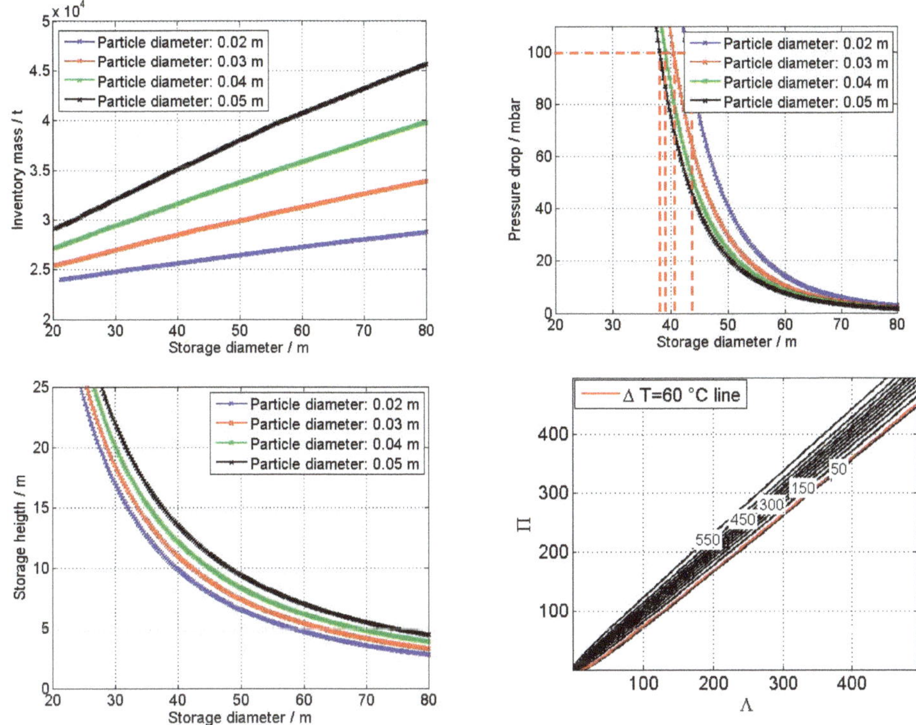

Figure 4. Simulation results of designing the TES.

3.4. Comparative Assessment and Definition of a Lead Concept

Quality Function Deployment (QFD) and Failure Mode and Effect Analysis (FMEA)

For the identification of the lead concept, an evaluation method was developed, which is essentially based on QFD and FMEA, and includes an aptitude analysis and a risk analysis. Both methods, QFD and FMEA, are established management tools used in product development. While the aim of QFD is to successfully fulfil the wishes of the customers in their products and services, the use of FMEA attempts to minimize potential errors and risks. Extensive information on QFD and FMEA can be found elsewhere [20,21]. The two methods have been simplified for the present usage for deciding which design of slag-based TES is the best and has the lowest risks. The analysis approach used is described by Figure 5.

Firstly, criteria of the aptitude analysis were collected and divided into two different groups, economical and technical demands, with the first group weighted by 0.6 and the second by 0.4. This is due to the fact that the main focus is on cost reduction by using a very inexpensive inventory material—provided that the functionality is performed. All points considered in the aptitude analysis are listed in Table 4. Secondly, these criteria were prioritized among each other by pair-by-pair comparisons and the listed weighting factor of each criterion was generated accordingly.

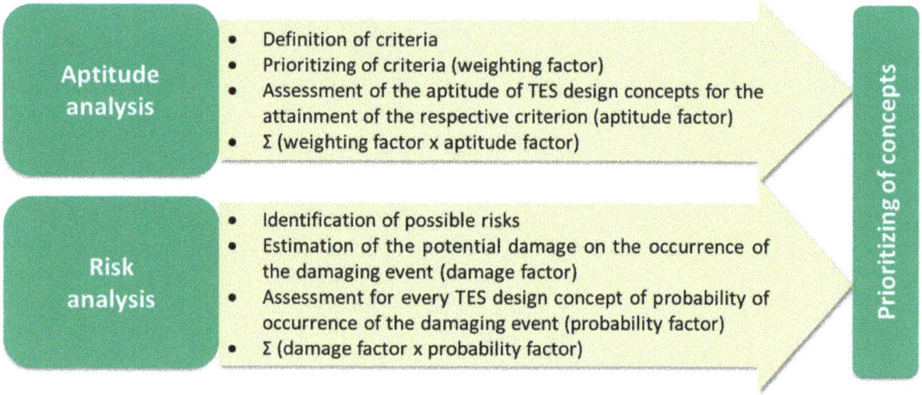

Figure 5. Quality function deployment (QFD)- and failure mode and effect analysis (FMEA)-based analysis approach.

Table 4. Criteria for aptitude analysis.

Group	Criterion	Description	Weighting Factor
Economical demand	Low investment costs	Will be calculated in a simplified way; includes inventory, containment and liner (protects insulation) costs. Simple designs without internals for flow guidance and distribution have advantages here.	7.1
	Low operating costs	Will be calculated in a simplified and qualitative way; includes, e.g., auxiliary power for ventilation, costs for inventory change etc.	5.9
	High storage degree of utilization	Is calculated by thermal simulation; indicates the utilization of the inventory, that means how much of the inventory undergoes the full temperature increase.	0.6
	Low space need	Base area of container is considered as well as the needed container number.	1.8
	High operational availability	Is reduced, for example, by out of order or maintenance times of the TES. Possible bypass flows due to settlement effects and hazards to thermal insulation at hot points due to high loads are taken into account.	10.0
	Long lifetime	Of the TES and its subcomponents. The higher the inventory level, the higher the forces on inventory, insulation and container. The lower the inventory level, the more gentle on the materials and the higher the potential lifetime.	10.0

Table 4. *Cont.*

Group	Criterion	Description	Weighting Factor
Technical demand	Low expense of protection for insulation	The basis is an inner liner which is used from a storage height of 10 m.	4.1
	High storage degree of uniformity	Ratio of effective emitted energy over discharge duration to maximal possible energy withdrawal over discharge duration.	9.0
	Low complexity	Of the construction and connection.	7.2
	High degree of maturity	Commercial availability of components. Concepts already available on the market must be evaluated as better than those for which only test setups or even only drawings exist.	10.0
	Good scalability	To larger or smaller storage	8.7
	Low expense of system integration	Of the TES.	6.7
	Low expense of fluid distribution	Good and even fluid distribution.	3.8
	Low expense of insulation	Inner and outer insulation of the TES.	3.6
	Low expense of filling	Filling the TES with slag pebbles.	2.1
	Low maintenance effort	Level of access. In particular, good accessibility at all points is crucial here.	6.7
	Low cleaning effort of working fluid	Filter mandatory? Cleaning amount of filter.	4.6
	Low safety effort	Safety must be ensured but at what expense?	1.3

Each concept was then evaluated according to the criteria developed previously. For this purpose, a so-called aptitude factor was introduced. The design results presented in the previous section are also taken into account here. The multiplication of these two factors and the subsequent addition of the individual criterion values results in an aptitude value for each concept. The results are shown in Figure 6.

The highest value is achieved by the vertical TES with axial flow direction (83%), followed by the horizontal TES with axial flow direction (75%) and the vertical TES with radial flow direction (72%). Mainly, this is caused by high aptitude values in the areas that have high weighting factors. In particular, operational availability, lifetime, investment costs and degree of maturity should be mentioned here. While horizontal setups promise longer lifetimes due to less inventory material failure, the more or less state-of-the-art vertical axial design achieves better scores with the degree of maturity, operational availability and investment costs criteria. Radially flowed variants have significant advantages in terms of operating costs due to low pressure losses along the flow path and the storage degree of uniformity. In the overall assessment, however, the vertical axial concept is slightly ahead of the horizontal axial and vertical radial concept.

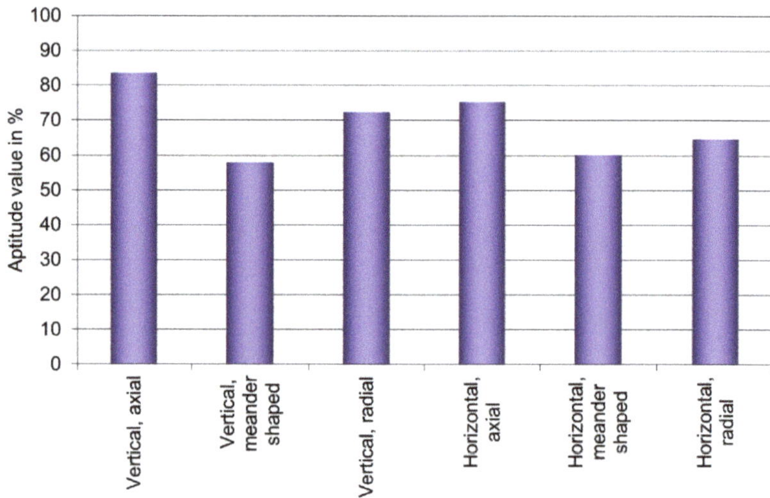

Figure 6. Aptitude values for each concept.

As a second step, the risk analysis was implemented. As with the aptitude analysis mentioned above, this was done in several steps. First, possible risks were identified (see Table 5) and estimated in relation to the potential damage when the damaging event occurred. This leads to a so-called damage factor (see Table 5). Secondly, each concept was also evaluated according to the probability of occurrence of the damaging events. A probability factor was therefore generated. The third step was carried out as in the aptitude analysis: the factors were multiplied and each risk value was summed. This results in a risk value for each concept, as shown in Figure 7.

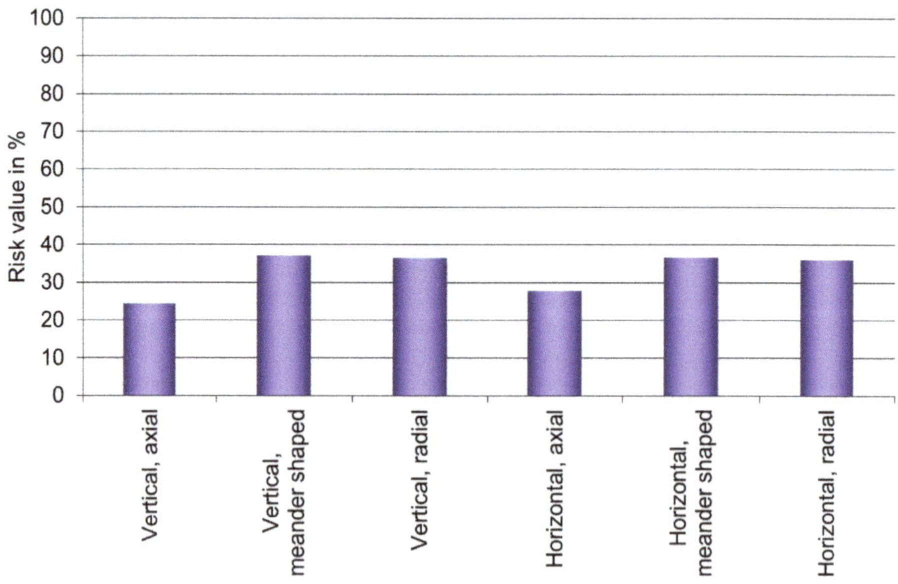

Figure 7. Risk value for each concept.

Table 5. Criteria for risk analysis and damage factors.

Criteria	Description	Damage Factor *
Thermal design uncertainties	Uncertainties in material parameters, model uncertainties	5
Thermomechanical design uncertainties	Uncertainties in material parameters, model uncertainties	3
Fluid mechanical design uncertainties	Model uncertainties	5
Material failure	Inventory, insulation	4
Corrosion	Inventory, insulation, container, piping (everything that is in contact with the high temperature fluid (HTF))	7
Operating restrictions due to the complexity of the storage system	Piping, amount of container, storage installations, isolation equipment	4
Operational safety	Outer damages which can influence the operational safety, e.g., fluid leaking	9
Economic uncertainties	False estimation of investment and operational costs	2
Lack of competition for plant components	Low amount of providers or no provider for essential components	3

* High damage factor (7–10): Hazard to human life; effect on whole plant (required changes); large effect on plant performance, lifetime and availability. Medium damage factor (4–6): derated operation possible; effect on related components (required changes); medium effect on plant performance, lifetime and availability. Low damage factor (1–3): effect on TES only (required changes); low effect on plant performance, lifetime and availability.

The lowest value is achieved by the vertical TES with axial flow direction (24%) followed by the horizontal TES with axial flow direction (28%). As there are no different probabilities of occurrence for the risks with a high damage factor (operational safety and corrosion), the results are mainly influenced by criteria with medium damage factors. In particular, thermal and fluid mechanical design uncertainties as well as material failure should be mentioned here. While horizontal setups promise less inventory material failure, the more or less state-of-the-art vertical axial design achieves lower probability factors with the design uncertainties criteria. Radially flowed and meander-shaped variants have significant disadvantages in terms of design uncertainties. Although the thermal design uncertainties are small in themselves, the fluid mechanical design uncertainties have to be assigned a higher probability of occurrence due to possible bypass flows in the case of horizontal axial flow TES. In the overall assessment, however, the vertical axial concept is slightly ahead of the horizontal axial concept.

At the end of the QFD analysis, both aptitude values and risk values were plotted in a diagram, as shown in Figure 8. It is important to note that a high value of risk is not necessarily problematic for a research project because research can reduce risks.

Three of the six concepts have aptitude values above 70%. The highest value is achieved by the vertical TES with axial flow direction, which also indicates the lowest value of risk. In summary, this is an expression of the techno-economic optimum and the highest degree of maturity. It is therefore defined as the lead concept and will be fully taken into account in the further course of the project. Also, a closer look, but not as detailed as the lead concept, is taken at the horizontal TES with axial flow and the vertical TES with radial flow direction. This is done due to the fact that these two concepts also achieved a high aptitude value and thus have high potential.

In order to demonstrate the competitiveness of slag-based TES and to further reduce technical uncertainties, detailed research on materials and design as well as pilot plant trials are being conducted as part of the REslag project.

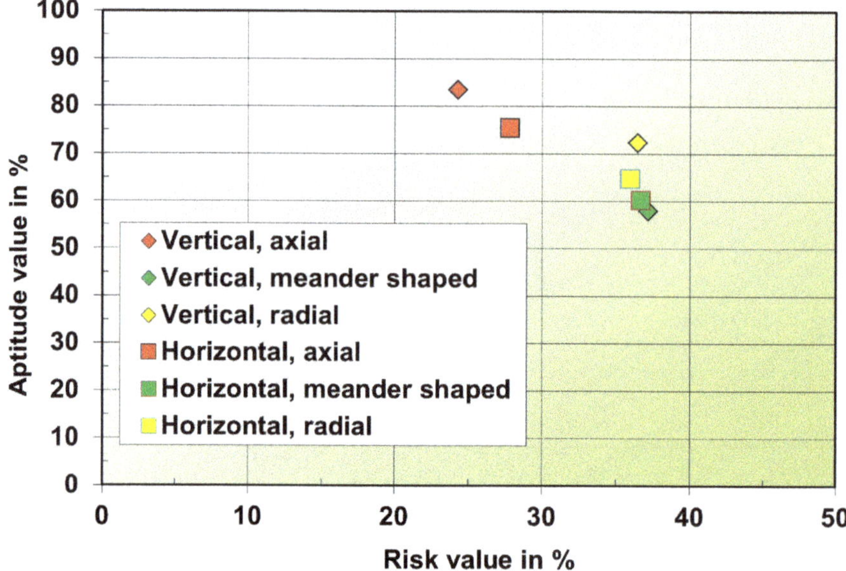

Figure 8. Results of the comparative assessment.

4. Summary and Conclusions

The use of metallurgical slags from EAF as inventory material for TES for CSP plants has not yet been intensively investigated and offers the opportunity to save raw materials and energy at the same time. Economically, the material offers the advantage of being a very cost-effective alternative to conventional inventory materials for solid heat storage. Since, however, the thermophysical properties of the slags under consideration are very similar to those of conventional refractory materials and ceramics, the TES concepts which are structurally simple and functional and pose the lowest technical risks will prevail.

In this paper, different slag-specific design concepts are systematically evaluated. For this purpose, an evaluation procedure based on QFD and FMEA was developed, which makes it possible to objectively evaluate the suitability and the risk of different concepts and thus to define the most suitable concepts. In the concepts examined here, which differ in their different installation methods and flow directions, the vertical TES with axial flow direction performed best, as it has both the best aptitude value and the lowest risk value. It forms the techno-economic optimum. The partly better functionalities of the other concepts are not so decisive here. It is therefore defined as a lead concept and will be further developed in the course of the REslag project. The horizontal TES with axial flow and the vertical TES with radial flow direction will also be investigated in less detail.

Author Contributions: The paper was written by M.K. under the guidance of S.Z. The pre-design of storage was carried out by J.H. The comparative assessment and definition of a lead concept were carried out by M.K., J.H. and P.K.

Funding: This project has received funding from the European Union's Horizon 2020 research and innovation programme under grant agreement No. 642067.

References

1. Pitz-Paal, R. Solar Energy-Concentrating Solar Power. In *Future Energy*; Elsevier Science: New York, NY, USA, 2014; pp. 405–431.
2. European Association for Storage of Energy (EASE); European Energy Research Alliance (EERA). *European Energy Storage Technology Development Roadmap-2017 Update*; EASE/EERA: Brussels, Belgium, 2017.
3. Temporärer ProcessNet-Arbeitskreis Rohstoffe und Kreislaufwirtschaft. *Anorganische Rohstoffe - Sicherung der Rohstoffbasis von Morgen*; DECHEMA, e.V.: Frankfurt, Germany, 2015.
4. Buchert, M. *Press Release: Transition to Sustainable Raw Materials Management Needed in Germany*; Oeko-Institut: Freiburg/Berlin, Germany, 2015.
5. Öko-Institut, e.V. *Rohstoffwende 2049: Zur Zukunft der Nationalen und Internationalen Rohstoffpolitik. Jahrestagung*; Öko-Institut e.V.: Freiburg, Germany, 2016.
6. Federal Ministry for the Environment, Nature Conservation and Nuclear Safety. *German Resource Efficiency Programme II-Programme for the Sustainable Use and Conservation of Natural Resources*; Federal Ministry for the Environment, Nature Conservation and Nuclear Safety: Berlin, Germany, 2016.
7. European Commission. *Communication from the Commission to the European Parliament, the Council, the European Economic and Social Committee and the Committee of the Regions: Roadmap to a Resource Efficient Europe*; European Commission: Brussels, Belgium, 2011.
8. Zunft, S.; Hänel, M.; Krüger, M.; Dreißigacker, V.; Göhring, F.; Wahl, E. Jülich Solar Power Tower-Experimental Evaluation of the Storage Subsystem and Performance Calculation. *J. Sol. Energy Eng.* **2011**, *133*, 1019–1023. [CrossRef]
9. Zunft, S.; Hänel, M.; Krüger, M.; Dreißigacker, V. A design study for regenerator-type heat storage in solar tower plants-Results and conclusions of the HOTSPOT project. *Energy Procedia* **2014**, *49*, 1088–1096. [CrossRef]
10. Laing, D.; Zunft, S. Using concrete and other solid storage media in thermal energy storage (TES) systems. In *Advances in Thermal Energy Storage Systems: Methods and Applications*; Woodhead Publishing: Cambridge, UK, 2015; pp 65–86.
11. Stahl, K.; Moser, P.; Marquardt, R.; Siebert, M.; Kesser, S.; Maier, F.; Krüger, M.; Zunft, S.; Dreißigacker, V.; Hahn, J. *Flexibilisierung von Gas- und Dampfturbinenkraftwerken durch den Einsatz von Hochtemperatur-Wärmespeichern (FleGs): F&F Vorhaben zur Vorbereitung von Hochtemperatur-Wärmespeichern und Deren Integration in den Gas- und Dampfturbinenprozess*; RWE Power AG: Essen, Germany, 2013.
12. Stahl, K.; Zunft, S.; Kessler, S.; Siebert, M. Flexibilisierung von GuD-Kraftwerken durch den Einsatz von Hochtemperatur-Wärmespeichern. In *VDI-Konferenz "GuD-Kraftwerke im Dynamischen Netzbetrieb", 29–30 November, Köln Germany*; Verein Deutscher Ingenieure (VDI): Düsseldorf, Germany, 2011.
13. Zunft, S.; Freund, S. Large-scale Electricity Storage with Adiabatic CAES-the ADELE-ING project. In Proceedings of the Energy Storage Global Conference, Paris, France, 19–21 November; Energy Storage Association: New York, NY, USA, 2014.
14. Zunft, S. *ADELE-ING: Engineering-Vorhaben für die Errichtung der Ersten Demonstrationsanlage zur adiabaten Druckluftspeichertechnik*; German Aerospace Center: Stuttgart, Germany, 2017.
15. Krüger, M.; Haunstetter, J.; Zunft, S. Concentrated solar tower power plant using slag as inventory material for a thermal energy storage (TES). In *AIP Conference Proceedings*; AIP Publishing: Melville, NY, USA, 2018; Volume 2033, pp. 090018-1–090018-8.
16. RESLAG. Available online: http://www.reslag.eu (accessed on 18 February 2019).
17. Grosu, Y.; Ortega-Fernández, I.; del Amo, J.M.L.; Faik, A. Natural and by-product materials for thermocline-based thermal energy storage system at CSP plant: Compatibility with mineral oil and molten nitrate salt. *Appl. Therm. Eng.* **2018**, *136*, 657–665. [CrossRef]
18. Dreißigacker, V.; Müller-Steinhagen, H.; Zunft, S. Thermo-mechanical analysis of packed beds for large-scale storage of high temperature heat. *Heat Mass Transf.* **2010**, *46*, 1199–1207. [CrossRef]
19. Ortega-Fernández, I.; Calvet, N.; Gil, A.; Rodríguez-Aseguinolaza, J.; Faik, A.; D'Aguanno, B. Thermophysical characterization of a by-product from the steel industry to be used as a sustainable and low-cost thermal energy storage material. *Energy* **2015**, *89*, 601–609. [CrossRef]

20. Akao, Y. *Quality Function Deployment: Integrating Customer Requirements into Product Design*; Productivity Press: Cambridge, UK, 1990.
21. Akao, Y.; Mizuno, S. *QFD: The Customer-Driven Approach to Quality Planning and Deployment*; Asian Productivity Organisation: Tokyo, Japan, 1994.

 *applied sciences*

Article

Numerical Investigation of Energy Saving Characteristic in Building Roof Coupled with PCM Using Lattice Boltzmann Method with Economic Analysis

Bilin Shao [1], Xingxuan Du [1] and Qinlong Ren [2,*

[1] School of Management, Xi'an University of Architecture and Technology, Xi'an 710055, Shaanxi, China; sblin0462@163.com (B.S.); Duxx0312@163.com (X.D.)
[2] Key Laboratory of Thermo-Fluid Science and Engineering of MOE, School of Energy and Power Engineering, Xi'an Jiaotong University, Xi'an 710049, Shaanxi, China
* Correspondence: qinlongren@xjtu.edu.cn

Received: 10 September 2018; Accepted: 25 September 2018; Published: 26 September 2018

Featured Application: The current work has potential applications in economically designing green buildings in the summer hot and winter cold region of China.

Abstract: Due to their characteristics of high energy storage density and a nearly constant melting temperature, phase change materials (PCMs) could be inserted into the roof of green buildings in order to reduce the energy consumption and ameliorate the room thermal comfort. In this paper, an enthalpy based multiple-relaxation-time (MRT) lattice Boltzmann method (LBM) was developed to calculate the transient phase change conjugate heat transfer with solar radiation inside the green building's PCM roof in the hot summer and cold winter areas of China. The effect of the PCM melting temperature on the variation of the roof internal temperature was investigated and the energy saving characteristic of the PCM roof under an intermittent energy utilization condition was also analyzed by comparing with the performance of the roof filled with sensible insulating materials (SIMs). Then, the life cycle incremental costs and incremental benefits of a PCM roof and SIM roof were studied by using the comprehensive incremental benefit model so that the green building roof could be economically evaluated. The results indicate that a temperature rise inside the roof during summer cooling time could be delayed due to the latent heat of the PCMs. It was also found that the melting temperature and the thickness of the PCM layer should be chosen appropriately for enhancing the energy saving amount of a PCM roof. Based on this, the PCM roof could have a better energy saving capability than the SIM roof. During the winter heating time, as the environment temperature and the room temperature are both below the PCM melting temperature, the PCM roof does not have a latent heat characteristic so that it performs like a SIM roof. Furthermore, due to the high price of PCMs, the incremental cost of green building is increased, which makes the PCM roof have a negative comprehensive incremental benefit. Under this circumstance, developing PCMs with a low price and stable chemical properties is a key scientific bottleneck for a wider application of PCM roofs in the architecture engineering field.

Keywords: hot summer and cold winter area; PCM roof; comprehensive incremental benefit; conjugate phase change heat transfer; lattice Boltzmann method

---

1. Introduction

Energy and environment are the two of the most significant basic factors for the development of the human society. The combustion of fossil fuels, which emit a great deal of carbon dioxide contributes to the global warming issue and effective energy saving techniques are essential for solving the energy crisis and environmental issues [1]. Building energy consumption was responsible for more than 40% of the total energy cost in China during the past seven years [2,3]. Furthermore, the energy consumption of a building is still increasing as people's living standards improve. For this reason, developing green buildings with high energy saving characteristics has been an indispensable task for local government in order to protect the living environment and realize sustainable development. Reducing the energy release through the building envelope is an important approach for decreasing the buildings energy consumption. Due to their low thermal conductivities and latent heat characteristics, the PCMs could be inserted into the building envelope as a thermal insulation layer to decrease the building energy cost, which has attracted a lot of research attention during recent years [4–6]. As the PCMs absorb and release heat during a solid-liquid phase change process at a nearly constant temperature, it could not only increase the thermal resistance of a building's envelope but also attenuate the temperature oscillation so that the thermal comfort of the building's rooms is improved.

The research methodologies for building a PCM envelope can be generally categorized as follows: Theoretical method, experimental method, and numerical method. With the fast progress of computer science and numerical modelling schemes, numerical methods have been widely used to analyze the energy saving characteristics of buildings [7]. Barrientos applied a one-dimensional finite difference method to investigate the transient conjugate heat transfer inside the building walls contained with PCMs [8]. It was found that the integration of a PCM layer into a building wall could diminish the instantaneous heat flux magnitude through the wall when the PCM melting temperature was appropriately chosen. Besides, due to the elevated solar radiation, the wall orientation has a more obvious effect on the energy consumption of a building during the summertime. Zwanzig solved the one-dimensional transient heat equation in the multilayered wallboard using the Crank-Nicolson discretization scheme [9]. It was found that an optimum location for PCM placement existed in terms of different thermal resistances between the PCM layers and the external thermal boundary conditions. Jin numerically optimized the location of the PCM layer under different parametric conditions by validating their model with experiments conducted using a dynamic wall insulator [10]. They presented that the optimum PCM layer location approaches the exterior surface of the wall when the melting temperature and the latent heat of PCMs were increased. By validating the numerical model with a PCM wallboard heat storage experiment, Xie investigated the thermal performance of PCM wallboards for practical engineering [11]. Their results indicated that the PCM wallboard performance could be adverse in different seasons and the thermal analysis through an entire year is also necessary. The above research was mainly related to the heat transfer properties of PCM walls. However, the thermal loss of roofs contributes almost 70% of the total heat loss in building envelopes due to direct solar radiation and heat transfer between the roof and the exterior environment. Under this situation, the thermal performance of roofs imbedded with PCM layers deserves more scientific investigations. Li numerically analyzed the thermal performance of a roof that contained PCM in a single residential building with respect to the factors of solar radiation intensity, roof slope, PCM melting temperature and layer thickness [3]. They concluded that the roof slope has a more significant influence on the thermal performance of a PCM roof compared to the melting temperature and latent heat of PCM in the region of northeast China. Tokuc developed a one-dimensional model based on the first law of thermodynamics to carry out the time-dependent simulation of a PCM roof in Istanbul under summer conditions [12]. The results demonstrated that a PCM thickness of 2 cm is suitable for use in flat roofs in this specific area and their experimental and numerical results were consistent. Liu analyzed the thermal performance of a PCM-filled double glazing roof using a numerical method [13]. It was indicated that the semi-transparent property and zenith angle had a big effect on the thermal performance. In addition, the temperature lag time of a PCM roof increases with the increment of the PCM layer thickness. Although a lot of research related to PCM inserted into a building envelope exists, little work has been carried out for the hot summer and cold winter region in China where the climate conditions are quite different from other areas. In China's hot summer and cold winter region, an air conditioner is indispensable for both the summer cooling season and the winter heating season, which consumes a huge amount of electricity. Based on this, the current paper aims to investigate the conjugate phase change heat transfer and the energy saving benefit in the PCM roof of the hot summer and cold winter region in China.

During the past two decades, the lattice Boltzmann method (LBM) has been developed as a powerful numerical method for solving complicated heat transfer problems [14,15]. For the solid-liquid phase change phenomenon, the existing LBM schemes could be classified into three categories: (1) The phase-field method [16,17]; (2) the immersed boundary method [18]; (3) the enthalpy-based method [19–24]. Due to its simplicity and numerical robustness, the enthalpy-based method is the most widely used scheme for simulating solidification and melting problems in scientific and engineering fields. Jiaung firstly investigated the solid-liquid phase change problem using an enthalpy-based lattice Boltzmann method [19]. However, the iteration of latent heat source term is required in their scheme, which increases the computational load. Eshraghi developed an implicit LBM scheme for conduction with a solid-liquid phase change [20]. By solving a linear system of equations, the numerical iteration process for the latent heat source term could be avoided. To further improve the computational efficiency, Huang modified the equilibrium function for temperature for which the circumstances for the iteration of the latent heat source or solving a linear system of equations are not indispensable [21]. In order to ameliorate the numerical stability and reduce the numerical diffusion during the phase change, Huang further developed the multiple-relaxation-time (MRT) LBM for a solid-liquid phase change using enthalpy formulation [22]. Besides, by decoupling the thermal conductivity and the specific heat from the relaxation time and the equilibrium distribution function, this model is demonstrated to be appropriate for modelling the conjugate heat transfer. Recently, Li also developed MRT LBM models for axisymmetric and three-dimensional solid-liquid phase change problems [23,24]. On the other hand, due to its highly parallel nature, the lattice Boltzmann method has been successfully installed into graphics processor units (GPU) to achieve parallel computing for several different heat transfer and fluid flow applications [25–30].

In this paper, the MRT enthalpy-based LBM with GPU acceleration was used to investigate the conjugate phase change heat transfer of a PCM roof in the hot summer and cold winter region of China. Then, by obtaining the energy loss through the roof during the summer cooling time and the winter heating time, the comprehensive incremental model was applied to evaluate the economic benefit of the PCM roof building during its life cycle. The remainder of the paper is organized as follows. In Section 2, the mathematical model for conjugate heat transfer with phase change and its thermal boundary conditions is presented. In Section 3, the details of the enthalpy-based MRT lattice Boltzmann method and the comprehensive incremental benefit model are shown. The results and discussion are presented in Section 4. A conclusion is finally drawn in Section 5.

2. Mathematical Model for Conjugate Heat Transfer with Phase Change

According to the configuration standard of a building envelope in the hot summer and cold winter region of China, the current research focuses on the heat transfer capability and energy saving characteristics of the following three different roofs: (1) Ordinary roof, its configuration from the top to the bottom is: A 20 mm thick cement layer, a 100 mm thick reinforced concrete layer, and a 20 mm thick lime layer; (2) a PCM roof, its configuration from the top to the bottom is: A 20 mm thick cement layer, a 100 mm thick reinforced concrete layer, a 30 mm thick phase change material (PCM) layer, and a 20 mm thick lime layer; (3) a SIM roof, its configuration from the top to the bottom is: A 20 mm thick cement layer, a 100 mm thick reinforced concrete layer, a 30 mm thick sensible insulating material (SIM) layer, and a 20 mm thick lime layer. In order to simplify the mathematical complication, the following reasonable assumptions are made for the current modelling: (1) The thermophysical properties of PCM, SIM, and other materials are constant. (2) The volume expansion of PCM during solidification and melting is negligible, and its melting temperature is constant. (3) The contact thermal resistance between different material layers is neglected. (4) The effects of people, furniture, and other heat sources in the room are not considered. (5) The temperature distribution inside the room is uniform. Based on the above conditions, the transient conjugate heat transfer with phase change inside the PCM roof is governed by the following energy equation:

$$\rho_i \frac{\partial H_i}{\partial t} = k_i \frac{\partial^2 T_i}{\partial y_i^2} \tag{1}$$

where ρ_i is the density, k is the thermal conductivity, T is the temperature, t is the time, y is the Cartesian coordinate in the vertical direction, H is the enthalpy, and the index $i = 1 \sim n$ represents the ith layer material of a building roof. For the PCM layer, the enthalpy could be defined as:

$$H = c_p(T - T_r) + f_l h_s \tag{2}$$

where c_p is the specific heat, T_r is the reference temperature, f_l is the liquid fraction of the PCM, and h_s is the latent heat. For other sensible material layers, the enthalpy H is expressed as:

$$H = c_p(T - T_r). \tag{3}$$

At the external surfaces of a roof ($i = 1$ and $y_1 = 0$ mm), the solar radiation and convective heat transfer between the environment and the cement layer are taken into account, which could be given by the third kind of thermal boundary condition as:

$$- k_1 A \frac{\partial T_1}{\partial y_1} = h_{out} A (T_e - T_1) + Q_{rad} \tag{4}$$

where h_{out} is the convective heat transfer coefficient of the external roof surface, which is used as $h_{out} = 19$ W/(m·K) for summertime and $h_{out} = 23.3$ W/(m·K) for winter time [2]; A is the roof surface area, T_e is the environmental temperature, and Q_{rad} is the solar radiation. The environment temperature, T_e, and solar radiation, Q_{rad}, could be simultaneously considered by introducing the equivalent temperature T_{eq} [2]:

$$T_{eq} = T_e + \alpha I / h_{out} - I_w / h_{out} \tag{5}$$

α is the solar absorption coefficient of the roof surface, which was chosen to be $\alpha = 0.8$ for the cement material; I is the solar radiation intensity. For the building roof, the term I_w / h_{out} was 3.5–4 K [2], and $I_w / h_{out} = 3.5$ K was used in the current work. For the interior surface of the roof ($i = n$ and $y_n = L$ mm), where L is the roof thickness, the third kind of thermal boundary condition was applied:

$$- k_n A \frac{\partial T_n}{\partial y_n} = h_{in} A (T_n - T_{in}) \tag{6}$$

h_{in} is the convective heat transfer coefficient of the interior roof surface, which is $h_{in} = 8.7$ W/(m·K) [2]. For the interfaces between different materials in the building roof, the Dirichlet-Neumann boundary condition for conjugate heat transfer should be satisfied:

$$T_i = T_{i\pm1} \tag{7}$$

$$- k_i \frac{\partial T_i}{\partial y_i} = -k_{i\pm1} \frac{\partial T_{i\pm1}}{\partial y_{i\pm1}}. \tag{8}$$

To characterize the energy consumption of a building's roof, the heat flux magnitude Q through the interior roof surface per meter square is defined as:

$$Q(t) = |h_{in}(T_n(t) - T_{in})|. \tag{9}$$

Using the ordinary roof (without a PCM layer and SIM layer) as a reference, the energy saving amount, ΔW, during the air conditioner (AC) working period for a PCM roof and SIM roof was calculated as:

$$\Delta W = \int_{t1}^{t2} \left[Q_{ordinary\ roof}(t) - Q_{PCM\ or\ SIM\ roof}(t) \right] dt \tag{10}$$

where t_1 and t_2 are the time limits of the AC working period. The current paper aims to investigate the energy saving benefit of a PCM roof building during the working daytime in the hot summer and cold winter region of China. Hence, the time limit of the AC working period was chosen to be $t_1 = 8 : 00$ a.m. and $t_2 = 18 : 00$ p.m. Besides, the windows of the building room were assumed to be open during the other times so that the internal room temperature, T_{in}, was equal to the environment temperature, T_e, when the air conditioner was off.

3. Lattice Boltzmann Method and Comprehensive Incremental Benefit Model

3.1. Lattice Boltzmann Method

The enthalpy-based multiple-relaxation-time lattice Boltzmann method derived by Huang was used to simulate the solid-liquid phase change and the conjugate heat transfer in the building's roof [22]. The evolution equation for the distribution function is given as:

$$g_i(x + e_i \Delta t, t + \Delta t) = g_i(x, t) - M^{-1} S[m(x, t) - m^{eq}(x, t)] \tag{11}$$

where the distribution function for the momentum space is given by:

$$m = M g_i = (m_0, m_1, \ldots, m_8)^T, \ m^{eq} = M g_i^{eq} = \left(m_0^{eq}, m_1^{eq}, \ldots, m_8^{eq} \right)^T. \tag{12}$$

The matrix for transforming the distribution function between momentum space and velocity space is given by:

$$M = \left\{ \begin{array}{rrrrrrrrr} 1 & 1 & 1 & 1 & 1 & 1 & 1 & 1 & 1 \\ -4 & -1 & -1 & -1 & -1 & 2 & 2 & 2 & 2 \\ 4 & -2 & -2 & -2 & -2 & 1 & 1 & 1 & 1 \\ 0 & 1 & 0 & -1 & 0 & 1 & -1 & -1 & 1 \\ 0 & -2 & 0 & 2 & 0 & 1 & -1 & -1 & 1 \\ 0 & 0 & 1 & 0 & -1 & 1 & 1 & -1 & -1 \\ 0 & 0 & -2 & 0 & 2 & 1 & 1 & -1 & -1 \\ 0 & 1 & -1 & 1 & -1 & 0 & 0 & 0 & 0 \\ 0 & 0 & 0 & 0 & 0 & 1 & -1 & 1 & -1 \end{array} \right\}. \tag{13}$$

The equilibrium distribution function m^{eq} in momentum space is given by:

$$m^{eq} = (H, \ -4H + 2c_{p,ref}T, \ 4H - 3c_{p,ref}T, 0, \ 0, \ 0, \ 0, \ 0, \ 0)^T. \tag{14}$$

In the current work, the specific heat of PCM or SIM was chosen to be the reference specific heat $c_{p,ref}$. The relaxation time matrix, S, in momentum space is given as:

$$S = diag(s_0, \ s_e, \ s_\varepsilon, \ s_j, \ s_q, \ s_j, s_q, \ s_e, \ s_e). \tag{15}$$

To reduce the numerical diffusion, Huang pointed out that the following relation should be satisfied [22]:

$$\left(\frac{1}{s_e} - \frac{1}{2} \right) \left(\frac{1}{s_j} - \frac{1}{2} \right) = \frac{1}{4}. \tag{16}$$

When the collision step was completed in the momentum space, the post-collision distribution function in velocity space could be then calculated using an inverse transformation:

$$g_i(x, t + \Delta t) = M^{-1} m(x, t + \Delta t). \tag{17}$$

Then, the streaming process is completed as:

$$g_i(x + e_i \Delta t, t + \Delta t) = g_i(x, t + \Delta t). \tag{18}$$

The thermal boundary condition scheme, derived by Eshraghi and Felicelli, was used in this paper [20], and the enthalpy, H, was computed as:

$$H = \sum_{i=0}^{8} g_i. \tag{19}$$

The corresponding temperature, T, is given as:

$$T = \begin{cases} T_m - \frac{H_s - H}{C_p} & H \leq H_s \\ T_m & H_s < H < H_l \\ T_m + \frac{H - H_l}{C_p} & H \geq H_l \end{cases} \tag{20}$$

where T_m is the PCM melting temperature, H_s and H_l are the enthalpies of the solid and liquid state PCM respectively. The liquid fraction f_l of PCM is computed by:

$$f_l = \begin{cases} 0 & H \leq H_s \\ \frac{H - H_s}{H_l - H_s} & H_s < H < H_l. \\ 1 & H \geq H_l \end{cases} \tag{21}$$

3.2. Code Validation

The Compute Unified Device Architecture (CUDA) Fortran code accelerated by GPU was developed in the current work. The details of CUDA implementation are presented in detail in our previous work [31]. Firstly, the code was validated using the one-dimensional conjugate heat transfer without a phase change between two materials. Initially, the temperature was set to be $T = 1$ in the region A at $x > 0$ and $T = 0$ in the region B at $x < 0$. The analytical solution for this problem is given by [32]:

$$T^A(x, t) = \frac{1}{1 + \sqrt{(\rho C_p)^B k^B / (\rho C_p)^A k^A}} \left[1 + \sqrt{(\rho C_p)^B k^B / (\rho C_p)^A k^A} erf \left(\frac{x}{2\sqrt{k^A t / (\rho C_p)^A}} \right) \right] \tag{22}$$

$$T^B(x, t) = \frac{1}{1 + \sqrt{(\rho C_p)^B k^B / (\rho C_p)^A k^A}} erfc \left(-\frac{x}{2\sqrt{k^B t / (\rho C_p)^B}} \right). \tag{23}$$

The thermophysical properties and prices of materials used in this work are presented in Table 1. For this calibration, the reinforced concrete was used for region A and paraffin (solid state) was chosen as the material for region B with the characteristic length of 1 m. The conjugate heat transfer problem was simulated by MRT LBM on GPU with the grid number of 100×100. As displayed in Figure 1a, the current results agree well with the analytical solutions, which demonstrate the accuracy of the current code for solving conjugate heat transfer problems. To calibrate the current program for the solid-liquid phase change problem, the code was coupled with a D2Q9 LBM solver for fluid flow and then the natural convection with melting at $Pr > 1$ was simulated. Jany and Bejan applied the scaling laws to develop the correlation of an average Nusselt number, Nu_{ave}, of a hot wall for this problem, which is given as [33]:

$$Nu_{ave} = (2FoSte)^{-1/2} + \left[0.33Ra^{1/4} - (2FoSte)^{-1/2}\right]\left\{1 + \left[0.0175Ra^{3/4}(FoSte)^{3/2}\right]^{-2}\right\}^{-1/2}. \quad (24)$$

As presented in Figure 1b, the current LBM code matched the results from the scaling laws at the Stefan number $Ste = 0.1$ with different Rayleigh numbers when 400×400 grids were used. It indicated that the current code was accurate for modelling the solidification and melting processes.

Table 1. Thermophysical properties and prices of materials in a building's roof (Data resource: China Chemical Engineering Website).

Material	ρ (kg/m^3)	c_p (J/kg·K)	k (W/m·K)	C_u (yuan/m^3)
Cement	1807	840	0.854	850
Concrete	2500	920	1.74	-
Paraffin	775	534	0.148	8800
Perlite	400	1170	0.14	350
Lime	1600	1050	0.81	850

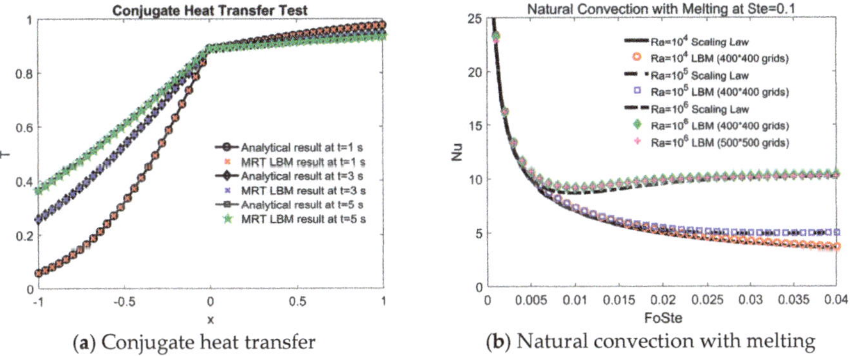

(a) Conjugate heat transfer   (b) Natural convection with melting

Figure 1. CUDA code calibration for conjugate heat transfer and solid-liquid phase change.

3.3. Comprehensive Incremental Benefit Model

Compared to the traditional buildings, the green buildings with PCM or SIM have a larger incremental cost during the construction period. However, with their better energy saving characteristics, the green buildings will generate an economic benefit, an environmental benefit, and a social benefit during their lifetime. The comprehensive incremental benefit is defined as [34]:

$$\Delta E = \Delta S - \Delta C \quad (25)$$

where ΔE is the comprehensive incremental benefit, ΔS is the incremental benefit, and ΔC is the incremental cost for the extra expense of constructing the green building compared with the traditional building, which is expressed as:

$$\Delta C = \Delta C_i + \Delta C_p \quad (26)$$

ΔC_i is the cost of integrating extra materials such as PCM or SIM:

$$\Delta C_i = (C_u + C_o + C_l) \times L \times H \times D \quad (27)$$

where C_u is the price of PCM or SIM as shown in Table 1, C_o is the price of other adhesive materials, which was chosen to be $C_o = 25$ yuan/m^3 and $C_l = 15$ yuan/m^3 was the labor fee. L was the length of the roof set to be 6 m, H is the width of the roof set to be 6 m, and D is the thickness of PCM or SIM layer. ΔC_p is the cost for numerical modelling and consultation for the building design, which is $\Delta C_p = 0.1\% \times \Delta C_i$. The incremental benefit ΔS is calculated as:

$$\Delta S = \Delta S_y \cdot P(\Delta S_y, i_0, n_y) \tag{28}$$

P is the annual income discount rate coefficient given as:

$$P = 1/i_0 - 1/\left[i_0 \times (1 + i_0)^{n_y}\right] \tag{29}$$

n_y is the year of the building life cycle, which is 50 years according to the standard of the Chinese government; i_0 is the discount rate, which is set to be $i_0 = 11\%$ for the Shaanxi province in China. The annual benefit ΔS_y is defined as:

$$\Delta S_y = \Delta S_e + \Delta S_h + \Delta S_s \tag{30}$$

ΔS_e is the economic benefit of green building:

$$\Delta S_e = \frac{(\Delta W_s * 70 + \Delta W_w * 70) \times L \times H}{3600 \times \eta} \times EP_{unit} \tag{31}$$

where ΔW_s (kJ/m^2) is the energy saving amount during the summer AC cooling time, and ΔW_w (kJ/m^2) is the energy saving amount during the winter AC heating time. In this paper, the period of summer cooling and winter heating was chosen to be 70 days for each. η is the Energy Efficiency Ratio (EER) or Coefficient of Performance (COP) of an air conditioner, which was set as 3.3 in this work. EP_{unit} is the electricity price in the city of Hanzhong, Shaanxi Province which is $EP_{unit} = 0.4983$ (yuan/kW·h).

ΔS_h is the environmental benefit of a green building and it includes the benefits generated by the reduction of CO_2, SO_2, and NO_x emissions. Using an effective equivalent method, the coefficient for transferring the saved electricity of the green building to the reduced combustion amount of coal was 0.0004 ton/(kW·h). The emission coefficients of CO_2, SO_2, NO_x, and smoke dust were 2.4567, 0.0165, 0.0156, 0.096 respectively. Furthermore, the economic value for the reduced emissions of CO_2, SO_2, NO_x, and smoke dust were 390 (yuan/ton), 4344.93 (yuan/ton), 632 (yuan/ton), and 275 (yuan/ton). Hence, the environmental benefit of a green building's roof, ΔS_h, is given as:

$$\begin{aligned} \Delta S_h = &\frac{(\Delta W_s * 70 + \Delta W_w * 70) \times L \times H}{3600 \times \eta} \times 0.0004 \\ &\times (2.4567 \times 390 + 0.0165 \times 4344.93 + 0.0156 \times 632 + 0.096 \times 275) \end{aligned} \tag{32}$$

ΔS_s is the social benefit of the green building's roof and it includes the reduced investment for electrical power installation and the economic loss due to the electricity consumption peak. From an investigation in China, when the electricity consumption decreases by 1 kW·h, the electrical investment is reduced by 0.2 yuan/year and the economic loss decreases by 0.22 yuan/(kW·h). Then, the social benefit of a green building's roof, ΔS_s, is computed by:

$$\Delta S_s = \frac{(\Delta W_s * 70 + \Delta W_w * 70) \times L \times H}{3600 \times \eta} \times (0.2 + 0.22). \tag{33}$$

4. Results and Discussions

The city of Hanzhong in Shaanxi province was chosen to be a representative of a hot summer and cold winter region in China. The temperature of Hanzhong during July 2017 and January 2018 is presented in Figure 2. The average maximum temperature and minimum temperature for summertime in Hanzhong is 33.97 and 24.13 respectively while it is 6.26 and −0.67 respectively for winter time. During the summertime, the average solar radiation intensity in the daytime is $I_{ave} = 755.40\ W$ and the sunrise time is 6:00 a.m. while the sunset time is 20:00 p.m. During the winter time, the average solar radiation intensity in the daytime is $I_{ave} = 646.75$, and the sunrise time is 8:00 a.m. while the sunset time is 17:00 p.m. The environmental temperature, T_e, is expressed by using a sinusoidal function as follows [35]:

$$T_e = -sin\left[\frac{2\pi(t + 7200)}{86400}\right]\left(\frac{T_{max} - T_{min}}{2}\right) + \left(\frac{T_{max} + T_{min}}{2}\right) \tag{34}$$

where T_{max} is the maximum temperature during a day, T_{min} is the minimum temperature during a day, and t is the time. Similarly, the solar radiation intensity I during daytime is given as:

$$I = \sqrt{2}I_{ave}sin\left[\frac{2\pi(t - 21600)}{100800}\right] \text{ (summertime)} \tag{35}$$

$$I = \sqrt{2}I_{ave}sin\left[\frac{2\pi(t - 28800)}{64800}\right] \text{ (winter time).} \tag{36}$$

The current research investigated the energy saving characteristics of a PCM roof under the influence of solar radiation with intermittent air conditioner (AC) working conditions. During the AC working time, the internal room temperature, T_{in}, was set to be 26 for the summer cooling season while it was 18 for the winter heating season. During the AC off time, the windows of the building were open so that the internal room temperature, T_{in}, was equal to the environmental temperature T_e. In order to eliminate the effect of the initial roof temperature on the energy consumption results, the computation was iterated for 10 days (240 h) until the temperature of the roof exhibited a periodic variation as shown in Figure 3.

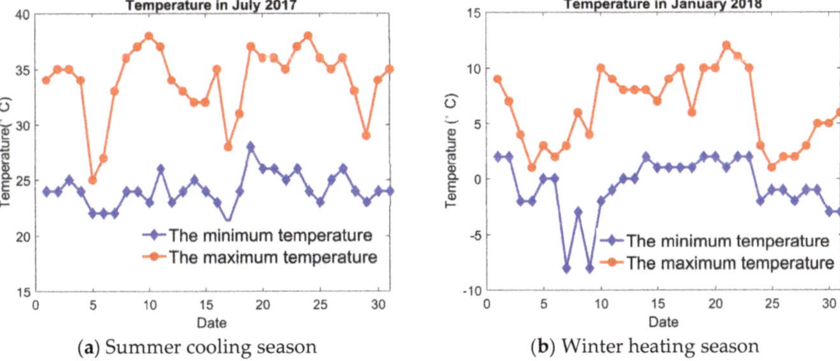

(**a**) Summer cooling season

(**b**) Winter heating season

Figure 2. The temperature variation of Hanzhong city.

4.1. The Influence of Roof Type

Paraffin was used as the phase change material (PCM) while perlite was chosen to be the sensible insulation material (SIM) for the roofs in the current work because of their similar thermophysical properties for a fair comparison as shown in Table 1. The energy saving characteristics of an ordinary roof (without PCM or SIM), SIM roof, and PCM roof during the summer cooling season were compared. As shown in Figure 4, the magnitudes of the heat flux, Q, for different types of roofs during the air conditioner cooling period are presented. The results indicate that the heat flux, Q, of an ordinary roof increases much more dramatically during the daytime compared with a SIM roof and a PCM roof, which indicates its larger energy consumption. By integrating the heat flux magnitude, Q, with respect to time, t, during the AC working period, the energy consumption of an ordinary roof was

2861.3 kJ/m^2. As a comparison, the heat flux magnitude, Q, of a SIM roof increased more smoothly due to the existence of a SIM layer with a low thermal conductivity, as displayed in Figure 4. For a SIM layer thickness of 30 mm, the energy consumption of a SIM roof during the AC cooling time was 1294.0 kJ/m^2. It means that the energy consumption of the roof decreased by 54.78% when the SIM layer was inserted into the roof compared with the ordinary roof. In contrast to the ordinary roof and the SIM roof, as presented in Figure 4, the heat flux magnitude, Q, of a PCM roof remained constant for a long time under the influence of environmental temperature and solar radiation. As a consequence, when the PCM layer thickness was 30 mm, with a melting temperature of 31 , the energy consumption of the PCM roof during the AC cooling time was 1048.3 kJ/m^2, which is 63.36 and 18.99% lower than those of the ordinary roof and the SIM roof respectively. Under this circumstance, it should be concluded that the PCM roof had the best energy saving characteristic during the summer AC cooling time among all the three type roofs.

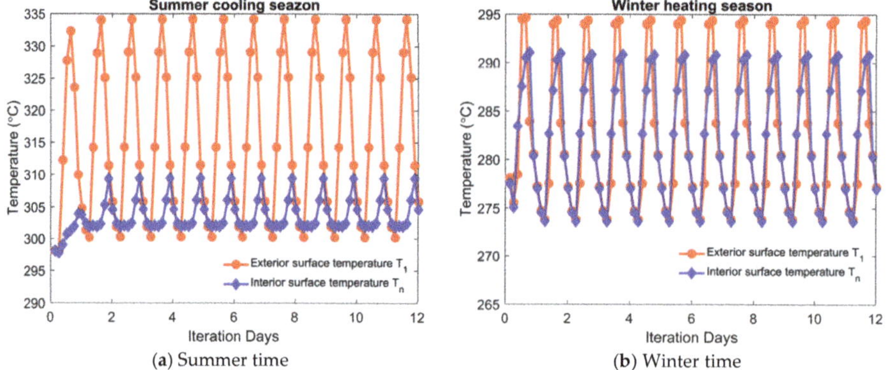

(a) Summer time   (b) Winter time

Figure 3. Iteration process of the roof temperature during the computation.

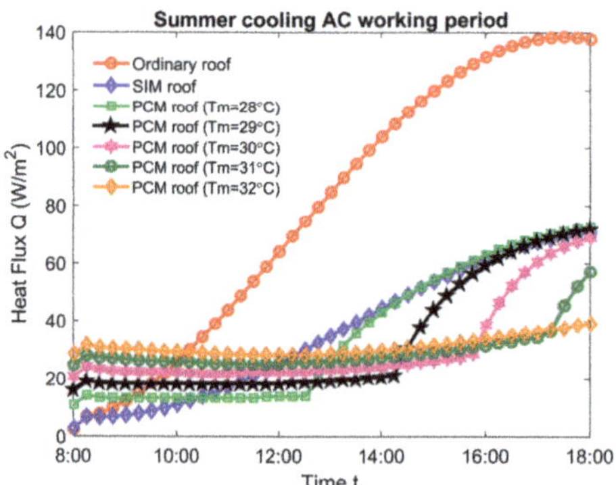

Figure 4. Heat flux magnitude of different types of roofs during the summer AC cooling time.

To further explain and understand the reason that a PCM roof has the minimum energy consumption during the summer cooling season, the temperature distributions of a PCM roof and a SIM roof ($D = 30$ mm and $T_m = 31$) during the AC working period are shown in Figure 5. As presented in Figure 5, under the combined influences of the increasing environmental temperature T_e and the solar radiation intensity, I, the exterior surface temperature of the building's roof T_1 increased during the daytime and achieved the maximum value at 14:00 p.m. Due to the latent heat characteristic of the PCM layer, the interior surface temperature of the PCM roof ($x = 170$ mm) remained almost unchanged until 16:00 p.m. However, when the PCM layer was fully melted and became a liquid, it behaved like the sensible insulation material and lost the latent heat characteristic. Hence, the temperature of the PCM roof interior surface could not be fixed at a relatively constant value and began to increase as shown by the curve at 17:00 p.m. On the other hand, although the low thermal conductivity of the SIM layer could prevent the heat loss to some extent compared with the ordinary roof, its internal surface temperature increased continuously, which actually consumed more energy for the air conditioner to cool down the room than the PCM roof.

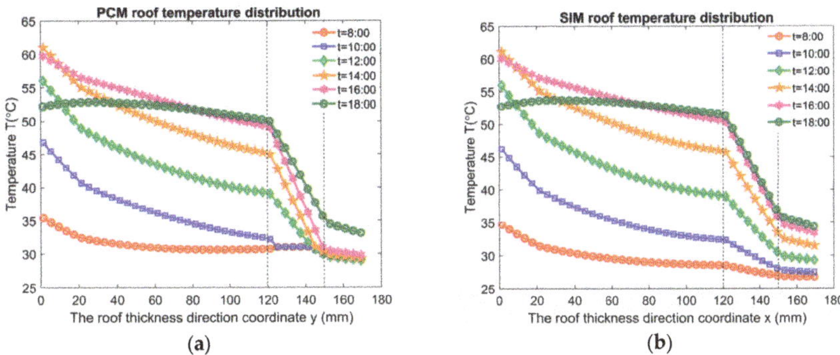

Figure 5. The PCM roof and SIM roof temperature distribution during the summer AC cooling time.

4.2. The Influence of PCM Melting Temperature

As one of the most significant factors of PCMs, which affects their thermal performance, the melting temperature, T_m, of the PCM layer plays an essential role in the energy saving characteristics of a PCM roof. The energy saving amount, ΔW, of a PCM roof in terms of the melting temperature, T_m, is plotted in Figure 6. It shows that the energy saving amount firstly increased with the increment of the melting temperature, T_m, and it reached the optimum value at $T_m = 31$. As the melting temperature, T_m, further increased to be higher than 31 , the energy saving amount value of the PCM roof became less. When the melting temperature, T_m, was 28 , 29 , 30 , 31 , and 32 , the PCM roof energy saving amount, ΔW, during the summer cooling season was 1603 kJ/m^2, 1685 kJ/m^2, 1762 kJ/m^2, 1813 kJ/m^2, and 1748 kJ/m^2 respectively. Compared to the roof of $T_m = 28$, the energy saving amount, ΔW, for the roof with $T_m = 31$ was increased by 13.10%. The above discussion indicates that using the PCMs with an appropriate melting temperature, according to climate conditions, is important for optimizing the energy consumption characteristics of PCM roof buildings.

To explain the mechanism, an optimum PCM melting temperature for reducing the roof energy consumption amount exists, the transient PCM average liquid fraction f_l during the summer cooling season is plotted in Figure 7. At a melting temperature of $T_m = 28$, it was found that only 4.8% of the amount of PCM solidified during the early morning time and all of the PCM became liquid again at 12:15 p.m. When the PCM layer became fully liquid, it lost the latent heat characteristic and behaved like the SIM layer so that the internal temperature inside the PCM roof could not be controlled and started to increase at this state. Contrary to the PCM roof with a melting temperature $T_m = 28$, the maximum solid state PCM amount was 31.65% during the early morning for the PCM roof with $T_m = 31$ under which situation the energy storage capacity of the PCM layer increased. For this reason, the time for a PCM layer of $T_m = 31$ being fully melted to a liquid state was delayed to 17:00 p.m. The means that the interior surface temperature of a PCM roof could be fixed at a nearly constant value until 17:00 p.m. so that the energy consumption from using the air conditioner was highly reduced. Unfortunately, as the melting temperature T_m was further increased to 32 , the PCM layer did not become fully liquid until 18:30 p.m. Although the PCM layer of $T_m = 32$ had the capability of modulating the temperature inside the PCM roof during the whole AC working period, this PCM layer also solidified more rapidly at a relatively higher solidification temperature during the nighttime and the early morning time from 0:00 a.m. to 9:00 a.m., as shown in Figure 7. This phenomenon makes the PCM layer have a trade-off effect on the energy saving characteristics of a PCM roof: (1) A relatively higher melting temperature, T_m, could make the PCM layer have a longer latent heat characteristic, which delays the temperature rise in the PCM roof during the daytime so that the energy consumption of the air conditioner is reduced. (2) The PCM layer with a higher melting temperature, T_m, solidified more rapidly during the nighttime at a nearly constant temperature. Hence, its temperature could not be further cooled down by the lower environment temperature, T_e, under which the air conditioner energy cost during the daytime increased. The variations of the PCM roof interior surface temperature, T_n, and the corresponding environmental temperature, T_e, are displayed in Figure 8. Although the interior surface temperature rise of the PCM roof with a melting temperature $T_m = 32$ could be delayed more than the other PCM roofs, its interior surface temperature, T_n, during the nighttime was the highest. Furthermore, during the nighttime, the environmental temperature, T_e, was much lower than the melting temperature of the PCM roofs in the current work, which means that the environmental temperature could cool down the temperature inside the roof. Based on this, a higher melting temperature of a PCM roof at nighttime has a negative contribution to the energy saving amount of the PCM roof. Hence, by balancing the trade-off influence of the PCM layer melting temperature during the daytime and nighttime, an optimum melting temperature, T_m, exists, which was 31 in this work for a PCM roof to reduce the maximum amount of energy consumption. For the remainder of the paper, the melting temperature of $T_m = 31$ was used.

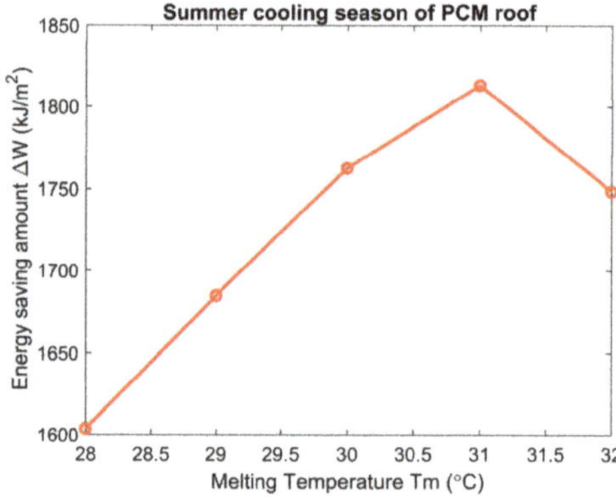

Figure 6. The effect of the melting temperature on the PCM roof energy saving amount.

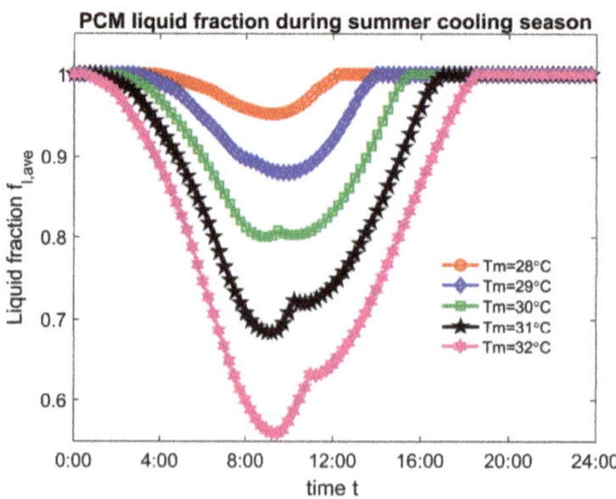

Figure 7. The PCM layer average liquid fraction during the summer cooling season at different melting temperatures.

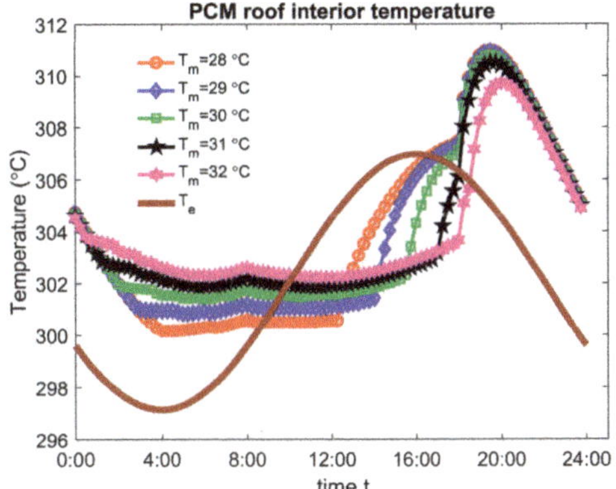

Figure 8. The PCM roof interior surface temperature during the summer cooling season at different melting temperatures.

4.3. The Influence of PCM or SIM Layer Thickness

The PCM or SIM layer thickness not only affected the energy saving characteristics of green building roofs but also determined its incremental cost during the construction period and the corresponding building's comprehensive incremental benefit. In Figure 9, the effect of PCM or SIM layer thickness on the energy saving amount of a green building's roof is plotted. The energy saving characteristics of both PCM roof and SIM roof improve with the increment of their PCM or SIM layer thicknesses. The results indicate that the energy saving amount, ΔW, of a SIM roof increases almost linearly with respect to the increasing SIM layer thickness because of the linearly increased thermal resistance. contrary to the SIM layer, the increasing the rate of the energy saving amount, ΔW, of a PCM roof decreased in terms of the PCM layer thickness and its energy saving amount almost approached a constant value when the PCM layer thickness is large. To clarify the underlying mechanism of this situation, the average liquid fraction of a PCM layer during the summer cooling time is shown in Figure 10. It was found that the energy storage capacity of a latent heat PCM layer increased with its increasing layer thickness. For the PCM layer of $D = 20$ mm, the PCM became fully melted and lost the capability of controlling the roof temperature at 15:15 p.m. As a consequence, the interior roof surface temperature began to rise after 15:15 p.m., which increased the energy consumption of the PCM roof. In comparison, when the PCM layer of $D = 60$ mm was used, the inside roof temperature was modulated by the PCM layer until 21:00 p.m., which guaranteed that the PCM layer had the latent heat characteristic during the air conditioner working time so that the maximum energy saving amount was achieved. However, similar to the discussion in Section 4.2 for the effect of PCM melting temperature, the increment of the PCM layer thickness also had a negative influence on the energy saving capability of the PCM roof. As the environmental temperature became low during the nighttime, the latent heat function of the PCM layer at nighttime, after 18:00 p.m., maintained its temperature nearly at the solidification point, which prevented heat transfer from the internal roof region into the environment. As a consequence, due to the trade-off effect of the PCM layer thickness, the increasing rate of the energy saving amount of a PCM roof decreases. Besides, when the PCM or SIM layer thickness was $D \leq 45$ mm, the energy saving amount of the PCM roof was more than that of the SIM roof because of the latent heat characteristic of the PCM layer, which attenuated the temperature oscillation inside the roof. However, it is interesting to note that the SIM roof had a better energy saving capability than the PCM roof when their layer thickness, D, was more than 50 mm. The reason for this phenomenon is that a SIM roof with larger thickness could prevent the heat transfer from the environment into the interior room in the daytime and the interior SIM roof temperature could also be efficiently cooled down without the effect of latent heat during the nighttime, which makes the SIM roof have a different thermal performance compared to the PCM roof.

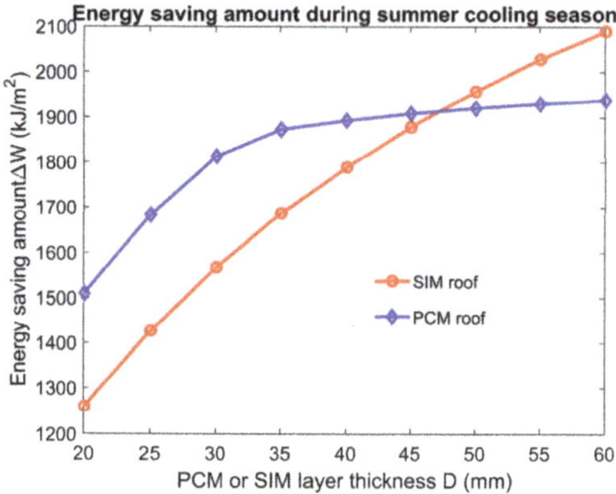

Figure 9. The effect of layer thickness on the energy saving amount of PCM or SIM roofs during the summer cooling season.

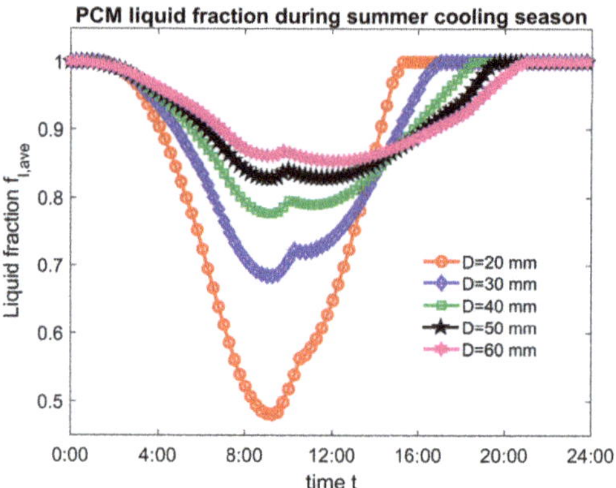

Figure 10. The PCM layer average liquid fraction during the summer cooling season with different layer thicknesses.

In order to evaluate the comprehensive incremental benefit of a green roof building during the life cycle in a hot summer and cold winter region of China, the energy saving amount of a PCM roof and a SIM roof during the winter AC heating season should also be investigated. When the PCM layer with a melting temperature $T_m = 31$ was inserted into the green building roof, it was kept in the solid state without melting and solidification processes in Hanzhong city during the winter AC heating time. As a result, the PCM layer did not have the capability of modulating the temperature change inside the roof with latent heat and its function was similar to the sensible insulation material with a low thermal conductivity. As the thermal conductivities of paraffin (PCM) and perlite (SIM) used in this work were similar, the energy saving amounts of a PCM roof and SIM roof were almost the same when the layer thickness, D, was less than 35 as shown in Figure 11. However, the specific heat, c_p, of perlite is much larger than that of paraffin. When the thicknesses of a PCM roof and a SIM roof were large, the interior surface temperature rise inside the SIM perlite roof was much slower than that inside the PCM paraffin roof during the daytime under the influence of solar radiation. Hence, as displayed in Figure 11, it costs more energy for heating the room of a SIM roof using the air conditioner than that of a PCM roof.

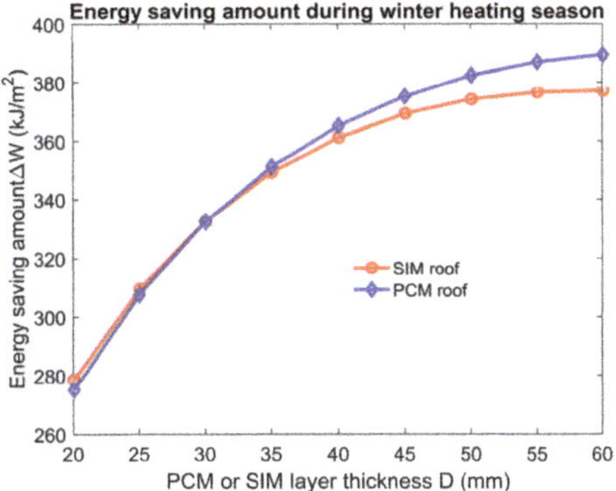

Figure 11. The effect of layer thickness on the energy saving amount of PCM or SIM roofs during the winter heating season.

From the above discussions, due to the effect of latent heat, the PCM layer could modulate the temperature inside the building roof and improve the thermal comfort in the room and its energy saving capability was also better than the SIM roof if the appropriate PCM melting temperature and PCM layer thickness were chosen. However, besides the energy saving capability of green building roofs, its incremental cost in the construction period and the incremental benefit during the operational period are significant in practical engineering. The comprehensive incremental benefits of a paraffin PCM roof and a perlite SIM roof during life cycle with different PCM or SIM layer thicknesses are shown in Figure 12. The data indicate that the perlite SIM roof has a positive comprehensive incremental benefit for all the different SIM layer thicknesses. When the SIM layer with a thickness of $D = 60$ mm was used, the SIM roof achieved the maximum comprehensive incremental benefit of 3110 yuan (6 m × 6 m roof). On the contrary, the comprehensive incremental benefit of a paraffin PCM roof was always negative, which indicates that the paraffin PCM roof could not have an economic benefit so that it is not proper for the practical engineering application at the current stage. The reason for the huge difference between the comprehensive incremental benefit of a paraffin PCM roof and a perlite SIM roof was that the material price of paraffin is 24.14 times more than that of perlite as shown in Table 1. In this case, developing cheap PCMs with stable chemical properties is a key scientific bottleneck for wide engineering applications of PCM envelope building in recent future.

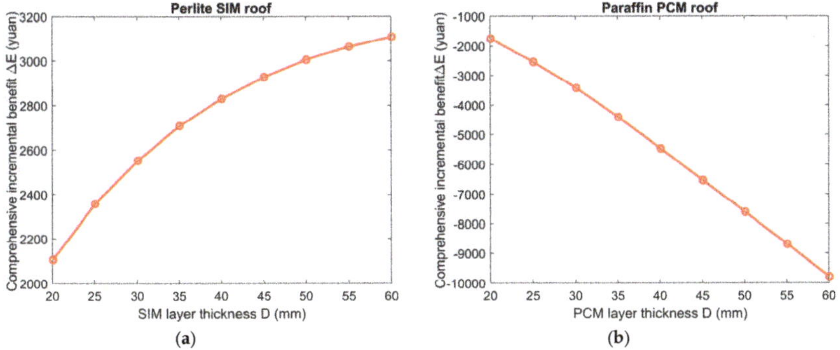

Figure 12. The comprehensive incremental benefit of a SIM roof and a PCM roof during the life cycle.

5. Conclusions

In this paper, an enthalpy-based multiple-relaxation-time (MRT) lattice Boltzmann method (LBM) was developed with GPU acceleration to investigate the conjugate heat transfer and solid-liquid phase change inside the PCM roof. To demonstrate the advantage of a PCM roof with latent heat properties for improving the building's thermal comfort, the energy saving characteristics of a PCM roof and a SIM roof under different conditions were compared. It was found that the latent heat of PCM had a trade-off effect on the energy saving characteristics of a building's roof. Under this circumstance, an optimum PCM melting temperature for the PCM roof exists to achieve the maximum energy saving amount. Besides, the PCM or SIM layer thickness also had obvious effects on the thermal performance of the PCM or SIM roof. The energy saving amount of the SIM roof increased linearly with the increasing SIM layer thickness due to the enhanced thermal resistance. However, the increasing rate for the PCM roof energy saving amount decreased with the increment of the PCM layer thickness because of the trade-off influences of latent heat during the daytime and nighttime. In general, a PCM roof could exhibit a better thermal performance than the SIM roof for reducing energy consumption when the appropriate melting temperature and layer thickness are chosen. Unfortunately, due to the high price of PCMs, the comprehensive incremental benefit of a PCM roof is much lower than that of a SIM roof. To make the wide application of PCM envelope building in practical engineering, the PCMs with a low price and stable chemical properties should be developed in future scientific research.

Author Contributions: Conceptualization, B.S. and Q.R.; Methodology, X.D.; Validation, X.D.; Formal Analysis, X.D.; Investigation, X.D.; Writing-Original Draft Preparation, X.D.; Supervision, B.S.; Funding Acquisition, B.S. and Q.R.

Funding: This research was funded by the National Natural Science Foundation of China (No. 51806168) and China Post-doctoral Science Foundation (No. 2017M623169).

Conflicts of Interest: The authors declare no conflict of interest.

References

1. Erdem, C.; Young, C.H.; Saffa, B.R. Performance investigation of heat insulation solar glass for low-carbon buildings. *Energy Convers. Manag.* **2014**, *88*, 834–841.
2. Kong, X.; Lu, S.; Li, Y.; Huang, J.; Liu, S. Numerical study on the thermal performance of building wall and roof incorporating phase change material panel for passive cooling application. *Energy Build.* **2014**, *81*, 404–415. [CrossRef]
3. Li, D.; Zheng, Y.; Liu, C.; Wu, G. Numerical analysis on thermal performance of roof contained PCM of a single residential building. *Energy Convers. Manag.* **2015**, *100*, 147–156. [CrossRef]
4. Baetens, R.; Jelle, B.P.; Gustavsen, A. Phase change materials for building applications: A state-of-the-art review. *Energy Build.* **2010**, *42*, 1361–1368.
5. Zhou, D.; Zhao, C.Y.; Tian, Y. Review on thermal energy storage with phase change materials (PCMs) in building applications. *Appl. Energy* **2012**, *92*, 593–605.
6. Akeiber, H.; Nejat, P.; Majid, M.Z.A.; Wahid, M.A.; Jomehzadeh, F.; Famileh, I.Z.; Calautit, J.K.; Hughes, B.R.; Zaki, S.A. A review on phase change material (PCM) for sustainable passive cooling in building envelopes. *Renew. Sustain. Energy Rev.* **2016**, *60*, 1470–1497. [CrossRef]
7. Al-Saadi, S.N.; Zhai, Z.J. Modeling phase change materials embedded in building enclosure: A review. *Renew. Sustain. Energy Rev.* **2013**, *21*, 659–673. [CrossRef]
8. Izquierdo-Barrientos, M.A.; Belmonte, J.F.; Rodríguez-Sánchez, D.; Molina, A.E.; Almendros-Ibáñez, J.A. A numerical study of external building walls containing phase change materials (PCMs). *Appl. Therm. Eng.* **2012**, *47*, 73–85. [CrossRef]
9. Zwanzig, S.D.; Lian, Y.; Brehob, E.G. Numerical simulation of phase change material composite wallboard in a multi-layered building envelope. *Energy Convers. Manag.* **2013**, *69*, 27–40. [CrossRef]
10. Jin, X.; Medina, M.A.; Zhang, X. Numerical analysis for the optimal location of a thin PCM layer in frame walls. *Appl. Therm. Eng.* **2016**, *103*, 1057–1063. [CrossRef]
11. Xie, J.; Wang, W.; Liu, J.; Pan, S. Thermal performance analysis of PCM wallboards for building application base on numerical simulation. *Sol. Energy* **2018**, *162*, 533–540. [CrossRef]
12. Tokuç, A.; Başaran, T.; Yesügey, S.C. An experimental and numerical investigation on the use of phase change materials in building elements: The case of a flat roof in Istanbul. *Energy Build.* **2015**, *102*, 91–104. [CrossRef]

13. Liu, C.; Zhou, Y.; Li, D.; Meng, F.; Zheng, Y.; Liu, X. Numerical analysis on thermal performance of a PCM-filled double glazing roof. *Energy Build.* **2016**, *125*, 267–275. [CrossRef]

14. Chen, S.; Doolen, G.D. Lattice Boltzmann method for fluid flows. *Annu. Rev. Fluid Mech.* **1998**, *30*, 329–364. [CrossRef]

15. Aidun, C.K.; Clausen, J.R. Lattice-Boltzmann method for complex flows. *Annu. Rev. Fluid Mech.* **2010**, *42*, 439–472. [CrossRef]

16. Miller, W. The lattice Boltzmann method: A new tool for numerical simulation of the interaction of growth kinetics and melt flow. *J. Cryst. Growth* **2001**, *230*, 263–269. [CrossRef]

17. Rasin, I.; Miller, W.; Succi, S. Phase-field lattice kinetic scheme for the numerical simulation of dendritic growth. *Phys. Rev. E* **2005**, *72*, 066705. [CrossRef] [PubMed]

18. Huang, R.; Wu, H. An immersed boundary-thermal lattice Boltzmann method for solid-liquid phase change. *J. Comput. Phys.* **2014**, *277*, 305–319. [CrossRef]

19. Jiaung, W.S.; Ho, J.R.; Kuo, C.P. Lattice Boltzmann method for the heat conduction problem with phase change. *Numer. Heat Transf. Part B Fundam.* **2001**, *39*, 167–187.

20. Eshraghi, M.; Felicelli, S.D. An implicit lattice Boltzmann model for heat conduction with phase change. *Int. J. Heat Mass Transf.* **2012**, *55*, 2420–2428. [CrossRef]

21. Huang, R.; Wu, H.; Cheng, P. A new lattice Boltzmann model for solid-liquid phase change. *Int. J. Heat Mass Transf.* **2013**, *59*, 295–301. [CrossRef]

22. Huang, R.; Wu, H. Phase interface effects in the total enthalpy-based lattice Boltzmann model for solid-liquid phase change. *J. Comput. Phys.* **2015**, *294*, 346–362. [CrossRef]

23. Li, D.; Ren, Q.; Tong, Z.X.; He, Y.L. Lattice Boltzmann models for axisymmetric solid-liquid phase change. *Int. J. Heat Mass Transf.* **2017**, *112*, 795–804. [CrossRef]

24. Li, D.; Tong, Z.X.; Ren, Q.; He, Y.L.; Tao, W.Q. Three-dimensional lattice Boltzmann models for solid-liquid phase change. *Int. J. Heat Mass Transf.* **2017**, *115*, 1334–1347. [CrossRef]

25. Ren, Q.; Chan, C.L. Numerical study of double-diffusive convection in a vertical cavity with Soret and Dufour effects by lattice Boltzmann method on GPU. *Int. J. Heat Mass Transf.* **2016**, *93*, 538–553. [CrossRef]

26. Ren, Q.; Meng, F.; Guo, P. A comparative study of PCM melting process in a heat pipe-assisted LHTES unit enhanced with nanoparticles and metal foams by immersed boundary-lattice Boltzmann method at pore-scale. *Int. J. Heat Mass Transf.* **2018**, *121*, 1214–1228. [CrossRef]

27. Ren, Q. Investigation of pumping mechanism for non-Newtonian blood flow with AC electrothermal forces in a microchannel by hybrid boundary element method and immersed boundary-lattice Boltzmann method. *Electrophoresis* **2018**, *39*, 1329–1338. [CrossRef] [PubMed]

28. Ren, Q.; He, Y.L.; Su, K.Z.; Chan, C.L. Investigation of the effect of metal foam chracteristics on the PCM melting performance in a latent heat thermal energy storage unit by pore-scale lattice Boltzmann modelling. *Numer. Heat Transf. Part A Appl.* **2017**, *72*, 745–764. [CrossRef]

29. Ren, Q. Bioparticle delivery in physiological conductivity solution using AC electrokinetic micropump with castellated electrodes. *J. Phys. D Appl. Phys.* **2018**. [CrossRef]

30. Ren, Q.; Meng, F.; Chan, C.L. Cell transport and suspension in high conductivity electrothermal flow with negative dielectrophoresis by immersed boundary-lattice Boltzmann method. *Int. J. Heat Mass Transf.* **2019**, *128*, 1129–1244. [CrossRef]

31. Ren, Q.; Chan, C.L. Natural convection with an array of solid obstacles in an enclosure by lattice Boltzmann method on a CUDA computation platform. *Int. J. Heat Mass Transf.* **2016**, *93*, 273–285. [CrossRef]

32. Ren, Q.; Chan, C.L. GPU accelerated numerical study of PCM melting process in an enclosure with internal fins using lattice Boltzmann method. *Int. J. Heat Mass Transf.* **2016**, *100*, 522–535. [CrossRef]

33. Jany, P.; Bejan, A. Scaling theory of melting with natural convection in an enclosure. *Int. J. Heat Mass Transf.* **1988**, *31*, 1221–1235. [CrossRef]

34. Li, J.; Tian, Z. Incremental cost-benefit of life cycle green buildings. *J. Eng. Manag.* **2011**, *5*, 004.

35. Ballim, Y. A numerical model and associated calorimeter for predicting temperature profiles in mass concrete. *Cem. Concr. Compos.* **2004**, *26*, 695–703. [CrossRef]

Article

Nucleation Triggering of Highly Undercooled Xylitol Using an Air Lift Reactor for Seasonal Thermal Energy Storage

Marie Duquesne [1,*], Elena Palomo Del Barrio [2] and Alexandre Godin [3]

[1] Bordeaux INP, CNRS, I2M Bordeaux, ENSCBP, 16 avenue Pey Berland, 33607 Pessac CEDEX, France
[2] CIC EnergiGUNE, Parque Tecnológico de Álava, Albert Einstein, 48. Edificio CIC, 01510 Miñano, Álava, Spain; epalomo@cicenergigune.com
[3] Université de Bordeaux, CNRS, I2M Bordeaux, Esplanade des Arts et Métiers, F-33405 Talence CEDEX, France; alexandregodin1@hotmail.com
* Correspondence: marie.duquesne@enscbp.fr; Tel.: +33-5-40-00-66-64

Received: 21 December 2018; Accepted: 11 January 2019; Published: 14 January 2019

Featured Application: A potential application of this work consists in using Xylitol as phase change material for latent heat thermal energy storage at temperatures varying from ambient up to slightly below 100 °C.

Abstract: Bio-based glass-forming materials are now considered for thermal energy storage in building applications. Among them, Xylitol appears as a biosourced seasonal thermal energy storage material with high potential. It has a high energy density and a high and stable undercooling, thus allowing storing solar energy at ambient temperature and reducing thermal losses and the risk of spontaneous nucleation (i.e., the risk of losing the stored energy). Generally when the energy is needed, the discharge triggering of the storage system is very difficult as well as reaching a sufficient power delivery. Both are indeed the main obstacles for the use of pure Xylitol in seasonal energy storage. Different techniques have been hence considered to crystallize highly undercooled Xylitol. Nucleation triggering of highly undercooled pure Xylitol by using an air lift reactor has been proven here. This method should allow reaching performances matching with building applications (i.e., at medium temperatures, below 100 °C). The advantages of this technique compared to other existing techniques to activate the crystallization are discussed. The mechanisms triggering the nucleation are investigated. The air bubble generation, transportation of nucleation sites and subsequent crystallization are discussed to improve the air injection operating conditions.

Keywords: energy discharge; bubbles burst; bubbles transportation; crystal growth rates; undercooling

---

1. Introduction

One of the key elements to optimize the use of renewable energies and to improve building performances is the development of thermal energy storage [1–5]. To this purpose, latent heat storage using phase change materials (PCMs) is known to provide a greater energy density than the sensible heat storage with a smaller temperature difference between the energy charge and the energy discharge. Cabeza et al. [6] reviewed many of PCM complying with this application as well as the eventual associated technical problems as separation, segregation, corrosion, and materials compatibility. Among inorganic systems such as salts and salt hydrates, organic compounds such as paraffin waxes, fatty acids or esters and polymeric materials [3–5], sugar alcohols (SAs) are promising phase change materials for seasonal energy storage applications at low to medium temperatures (below 100 °C) and allow avoiding these technical problems.

These bio-based glass-forming materials have been up to now widely studied in food engineering and pharmaceutics [7,8]. Recently, they have been also studied for thermal energy storage applications [9–14]. The idea consists in storing thermal energy into undercooled sugar alcohols at a temperature inferior to their melting one, as close as possible to the ambient temperature, to reduce thermal losses and to limit the risk of spontaneous energy discharge. When heating is needed, the storage system is discharged by triggering sugar alcohols nucleation, which is the first step of crystallization.

In spite of the fact that Xylitol has been identified as one of the most promising candidates among sugar alcohols for applications at temperature below 200 °C due to its high energy storage potential, few studies only have been conducted on it, as reported in [15,16]. Its melting point of 95 °C, its latent heat superior to 263 J·g^{-1} and its total energy density 4–5 times higher than that of water (110–150 kWh·m^{-3} whereas it is approximately 30 kWh·m^{-3} for water on a seasonal basis) are well suitable for their use in cheap solar collectors for instance. However, the strong undercooling behavior of Xylitol as well as its nucleation triggering which is very difficult due to the very high activation energy required for atomic rearrangements at the liquid–solid interface (i.e., the energy barrier to trigger the nucleation) are seen as the major obstacles to their employment as PCM. A decrease in temperature causes indeed a drastic increase in viscosity making the atom diffusion slow down or even stop. This is the reason why [15], among others, stipulated the impossibility to solidify pure Xylitol once at liquid state and thus, stopped their investigation about this material.

In the case of undercooled Xylitol, the crystallization step is thus a challenging process since it has a very high viscosity at ambient temperature (see the evolution of its viscosity versus temperature in [17]). Indeed, the transition from liquid to solid comes with a conformation change (from a linear structure in the liquid phase, to a bent one in the solid phase) which increases the required energy to overcome the activation energy barrier. These are the reasons why Xylitol crystal growth rates are low at ambient temperature. The rate values were measured and calculated according to temperature and presented in [18]. The maximum growth rates appear around 50–55°C and are of the order of magnitude of μm·s^{-1}.

In this work, it shown that these characteristics are, on the one part, an asset for a medium/long term storage. Indeed, the probability for spontaneous nucleation is negligible at high undercoolings which leads to a negligible risk of losing the stored energy and limited thermal losses. On the other part, the nucleation triggering, i.e., of the discharge triggering when the energy is needed, is very difficult and the crystal growth rates are too low for an efficient power delivery. Consequently, if these two main obstacles are overcome, Xylitol would be a very competitive phase change material for seasonal storage applications at low to medium temperatures, particularly in building field. To this end, an air lift reactor was designed and the proof of concept enabling to reach performances matching with building applications is described in [19].

This work aims at comparing the advantages of using an air lift reactor with the other potential techniques able to activate highly undercooled pure Xylitol nucleation and to accelerate its complete crystallization. These latter techniques includes local cooling, intentional seeding, ultrasonication, and mechanical agitation.

A preliminary parametric study in order to identify the key variables influencing these mechanisms is performed. To do so, air bubble generations, transportations of nucleation sites and subsequent crystallizations are observed and discussed. This understanding should help to identify the parameters triggering nucleation at any time (or temperature) and to crystallize the entire product in due time.

2. Attempts of Nucleation Triggering and of Discharge at an Acceptable Power

2.1. Materials

Xylitol was purchased from Roquette (batch E089X, purity 98.43%, Lestrem, France). In all the performed experiments, molten Xylitol is conditioned as follows: Xylitol is placed at 100 °C overnight in an oven to be fully melted, then, it is free-cooled down to the starting temperatures (ranging from 30 °C to 75 °C).

2.2. Xylitol-Conscious Nucleation and Crystallization: Commonly Used Techniques

The most common used techniques to crystallize Xylitol have been considered and the preliminary observations and analyses carried out are presented hereafter.

1. Local cooling. It fails to trigger nucleation in severely undercooled Xylitol due to the very high activation energy required for atoms diffusion and rearrangement at the solid–liquid interface.
2. Intentional seeding. It allows triggering Xylitol crystallization, even in cases with very high undercooled melts. However, the effect of seeding on nucleation is too local. The seed growth being too slow (4 days/10 mL), this technique will lead to very low heat release rates and too long discharge times for the aimed application.
3. Ultrasonication. High-power ultrasonification (450 W) allows also crystallizing the studied Xylitol. However, the crystallization rates are still too low for this technique to become appropriate at the storage system scale.
4. Solvents addition in Xylitol. It does not contribute to accelerate crystallization even in combination with ultrasonification.

2.3. Successful Xylitol Nucleation and Crystallization: Preliminary Tests

Two techniques only succeeds in activating the nucleation of severely undercooled melts in extended area with then, significantly increased crystallization rates of the studied systems.

The first one is the mechanical agitation and the second one consists in stirring undercooled Xylitol by bubbling. Both of them aim at triggering Xylitol nucleation and at speeding-up its crystallization.

2.3.1. Undercooled Xylitol Crystallization Induced by Mechanical Agitation

In industrial crystallization, mechanical agitation is often used in combination with seeding. The primary objectives of this method are to create and maintain a good dispersion of the seeds in the bulk and to enhance the rate of mass transfer between the solid and liquid phases [19]. Mechanical agitation without seeding has been here considered to crystallize undercooled Xylitol.

A glass beaker with a diameter of 85 mm and a height of 105 mm is filled with Xylitol initially in powder form and placed into an oven up to its molten liquid state. Small volumes (~400 mL) of undercooled Xylitol are used for testing. Then, the beaker is put out of the oven and remains at room temperature to make it naturally cool down to the selected starting temperature. Finally, the mixer is started with a power of 125 W at a constant and low speed and for a short duration of 5 s. This allows the agitation of the undercooled Xylitol.

The form and the position of the mixer within the undercooled Xylitol can be seen in Figure 1. The crystallization of the sample is observed with a high speed CCD camera (HM1024, Genie, Teledyne DALSA, Waterloo, ON, Canada). Simultaneously, a thermocouple glued to the wall of the beaker provides useful information regarding the sample temperature evolution with time. Several starting temperatures were investigated.

Figure 1 illustrates the device used for mechanical agitation and the snapshots in Figure 2 show an example of tests performed on highly undercooled melts at a starting temperature of 50 °C (T0 = 50 °C), i.e., with an undercooling of 45 °C (ΔT = 45 °C) regarding the melting temperature of 95 °C.

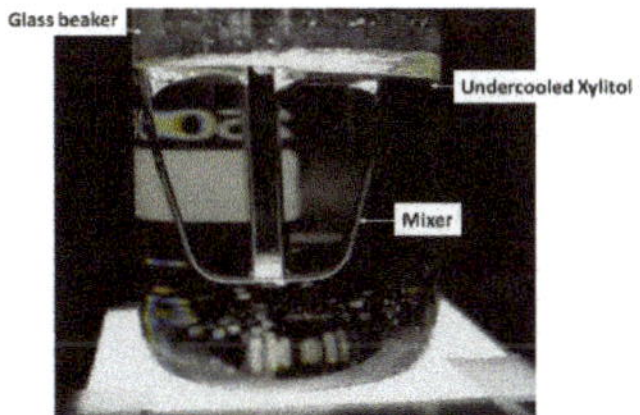

Figure 1. Picture of the mixer within a sample of undercooled Xylitol in a glass beaker at a starting temperature T0 = 50 °C (ΔT = 45 °C) before mechanical agitation.

It has been observed that mechanical agitation results in a very fast response. Indeed, nucleation is reached in only 5 s of mixing all over the agitated region. Then, the crystallization progresses downwards from the agitated region. As shown in Figure 2, crystal clusters detach from the crystallization front and fall by gravity, thus contributing to accelerate the process of crystallization. After 1 min 30 s an increase in temperature of the beaker exterior wall is measured (25 °C).

Figure 2. Snapshots of an undercooled Xylitol sample in a glass beaker at a starting temperature of T0 = 50 °C (ΔT = 45 °C) after mechanical agitation. Observation of its crystallization after 5, 92, and 215 s.

The obtained results suggest that mechanical agitation could be an efficient solution for discharging the storage system when necessary and at appropriate speeds. However, this technique shows a solidification which is less homogeneous from a spatial viewpoint than the bubbling one (Figure 2 versus Figure 3). Besides, the mechanical agitation is intrusive, and would require a specific reactors design and would probably lead to a significant extra-cost.

2.3.2. Stirring by Bubbling

As mechanical agitation, bubble stirring allows activating the nucleation of highly undercooled Xylitol in extended area. To this purpose, bubble columns are widely employed within chemical industry as gas–liquid contactors and multiphase reactors. Examples of applications of this reactor type include oxidations, hydrogenations, fermentations, and the production of synthetic fuels [20]. One of the main features of bubble column operation is that gas and liquid or suspended solid phases are brought in contact without the need for additional mechanical stirring equipment, making bubble column design and operation appear easier than that of other gas–liquid reactors. The gas distributor is usually located at the bottom of the column, while the liquid phase can either be distributed co-currently or counter-currently with respect to the flow direction of the gas phase.

Bubbling with air has been considered as an alternative to mechanical agitation to induce Xylitol crystallization. Contrary to mechanical agitation, bubbling is a low-intrusive technique, consequently, easy to implement in standard storage containers and hence cost effective [21].

The same previous conditions (glass beaker, Xylitol and protocol to prepare the undercooled Xylitol) have been used. Only the technique of nucleation triggering is different from the previous tests. The air distributor is a flexible pipe of 3 mm inner diameter which is connected to a small pump. The air inlet is located at the bottom of the column and the air flow is of 80 L·h^{-1}.

After 30 min, an increase in Xylitol temperature is measured (35 °C) within the beaker. The snapshots in Figure 3 show that air injection at the bottom of the beaker fosters Xylitol nucleation as well the crystallization progress over time in a complete different way than that from the one obtained using mechanical agitation. In this case, crystals form around air bubbles all over the stirred area.

Figure 3. Snapshots of the undercooled Xylitol crystallization induced by air bubbling in a glass beaker at different times (air flow rate = 80 L/h; Starting Temperature T0 = 50 °C, undercooling degree ΔT = 45 °C).

The possibility to induce the undercooled Xylitol crystallization by bubbling using a simple and cost effective technique has been proven through this experiment. The further researches carried out in order i/ to understand the effect of bubbling on crystallization, ii/ to identify the key related variables and iii/ to improve air injection operating conditions are presented in next section.

3. Undercooled Xylitol Crystallization Induced by Bubbling

A significant amount of research works on bubble columns regarding bubble formation and bubble rise velocity aspects, flow patterns characteristics and parameters, heat and mass transfers, columns design, and operating conditions has appear in literature in the last few decades (see e.g., recent reviews [22–24]), including experimental and numerical studies.

However, to the best of knowledge, the use of bubbling to trigger the nucleation of undercooled melts and to accelerate their crystallization rates has never been studied.

3.1. Air Lift Reactors

Two different liquid/gas reactors have been used to achieve an overall understanding of the Xylitol crystallization by bubbling, as can be seen in Figure 4:

- *Reactor n°1.* The reactor is a cylindrical glass beaker with a diameter of 85 mm and a height of 105 mm, filled with the molten sugar alcohol (SA) up to 60 mm of height. Air bubbles are generated at the bottom of the vessel by a single tube connected to a pump. The inner diameter of the tube is 3 mm. The pump can provide an air flow rate of 80 L·h^{-1}.
- *Reactor n°2.* The reactor is a cylindrical glass crystallizer with a diameter of 140 mm diameter and a height of 105 mm, filled with the molten SA up to 60 mm of height. As in reactor n°1, air bubbles are generated at the bottom of the vessel by a single tube (3 mm inner diameter) connected to a pump (80 L·h^{-1}).

Figure 4. Pictures of the 3 used air lift reactors (diameter × height mm): n°1 (85 × 105 mm) and n°2. (140 × 105 mm).

The same preparation protocol described in Section 2 is applied to Xylitol. Once the desired starting temperature is reached, the bubbling is started and its impact on the SA crystallization process is visualized by using a high speed CDD camera.

3.2. Bubbles Observation into Molten Xylitol

Numerous experiments have been performed in order to get an overall view of the crystallization process by bubbling. Most of them were carried out in reactor n°1. The air flow is still of 80 L·h^{-1} and the starting temperature of the melt (To) varies from 30 °C to 75 °C. The snapshots in Figure 3 show the different stages of bubbling from the side whereas the top views are shown in figure 4. The initiation and the progress of the crystallization in the experiment carried out at 50 °C (i.e., undercooling degree $\Delta T = 45$ °C) is the same whatever the applied initial thermal conditions.

The crystallization process follows the next steps:

- Bubbles are formed continuously at the submerged orifice. First they grow, detach from the orifice and rise in the melt up to the liquid free-surface. The bubbles generation is periodic and induces an oscillatory motion of the liquid, with a major circulation cell which fills the entire column. Snapshots in Figure 5 illustrates the beat of the free-surface of the liquid.
- At the free-surface of the liquid, the bubbles burst in a very short time and generate smaller bubbles (see Figure 6 (colored circles)). Part of these smaller bubbles are caught by the fluid and dragged by the flow cell as shown in Figure 6. As it can be seen in Figure 7, the small bubbles are then dispersed within the liquid and their amount increases with time.
- After an average stirring time of 5 min of the mixture "bubbles/Xylitol", Xylitol starts to crystallize. Crystallization begins on the surface of the bubble, at the interface air/Xylitol, then it progresses in the outward direction forming a spherical solid shell.

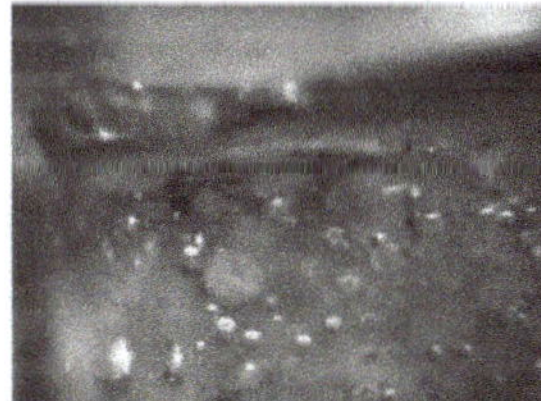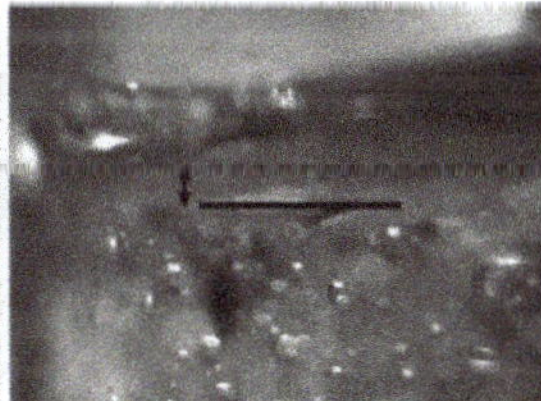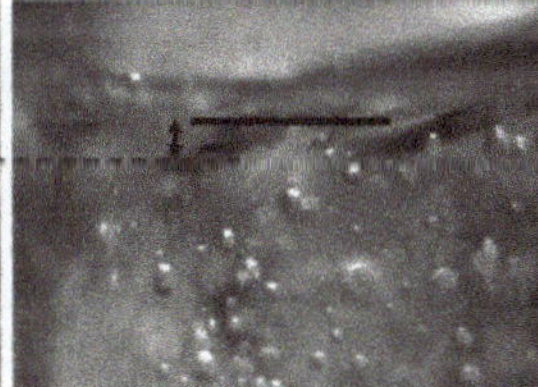

Figure 5. Oscillation (up and down motion) of the free-surface of the liquid with time produced by the periodic bubbling.

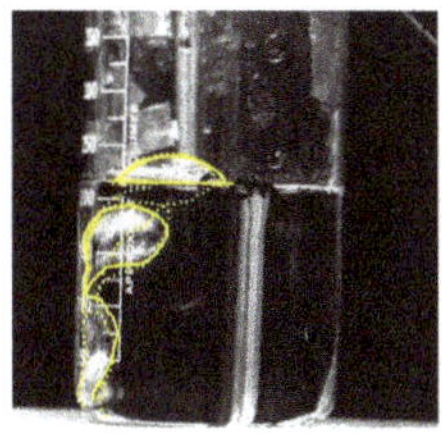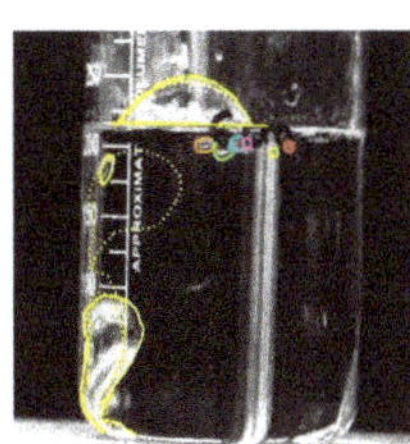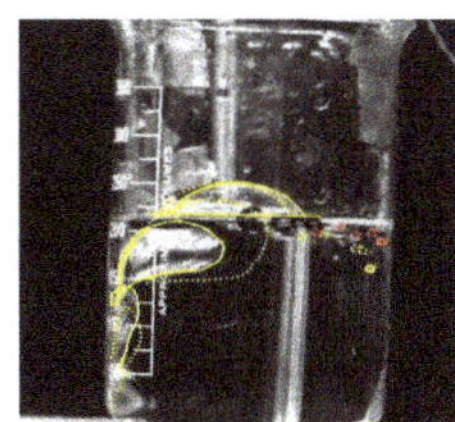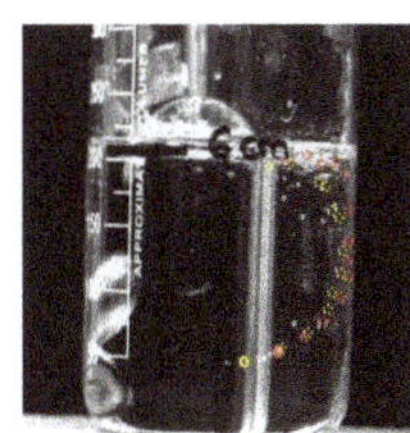

Figure 6. Pictures of the initial stages of bubbling when the small bubbles created at the free surface are captured by the liquid and entrained in its movement. Bubbles dragged by the liquid flow are highlighted by yellow and red circles.

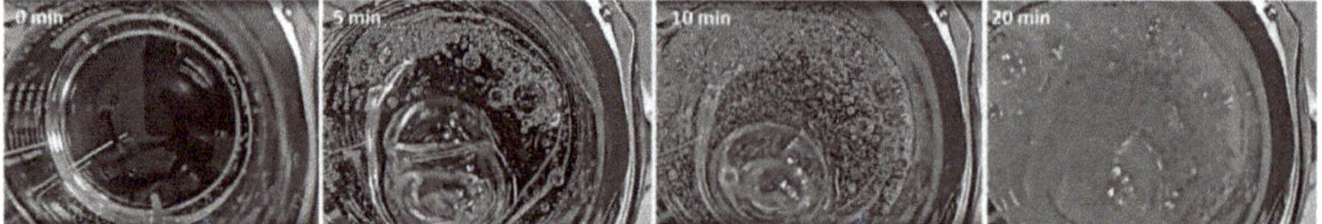

Figure 7. Top view pictures of bubbles reaching the free-surface of the liquid at different times of the crystallization process. The bubbles burst and generate small daughters.

Different stages of the crystallization process from the bubbling in the undercooled melt to a quite advanced stage of the crystallization process are presented in Figure 8a. Numerous spherical solid white particles can be observed in suspension.

Six of them with a diameter inferior to 0.5 mm have been extracted as can be seen in Figure 8b. Magnified pictures of the cross-section of two particles showing the presence of a hole in the center of each sphere are presented in Figure 8c. The hole corresponds to the air bubble on which crystallization has been initiated.

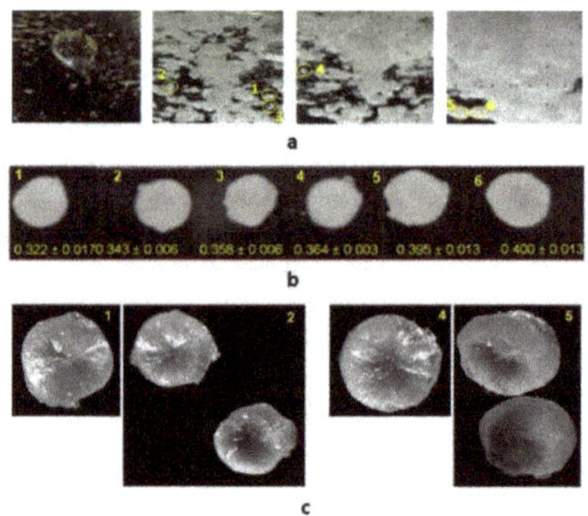

Figure 8. Xylitol crystallization pictures: (**a**) small air bubbles are dragged by the liquid circulation cell then white solid particles are formed by crystallization of Xylitol on the air bubbles surface. Finally, numerous solid particles can be observed in suspension; (**b**) Pictures of 6 particles extracted from the beaker and measured using a caliper (dimensions are given in millimeters); (**c**) magnified pictures of the cross-sections of particles cut in half.

An air injection allows the occurrence of air bubbles and of a flow cell into the molten Xylitol both observed for reactors 1 and 2. These phenomena participate to the formation and the spreading of crystallized Xylitol particles.

The bursting of the primitive air bubbles at the free liquid surface with production of small bubbles acting as nucleation sites appear, indeed, as a necessary condition for the crystallization of undercooled Xylitol.

The efficiency of the crystallization process depends thus on (i) the number of the produced nucleation sites; as well as (ii) the extent and the homogeneity of their dispersion within the melt.

Next section is consequently dedicated to the preliminary thermal study of the previous described experiments but the glass beaker filled with undercooled Xylitol is, this time, thermally insulated.

4. Preliminary Thermal Analysis as a Proof of Feasibility

Bubbling is started once the desired starting temperature is reached. The thermal response of Xylitol during bubbling is recorded by two thermocouples: one close to the tube injecting air and the other one placed on the opposite side. The air is injected at room temperature (20 °C).

A typical thermal response of high undercooled Xylitol during bubbling is in Figure 9 where three different periods can be clearly distinguished:

- *The induction period*, from the beginning of bubbling to the beginning of crystallization. In a perfectly adiabatic reactor, the temperature of the melt should be constant during this period. In practice, it slightly decreases due to thermal losses. Thermal losses take place both, by conduction through the reactor wall and by convection due to air bubbling at ambient temperature.
- *The recalescence period*, where Xylitol is crystallizing and its temperature raises. In a perfectly thermally insulated system, Xylitol should reach the melting point and the crystallization process should stop at that moment. However, due to thermal losses, the maximum temperature reached by Xylitol is inferior to its melting point and the crystallization can progress beyond this point. We consider in the following that the recalescence period ends when Xylitol reaches its maximum temperature.
- *The end-period*, where crystallization (if any) is driven by heat extraction due to thermal losses.

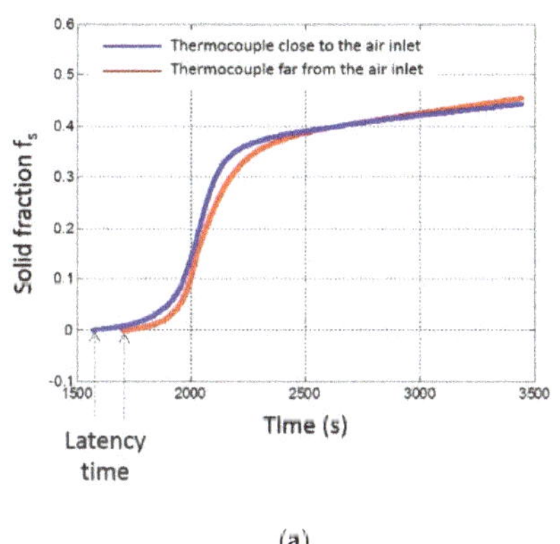

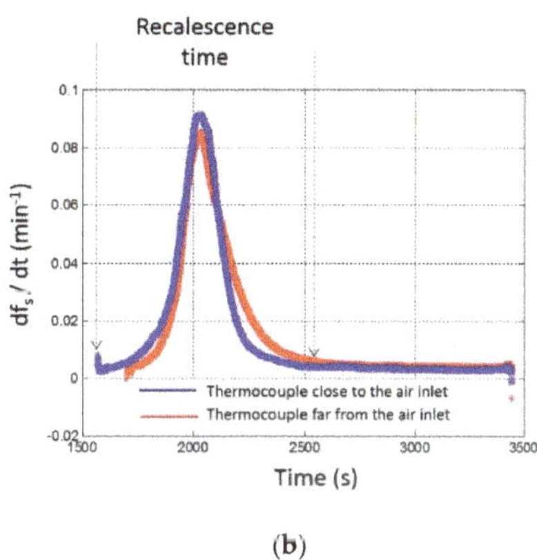

Figure 9. Time evolution of the solid fraction (**a**) and time behavior of the crystallization rate (**b**) calculated from the temperature data obtained using both thermocouples.

From the recorded temperature data, the evolution of the solid fraction over time as well as the total crystallization rate can be estimated. It has been shown in [19] that the solid fraction (f_s) can be estimated the following equation:

$$\frac{df_s}{dt} = \left(\frac{c_p}{L_m}\right)\left[\frac{dT}{dt} + k(T - T_\infty)\right] \tag{1}$$

where c_p is the specific heat of Xylitol, L_m is the latent heat and k the thermal exchange coefficient. Once, c_p, L_m and k parameters are known, the crystallization rate (df_s/dt) can be determined.

An example of the estimated time evolution of the crystallization rate as well the solid fraction corresponding to temperature data recorded by both thermocouples are depicted in Figure 9.

It can be seen that the crystallization progresses slowly at the beginning, then accelerates up to a maximum crystallization rate, and the rhythm of crystal production reduces afterwards. At the end of the recalescence period ($t \approx 2500s$ in Figure 9), the crystallization rate is such that power delivered by solidification equals the thermal losses.

These results confirm the possibility of using Xylitol as phase change material for seasonal thermal energy storage despite its high viscosity and by using a simple, non-intrusive and thus cost-effective bubbling technique.

Further works will consist in studying the bubbling performances for thermal storage applications. Quantitative analyses will be performed to estimate the thermal properties influencing the discharge process and to evaluate the reachable performances.

5. Conclusions

Xylitol high and stable undercooling allows long-term thermal energy storage at low to medium temperatures with reduced thermal losses and with negligible risk of spontaneous energy discharge. However, its nucleation triggering (energy discharge triggering) and hence, its subsequent crystallization rates (discharge power delivery) have been recognized as major obstacles to its use as PCM for applications below 200 °C such as building applications.

Using an air lift reactor is a very promising technique to discharge the storage system at the required power when needed.

In this paper, the proof of the concept was achieved by using pure Xylitol, i.e. without adding any additives. This paper provides a better understanding of the effect of primitive bubbles burst and small bubbles transportation on nucleation triggering of the whole highly undercooled Xylitol in a small container. Indeed, it has been shown that Xylitol nucleation occurs on the surface of the small bubbles generated at the liquid surface when primitive bubbles burst. Once dispersed within the reactor by the liquid flow, they act as nucleation sites. The bursting of the primitive bubbles at the free liquid surface with production of smaller bubbles is hence a necessary condition for Xylitol crystallization. As mechanical agitation, stirring by bubbling is able to trigger the nucleation of Xylitol in extended and highly spatially resolved areas. The efficiency of the crystallization process will be obviously dependent on the number of nucleation produced sites as well as the extent and the homogeneity of their dispersion within the melt. Besides, this technique is a low intrusive technique and, consequently, easy to implement in standard storage containers. The feasibility of nucleation triggering thanks to an air lift reactor bubble stirring being proven in this paper, further works will consist in providing an in-depth understanding of highly undercooled Xylitol crystallization induced by air injection, including bubbles formation and characteristics, nucleation sites generation, nucleation sites dispersion, and crystallization on the liquid/gas interfaces. Further works will also consists in studying the air injection performances for thermal energy storage applications to pave the way for bubbling conditions optimization.

Author Contributions: All authors contributed to this work. Conceptualization, M.D., A.G. and E.P.D.B.; methodology, M.D. and A.G.; investigation, M.D., A.G. and E.P.D.B.; data curation, M.D., A.G. and E.P.D.B.; writing—original draft preparation, E.P.D.B.; writing—review and editing, M.D. and A.G.; visualization, M.D. and A.G.; supervision, E.P.D.B.; project administration, E.P.D.B.

Funding: This research was funded by the European Community's Seventh Framework Program (FP7/2007-2013), grant number 296006.

Acknowledgments: The research leading to these results has received funding from the European Community's Seventh Framework Program (FP7/2007-2013) under grant agreement 296006. The authors acknowledge them as well the financial support of Region Nouvelle Aquitaine for subsidizing BioMCP project (Project-2017-1R10209-13023).

Conflicts of Interest: The authors declare no conflict of interest. The funders had no role in the design of the study; in the collection, analyses, or interpretation of data; in the writing of the manuscript, or in the decision to publish the results.

References

1. Agyenim, F.; Hewitt, N.; Eames, P.; Smyth, M. A review of materials, heat transfer and phase change problems formulation for latent heat thermal energy storage systems. *Renew. Sustain. Energy Rev.* **2010**, *14*, 615–628. [CrossRef]
2. Duquesne, M.; Toutain, J.; Sempey, A.; Ginestet, S.; del Barrio, E.P. Modeling of a non-linear thermochemical energy storage by adsorption on zeolites. *Appl. Therm. Eng.* **2014**, *71*, 469–480. [CrossRef]

3. Zalba, B.; Marin, J.M.; Cabeza, L.F.; Mehling, H. Review on thermal energy storage with phase change: Materials, heat transfer analysis and applications. *Appl. Therm. Eng.* **2003**, *23*, 251–283. [CrossRef]

4. Yuan, Y.; Zhang, N.; Tao, W.; Cao, X.; He, Y. Fatty acids as phase change materials: A review. *Renew. Sustain. Energy Rev.* **2014**, *29*, 482–498. [CrossRef]

5. D'Avignon, K.; Kummert, M. Experimental assessment as a phase change material storage tank. *Appl. Therm. Eng.* **2016**, *99*, 880–891. [CrossRef]

6. Cabeza, L.F.; Castell, A.; Barreneche, C.; de Garcia, A.; Fernandèz, A.I. Materials used as PCM in thermal energy storage in buildings: A review. *Renew. Sustain. Energy Rev.* **2011**, *15*, 1675–1695. [CrossRef]

7. Nguyen, P.T.N.; Ilrich, J. Sugar alcohols—Multifunctional agents in the freeze casting process of foods. *J. Food Eng.* **2015**, *153*, 1–7. [CrossRef]

8. Fujii, K.; Izutsu, K.I.; Kume, M.; Yoshino, T.; Yoshihashi, Y.; Sugano, K.; Terada, K. Physical characterization of meso-erythritol as a crystalline bulking agent for freeze-dried formulations. *Chem. Pharm. Bull.* **2015**, *63*, 311–317. [CrossRef]

9. SAM.SSA Project. *Sugar Alcohol based Materials for Seasonal Storage Applications*; ENER/FP7/296006; 2012–2015. Available online: https://cordis.europa.eu/project/rcn/103643/factsheet/en (accessed on 11 January 2019).

10. Solé, A.; Neumann, H.; Niedermaier, S.; Martorell, I.; Schossig, P.; Cabeza, L.F. Stability of sugar alcohols as PCM for thermal energy storage. *Sol. Energy Mater. Sol. Cells* **2014**, *126*, 125–134. [CrossRef]

11. Nomura, T.; Zhu, C.; Sagara, A.; Okinaka, N.; Akiyama, T. Estimation of thermal endurance of multicomponent sugar alcohols as phase change materials. *Appl. Therm. Eng.* **2015**, *75*, 481–486. [CrossRef]

12. Ona, E.P.; Zhang, X.; Kyaw, K.; Watanabe, F.; Matsuda, H.; Kakiuchi, H.; Yabe, M.; Chihara, S. Relaxation of Supercooling of Erythritol for Latent Heat Storage. *J. Chem. Eng. Jpn.* **2001**, *34*, 376–382. [CrossRef]

13. Gunasekara, S.N.; Pan, R.; Chiu, J.N.; Martin, V. Polyols as phase change materials for surplus thermal energy storage. *Appl. Energy* **2015**, *162*, 1439–1452. [CrossRef]

14. Biçer, A.; Sari, A. Synthesis and thermal energy storage properties of xylitol pentastearate and xylitol pentapalmitate as novel solid-liquid PCMs. *Sol. Energy Mater. Sol. Cells* **2012**, *102*, 215–226. [CrossRef]

15. Höhlein, S.; König-Haagen, A.; Brüggemann, D. Thermophysical characterization of $MgCl_2.H_2O$, Xylitol and Erythritol as Phase Change Materials (PCM) for Latent Heat Thermal Energy Storage (LHTES). *Materials* **2017**, *10*, 444. [CrossRef] [PubMed]

16. Guo, S.; Liu, Q.; Zhao, J.; Jui, G.; Wu, W.; Yan, J.; Li, H.; Jin, H. Mobilized thermal nergy storage: Materials, containers and economic evaluation. *Energy Convers. Manag.* **2018**, *177*, 315–329. [CrossRef]

17. Del Barrio, E.P.; Godin, A.; Duquesne, M.; Daranlot, J.; Jolly, J.; Alshaer, W.; Kouadio, T.; Sommier, A. Characterization of different sugar alcohols as phase change materials for thermal energy storage applications. *Sol. Energy Mater. Sol. Cells* **2017**, *159*, 560–569. [CrossRef]

18. Duquesne, M.; Godin, A.; del Barrio, E.P.; Achchaq, F. Crystal growth kinetics of sugar alcohols as phase change materials for thermal energy storage. *Energy Procedia* **2017**, *139*, 315–321. [CrossRef]

19. Godin, A.; Duquesne, M.; del Barrio, E.P.; Achchaq, F.; Monneyron, P. Bubble agitation as a new low intrusive method to crystallize glass-forming materials. *Energy Procedia* **2017**, *139*, 352–357. [CrossRef]

20. Vedantam, S.; Ranade, V. Crystallization: Key thermodynamic, kinetic and hydrodynamic aspects. *Sadhana* **2013**, *38*, 1287–1337. [CrossRef]

21. Gaddis, E.; Vogelpohl, A. Bubble formation in quiescent liquids under constant flow conditions. *Chem. Eng. Sci.* **1986**, *41*, 97–105. [CrossRef]

22. Kulkarni, A.A.; Joshi, J.B. Bubble formation and bubble rise velocity in gas-liquid systems: A review. *Ind. Eng. Chem. Res.* **2005**, *44*, 5873–5931. [CrossRef]

23. Abdulmouti, H. Bubbly two-phase flow: Part II—Characteristics and parameters. *Am. J. Fluids Dyn.* **2014**, *4*, 115–180.

24. Rollbusch, P.; Bothe, M.; Becker, M.; Ludwig, M. Bubble columns operated under industrially relevant conditions—Current understanding of design parameters. *Chem. Eng. Sci.* **2015**, *126*, 660–678. [CrossRef]

 *applied sciences*

Review

A Review on Nanocomposite Materials for Rechargeable Li-ion Batteries

Dervis Emre Demirocak [1,*], Sesha S. Srinivasan [2] and Elias K. Stefanakos [3]

[1] Department of Mechanical and Industrial Engineering, Texas A&M University-Kingsville, Kingsville, TX 78363, USA

[2] Department of Physics, Florida Polytechnic University, 4700 Research Way, Lakeland, FL 33805, USA; ssrinivasan@flpoly.org

[3] Clean Energy Research Center, College of Engineering, University of South Florida, Tampa, FL 33620, USA; estefana@usf.edu

* Correspondence: Dervis.Demirocak@tamuk.edu; Tel.: +1-361-5932-029

Academic Editor: Carlo Casari
Received: 15 June 2017; Accepted: 12 July 2017; Published: 17 July 2017

Abstract: Li-ion batteries are the key enabling technology in portable electronics applications, and such batteries are also getting a foothold in mobile platforms and stationary energy storage technologies recently. To accelerate the penetration of Li-ion batteries in these markets, safety, cost, cycle life, energy density and rate capability of the Li-ion batteries should be improved. The Li-ion batteries in use today take advantage of the composite materials already. For instance, cathode, anode and separator are all composite materials. However, there is still plenty of room for advancing the Li-ion batteries by utilizing nanocomposite materials. By manipulating the Li-ion battery materials at the nanoscale, it is possible to achieve unprecedented improvement in the material properties. After presenting the current status and the operating principles of the Li-ion batteries briefly, this review discusses the recent developments in nanocomposite materials for cathode, anode, binder and separator components of the Li-ion batteries.

Keywords: Li-ion batteries; nanocomposite materials; cathode; anode; binder; separator

1. Introduction

Rechargeable (rechargeable battery and secondary battery terms are used interchangeably throughout this article) batteries are one of the key technologies in today's information-rich and mobile society. For a long time, the most important application for rechargeable batteries has been the starter-ignition-lighting (SLI) batteries in motor vehicles. With the rapid market penetration of the portable electronics such as laptops, smartphones and tablets since 2000s, the rechargeable battery market significantly expanded. More recently, interest in plug-in hybrid electric vehicles (PHEV) and electric vehicles (EV) has been driving the further expansion of the rechargeable battery market, specifically Li-ion batteries. Additionally, large-scale utilization of rechargeable batteries for peak shaving and load leveling applications in the power grid has been envisioned by many, but is still in very early phases [1]. Apart from the aforementioned applications, rechargeable batteries are also critical for e-bikes and power tools [2]. Considering the importance of the rechargeable batteries in a variety of applications, the future of these batteries looks brighter than ever.

Rechargeable batteries comprise three quarters of the global battery market, and the most important rechargeable battery chemistries are lithium-ion (Li-ion), lead acid, nickel-metal-hydride (NiMH) and nickel-cadmium (Ni–Cd) as shown in Figure 1 [3].

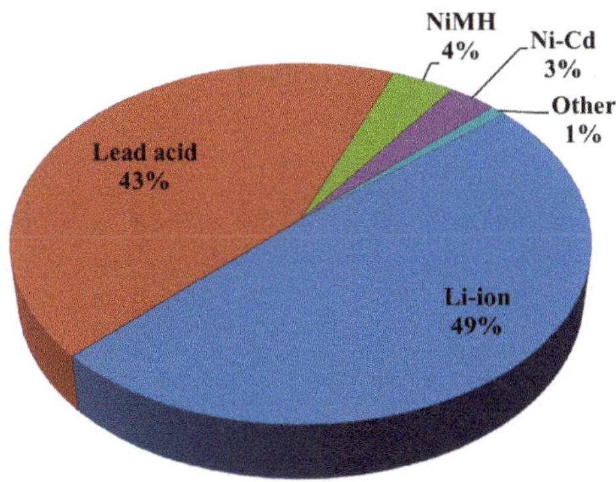

Figure 1. Market share of rechargeable battery chemistries in 2009 [3].

Li-ion and lead acid batteries dominate the market, and Li-ion is the fastest-growing market. The lead acid batteries are mostly utilized in the SLI, stationary (i.e., back-up power) and deep cycle applications, whereas Li-ion batteries are preferred in portable electronics. In recent years, Li-ion chemistry is making inroads to the lead acid market (i.e., SLI batteries), but the lead acid battery market still continues to grow due to its low cost and maturity [4]. The NiMH and Ni–Cd batteries have much smaller market share compared to lead acid and Li-ion batteries, and their share continues to shrink because Li ion batteries are rapidly replacing these types of batteries. The NiMH batteries are still utilized in some hybrid electric vehicles (HEV) such as the Toyota Prius, while Ni–Cd batteries are employed in power tools due to their high rate capability. The introduction of the Li-ion chemistries with high rate capability such as LiFePO$_4$ significantly dropped the market share of the Ni–Cd batteries. The popularity of Li-ion chemistry is associated with its high gravimetric and volumetric energy density as well as the significant cost reduction during the last decade. The main characteristics of the most common battery chemistries are summarized in Table 1 [5,6]

Table 1. Comparison of different battery chemistries [5,6].

Properties	Rechargeable Battery Chemistry			
	Li-ion	Lead Acid	NiMH	Ni–Cd
Specific energy density (Wh/kg)	90–190	30–55	60–120	45–80
Cycle life (80% of initial capacity)	500–2000	200–300	300–500	1000
Cell voltage (V)	3.3–3.8	2.0	1.2	1.2
Self-discharge/month at 25 °C	<5%	5–15%	30%	20%
Safety	Protection circuit is mandatory		Thermally stable	
Commercialized in	1991	1881	1990	1950
Toxicity	Low	High	Low	High

Despite the significant growth of the rechargeable battery market since 2000s, further improvements in energy density and cost reduction are required for EV and grid level energy storage applications. The progress in rechargeable battery technologies is rather slow, the energy density of the rechargeable batteries has only increased 3% per year on average during the last 60 years, and the rate of energy density increase has been slightly higher since the introduction of Li-ion batteries in 1991 as seen in Figure 2 [7]. However, the current rate of increase (i.e., 5–8%) in the energy density of secondary batteries is not sufficient to meet the targets set by the United States (U.S.) Department of Energy (DOE) [8].

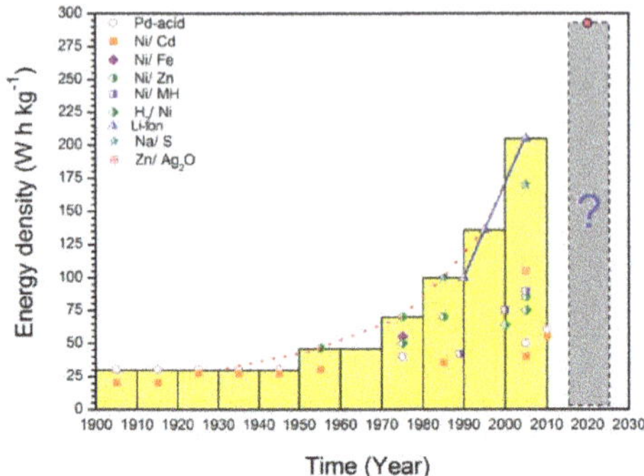

Figure 2. Increase in the energy density of the secondary batteries since 1900s. Adapted with permission from [7], © Royal Society of Chemistry, 2011.

There are two ways to improve the energy density of the secondary batteries: (i) by introducing new battery chemistries; and (ii) by employing enhanced battery designs and manufacturing processes (i.e., engineering improvements) to reduce the weight and volume of the inactive components in the existing battery chemistries. A closer look at Figure 2 shows that a quantum leap in energy density of secondary batteries was usually achieved when new battery chemistries were introduced. Therefore, it is critical to develop advanced chemistries and materials to facilitate the higher market penetration of secondary batteries.

The U.S. DOE "EV Everywhere" initiative was launched in 2012, and the energy storage targets for battery systems identified by this initiative are given in Table 2 [8]. Although significant progress has been made in improving the cost and energy density of batteries, it is estimated the targets can only be achieved by utilizing advanced Li-ion chemistries and materials.

Among the targets listed in Table 2, the cost target for the Li-ion batteries is critical to make PHEV and EV cost competitive with the gasoline powered cars. As shown in Figure 3, cost of Li-ion battery packs has significantly dropped during the last 10 years, mostly due to efforts by the EV market leaders such as Tesla (i.e., Model S) and Nissan (i.e., Leaf). The most recent cost estimate in 2016 for Li-ion battery packs is ~$190/kWh indicating that the U.S. DOE cost target for battery energy systems for 2022 is well within reach [9].

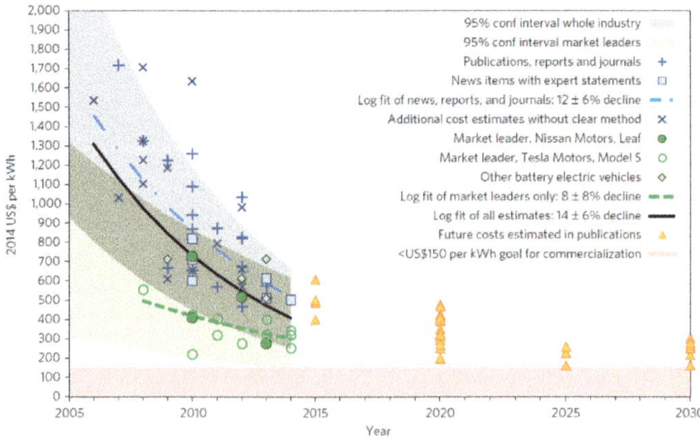

Figure 3. Cost of Li-ion battery packs in EV. Adapted with permission from [10], © Nature Publishing Group, 2015.

Table 2. The U.S. DOE "EV Everywhere" energy storage targets for battery systems [8].

Metrics	Status in 2012	Target for 2022
Battery cost ($/kWh)	500	125
Pack specific energy (Wh/kg)	80–100	250
Pack energy density (Wh/L)	200	400
Pack specific power (W/kg)	500	2000

The operation of a secondary battery relies on various interfacial processes at the nanoscale; therefore, manipulating the material properties at the nanoscale is essential for developing advanced materials. The most common strategies to improve the battery materials are particle size reduction, morphology control, composite formation, doping, functionalization, coating, encapsulation and electrolyte modification [11]. The battery electrodes (i.e., anode and cathode) that determine the overall performance of a battery are composite materials. A typical composite battery electrode includes active material, conductive carbon additive and a binder. Each of these components is vital to the operation of a rechargeable battery, and their performances can be improved by utilizing nanocomposites.

In this review article, the basics of rechargeable batteries and the key developments in nanocomposite materials are discussed with a focus on Li-ion batteries due to their technological significance. Specifically, the nanocomposite-based developments in cathode, anode, binder and separator materials for Li-ion batteries are discussed in detail. Readers can also benefit from the following review articles based exclusively on cathodes [12,13], anodes [14,15], binders [16,17] and separators [18].

2. Basics of Rechargeable Batteries

A rechargeable battery or a secondary battery is a type of battery that can be discharged and charged (i.e., cycled) multiple times over its lifetime as opposed to a primary battery, which can only be discharged one time only before being discarded. The smallest packaged unit of a battery is the cell. Most applications require higher voltages (i.e., PHEV/EV battery packs are in the range of 100–400 V, SLI batteries are ~12 V) than a single cell (i.e., ~3–4 V for Li-ion) can provide; therefore, multiple cells are assembled into modules (i.e., in a typical EV battery module there are 4 cells in a 2 series and 2 parallel formation) in parallel or in series configuration depending on the application. Battery pack is the final product formed by connecting multiple modules together. A battery pack is custom designed for a specific application and includes battery management system, sensors for voltage, current, and temperature measurements, cooling system in addition to battery modules.

A typical Li-ion rechargeable battery cell is composed of a cathode (positive electrode), an anode (negative electrode), a separator filled with an electrolyte and the current collectors for both anode and cathode. The electrodes (i.e., cathode and anode) in Li-ion batteries are composite materials, which include active materials (i.e., typically, metal oxides for cathode and graphite for anode) for storing Li ions, conductive carbon additive to improve the electronic conductivity and a binder to keep the active materials together and attached to the current collector. The cell components that are not involved in Li storage are called the inactive materials (i.e., current collector, separator, electrolyte, binder and carbon additive). The inactive components do not directly contribute to the cell capacity so it is critical to minimize their content; however, the inactive components play a remarkable role in the operation and safety of a battery. The cross-sectional view of a cylindrical Li-ion battery is given in Figure 4.

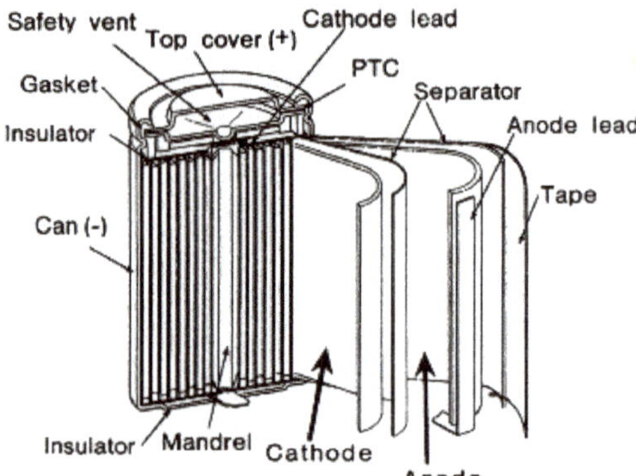

Figure 4. Schematic of a cylindrical Li-ion battery. Adapted with permission from [19], © Elsevier, 2001.

The energy and power characteristics of an electrochemical cell can be determined by utilizing thermodynamic and kinetics relations. The net useful energy of an electrochemical cell can be calculated by the following thermodynamic equation [20]:

$$\Delta G = -nFE \tag{1}$$

where ΔG is the Gibbs free energy, n is the number of electrons transferred per mole of reactants, F is the Faraday constant, and E is the cell potential. The Equation (1) implies that, the higher the cell potential (i.e., E) the higher the capacity of an electrochemical cell assuming number of electrons transferred per mole of reactants are the same (i.e., this is the case for different Li-ion chemistries such as $LiCoO_2$ and $LiFePO_4$). To maximize the useful energy of an electrochemical cell, the difference between the cathode and anode potentials (i.e., open circuit voltage, E_{OCV}) should be as large as possible, and this is the main reason behind the research efforts on developing high voltage batteries [21].

Although the thermodynamic relations are useful in understanding the limits of an electrochemical cell, the practical batteries do not operate under equilibrium conditions hence deviate from thermodynamic equilibrium condition when current is withdrawn due to kinetic limitations (i.e., polarization effects, η). The terminal cell potential (E_T) of an electrochemical cell is dependent on three different types of polarization effects: (a) activation polarization which is related to charge transfer reactions occurring at the electrode-electrolyte interface, (b) ohmic polarization which is the overall internal resistance of the cell components, and (c) concentration polarization which is related to mass transfer limitations during cell operation. The terminal cell potential is given by Equation (2) [20] as follows:

$$E_T = E_{OCV} - \eta \tag{2}$$

The polarization effects can be observed on a discharge curve of a battery as shown in Figure 5. The discharge curve can be utilized to calculate the capacity of the cell, and to understand the effects of the C-rate and temperature on the cell capacity. A C-rate is a measure of the rate at which a battery is charged or discharged relative to its maximum capacity. Charging a battery at 1C rate would take 1 h to completely charge the battery [22].

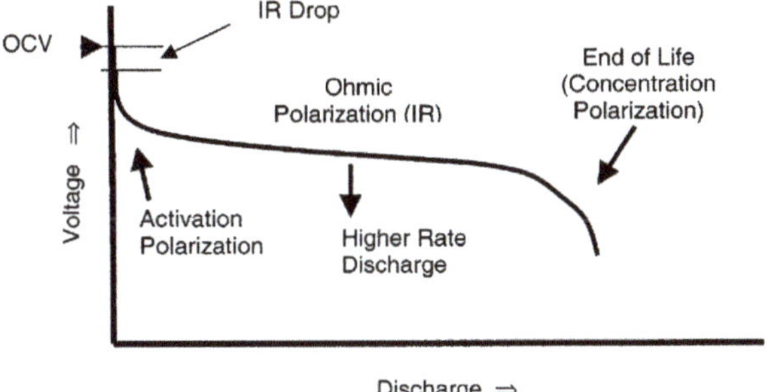

Figure 5. The discharge curve of a battery and the effects of different types of polarization. Adapted with permission from [20], © American Chemical Society, 2004.

Additional information regarding electrode kinetics can be obtained by electrochemical impedance spectroscopy (EIS) [23–25]. EIS is a non-destructive technique, and can reveal various reaction steps during Li-ion intercalation/deintercalation. Each electrode reaction has a distinctive impedance pattern, which can be modeled by an equivalent-circuit model. The depressed semicircle, as seen in Figure 6b, in an EIS curve indicates an inhomogeneous electrode surface. Therefore, constant phase element (CPE) is used in the equivalent circuit model rather than pure capacitance as in Figure 6c. The EIS of $InVO_4$ (indium vanadium oxide) anode is shown in Figure 6 where Re is the combined resistance (R) of electrolyte and cell components, Rsf+ct is resistance due to surface film (sf) and charge transfer (ct), CPEsf+dl is the capacitance due to surface film and double layer (dl), CPEb is the bulk (b) capacitance, Rb is the bulk resistance, Ws is the Warburg impedance that represents the solid state diffusion of Li-ions through the electrode lattice, and Cint is the intercalation (int) capacitance (C) [23]. Impedance values are highly dependent on voltage values during discharge/charge as shown in Figure 6a.

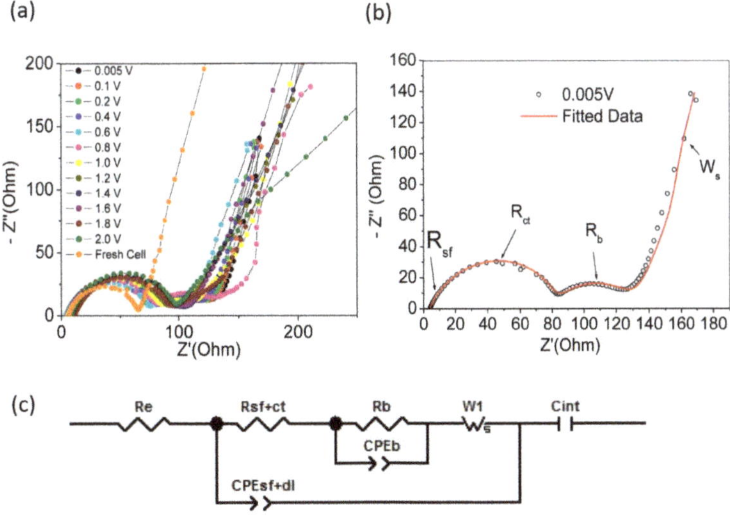

Figure 6. (**a**) EIS of InVO4 anode at different voltages; (**b**) EIS of InVO4 at 0.005 V and (**c**) the corresponding equivalent-circuit model. Adapted with permission from [23], © American Chemical Society, 2013.

3. Li-ion Batteries

Li-ion batteries are the most popular battery chemistry as of today due to their high gravimetric and volumetric energy density as compared to the other rechargeable battery chemistries. The main components of a Li-ion battery are the cathode, anode, electrolyte, separator and current collector. The schematic of a Li-ion battery is shown in Figure 7.

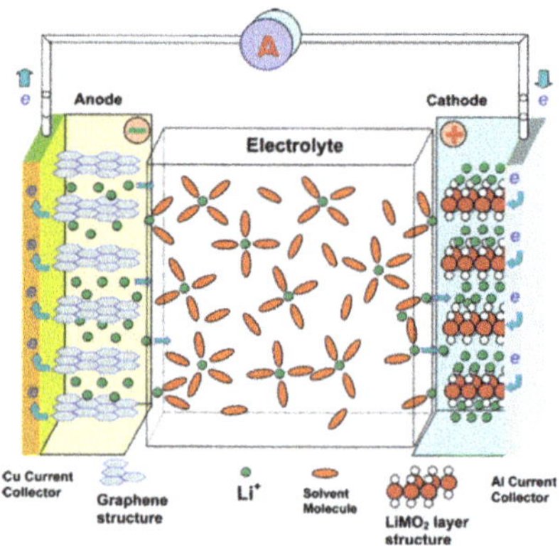

Figure 7. Schematic of a Li-ion battery. Adapted with permission from [26], © American Chemical Society, 2004.

During the operation of a Li-ion battery, Li ions shuttle back and forth in between cathode and anode through the electrolyte absorbed in the separator. Both cathode (i.e., transition metal oxides) and anode (i.e., graphite) are intercalation compounds with layered structures where Li ions can be reversibly inserted and removed. The primary reason behind using intercalation compounds instead of pure metallic Li as an anode is the poor safety and cycle life of the Li-ion batteries with metallic Li anode due to dendrite formation [27]. The cathode, anode, binder and separator are discussed in more detail in the following sections.

4. Cathode

Mostly, intercalation-type cathode materials based on transition metal oxides are utilized in the commercial Li-ion batteries. Basically, there are three types of cathodes: (i) Layered oxides such as $LiCoO_2$ which is the first Li-ion chemistry that was discovered in 1980, and commercialized by SONY in 1991 [27–29]. The $LiCoO_2$ chemistry still dominates the Li-ion battery market (i.e., ~37%) [30]. The layered oxides have two dimensional Li ion diffusion channels; (ii) Spinels such as $LiMn_2O_4$ has three dimensional Li ion diffusion channels, and was discovered in 1983 [31]; and (iii) Olivines such as $LiFePO_4$ has one dimensional Li ion diffusion channels, and was discovered in 1997 [32]. The crystal structures of these three main types of intercalation cathodes are shown in Figure 8.

Figure 8. The structure of the (a) layered oxide $LiCoO_2$; (b) spinel $LiMn_2O_4$; and (c) olivine $LiFePO_4$. Adapted with permission from [33], © Springer, 2013.

4.1. Layered Oxides, $LiMO_2$ (M = Co, Mn, Ni and Their Mixtures)

$LiCoO_2$ is the first layered oxide type cathode material with a practical capacity of 140 mAh/g and average voltage of 3.9 V vs. Li/Li^+. In practice, only half of the lithium (i.e., $1 > x > 0.5$) stored in Li_xCoO_2 is extracted because further Li extraction (i.e., $x < 0.5$ which corresponds to upper cut-off voltage of 4.2 V) triggers structural transition from a hexagonal to monoclinic phase as well as amplifies the side reactions, and results in quick degradation of the cycle life. Significant number of studies was conducted to understand the capacity fading mechanism over 4.2 V as well as to improve the capacity of $LiCoO_2$ by pushing the upper cut-off voltage above 4.2 V while keeping the cycle life at a reasonable level (i.e., ~500 cycles) [34–37]. It was shown by many that the cycle life of $LiCoO_2$ at higher voltages (i.e., >4.2 V) can be improved by metal oxide coatings such as Al_2O_3, SnO_2, ZrO_2, TiO_2, MgO, [34,38–42], metal phosphate coatings such as $AlPO_4$ [43–46], metal fluoride coatings such as AlF_3 and LaF_3 [47,48] and multicomponent metal fluoride coatings such as aluminum-tungsten-fluoride (AlW_xF_y) [49]. Although the underlying reasons for such an improvement in cycle life of $LiCoO_2$ at higher cut-off voltages due to coatings are still debatable [35,50,51], following mechanisms have been suggested: coating (i) inhibits the structural transformation [34,52], (ii) acts as a physical barrier, Figure 9, between the electrolyte and the active material, and prevents the trace amounts of hydrogen fluoride (HF) and water present in the electrolyte from reaching the active material thus effectively suppresses cobalt dissolution and the associated oxygen evolution [53,54], (iii) converts Lewis acids which in return corrode the insulating surface species and improves the electronic conductivity of the solid electrolyte interface (SEI) layer on $LiCoO_2$ [50].

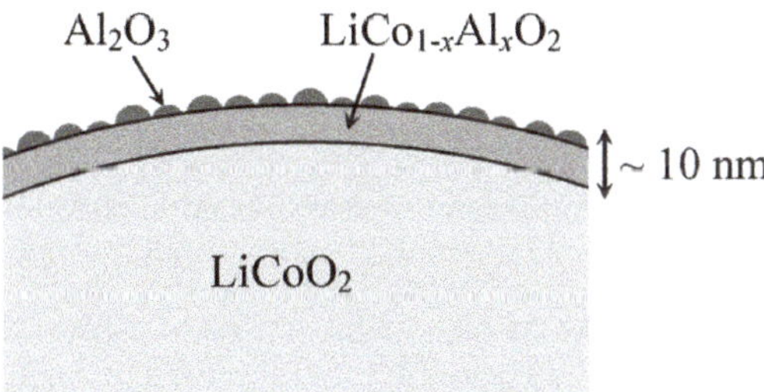

Figure 9. Schematic view of the formation of the solid solution at the coating/active material interface. Adapted with permission from [54], © American Chemical Society, 2009.

In addition to the cycle life, rate capability (i.e., electronic conductivity) can also be improved by metal oxide coatings [55]. Additionally, thermal stability of the $LiCoO_2$ can be improved by a metal oxide coating, which is desirable for safety purposes [37,43,55]. The sample preparation technique can also affect the cycle life of $LiCoO_2$. Tan et al. showed that $LiCoO_2$ prepared by the eutectic $LiNO_3$–LiCl molten salt can attain 80 charge–discharge cycles in the 2.5–4.4 V range without any capacity loss which was associated with the inhibited structural transitions (i.e., hexagonal $\leftrightarrow$ monoclinic $\leftrightarrow$ hexagonal) of $LiCoO_2$ [56].

One of the problems of the $LiCoO_2$ chemistry for large scale applications is its high cost due to limited Co sources; additionally, Co is a toxic substance, and environmentally friendly commercial scale recycling technologies are yet to be developed [57,58]. $LiNiO_2$ is a lower cost cathode compared to $LiCoO_2$, and has a similar structure to $LiCoO_2$; however, it is difficult to obtain stoichiometric $LiNiO_2$, and synthesis conditions significantly affect its compositional and structural properties which impede the widespread utilization of $LiNiO_2$ cathode [59]. $LiNiO_2$ exhibits similar structural transformations as in $LiCoO_2$; thus, the upper cut-off voltage is 4.2 V (i.e., x ~0.5), and beyond 4.2 V cycle-life quickly degrades. Several metal oxide coatings such as ZrO_2 have shown to improve the cycling stability of $LiNiO_2$ by inhibiting the structural transformation as shown by ex-situ X-ray diffraction (XRD) and cyclic voltammetry (CV) experiments [60]. The thermal stability of $Li_xNi_{1-x}O_2$ is also of great concern and it is function of Li content (i.e., x), the smaller the Li content (i.e., x < 0.5) the less thermally stable the $LiNiO_2$. In addition to the problems associated with the capacity fading, overcharging the $LiNiO_2$ also poses a safety risk; however, thermal stability of $LiNiO_2$ can be improved by Mg doping and coating [33,61]. $LiNiO_2$ can also be prepared using a single source precursor (i.e., $Li(H_2O)Ni(N_2H_3CO_2)_3]\cdot0.5H_2O$) which is a desired synthetic route to prepare this cathode material in large scale [62].

$LiMnO_2$ is another layered metal oxide which is isostructural to $LiCoO_2$; similar to $LiNiO_2$, $LiMnO_2$ is a low cost (i.e., Co is \$23.99/kg, Ni is \$8.94/kg and Mn is \$2.01/kg as of 9 May 2016 [63]) and environmentally friendly cathode material compared to $LiCoO_2$. $LiMnO_2$ irreversibly transforms from layered structure into spinel structure upon cycling and its capacity fades quickly [64,65]. The Al_2O_3 and CoO coatings improve the cycle life of $LiMnO_2$ by impeding Mn dissolution and lattice instability due to Jahn-Teller distortion [66,67].

Another strategy to lower the cost of $LiCoO_2$ is to partially or fully substitute Co with Ni, Mn and Al such as $LiNi_{1/3}Mn_{1/3}Co_{1/3}O_2$ (NMC), $LiNi_{0.8}Co_{0.15}Al_{0.05}O_2$ (NCA), $LiNi_xCo_{1-x}$ (0 < x < 1) and $Li(Ni_xMn_{1-x})O_2$ (0 < x < 1) [24,46,68–73]. These composite cathodes derived from $LiCoO_2$ have also higher specific capacity than the $LiCoO_2$, and are extensively used in portable electronics.

4.2. Spinel, LiM_2O_4 (M = Mn, and Mixtures of Co and Ni)

Spinel $LiMn_2O_4$ has three-dimensional Li ion conduction channels which translate to excellent rate capability; additionally, it is easy to synthesize $LiMn_2O_4$ from various precursors. Mn is a low cost and an earth abundant material. Spinel $LiMn_2O_4$ is also inherently safe due to its thermal stability [74]. On the other hand, the major problems associated with $LiMn_2O_4$ are Mn dissolution especially at high operating and storage temperatures (i.e., 55 °C) and the phase transition (i.e., Jahn-Teller distortion) that occurs if $LiMn_2O_4$ is over discharged. $Li_xMn_2O_4$ has two voltage plateaus at ~4 V (x = 0 to 1.0) and ~3 V (x = 1.0 to 2.0), and phase transition from cubic to tetragonal structure can be prevented if $LiMn_2O_4$ is not discharged below 3.5 V [75]. The over discharge of $LiMn_2O_4$ results in significant capacity fade and poor cycling hence should be avoided. The Mn dissolution is more problematic and occurs at both 3 V and 4 V voltage plateaus. It was shown by Tarascon et al. that Mn dissolution is a function of surface area of $LiMn_2O_4$ particles indicating Mn dissolution is originated at the active material-electrolyte interface [76]. The $LiMn_2O_4$ surface acts as a catalyst at high open circuit voltages and oxidizes the non-aqueous electrolyte producing harmful by-products, and higher temperatures accelerates these side reactions further. The dissolved Mn ions can also migrate to the anode side and get reduced and precipitate as Mn metal at the anode surface which can damage the structure of the solid electrolyte interface and increase the charge transfer resistance [77]. To circumvent the Mn dissolution problem, passivation layer needs to be formed between the active material and the electrolyte.

Various coatings and dopants such as lithium boron oxide (LBO) glass [76], SiO_2 [78], SnO_2 [79], MgO [80], Ag [81,82], Al [83], carbon nanotube (CNT) [84], $LiCoO_2$ [85] as well as mixing with other cathode materials such as NMC [86] have shown to suppress Mn dissolution hence improves the storage life, cycle life and/or capacity fading [87]. In particular, LBO glass was the first reported coating for $LiMn_2O_4$; it has a favorable ionic conductivity and can withstand oxidation at high potentials (i.e., 4 V) as well as molten LBO glass has good wetting properties. The LBO glass surface treatment improves the storage performance, but has a detrimental effect on cycling life which is associated with the reduction of average Mn valence state on the surface by boron containing solid species [76]. On the other hand, Al doping of $LiMn_2O_4$ increased the average valance of Mn, and showed enhanced cycling performance at high temperatures [83]. Li and Al co-doped manganese oxide spinel compounds prepared by polyvinyl pyrrolidone precursor method improved the initial discharge capacity and capacity retention [88]. $LiMn_2O_4$/CNT composites are also reported with improved capacity retention and high rate capability for flexible electronics applications [84]. More stable cathode materials such as $LiCoO_2$ are also utilized as a barrier coating for $LiMn_2O_4$ [85]. A 5 nm $LiCoO_2$ coating prepared by micro-emulsion technique improved the cycle life and reduced the self-discharge of $LiMn_2O_4$; however, resulted in slight decrease in capacity of pristine $LiMn_2O_4$ due to Li storage inability of the $LiCoO_2$ coating [85]. Synthesis of nano or submicron size $LiMn_2O_4$ particles and doping (i.e., partially substituting Mn site in $LiMn_2O_4$) $LiMn_2O_4$ with transition and nontransition metals were shown to reduce the Mn dissolution upon cycling hence improved the cycle life of $LiMn_2O_4$ [88–94]. $LiMn_2O_4$ prepared by the one-pot molten salt method resulted in nano sized (~50 nm) hollow spherical particles which showed improved capacity retention even at high C-rates (i.e., 96% retention at 5C rate) [89]. These nano sized spherical $LiMn_2O_4$ particles were also performed well at 50 °C which was associated with their high crystallinity and the presence of spherical particles [89]. In another study, Ru doped $LiMn_{2-x}Ru_xO_4$ (x = 0.1 and 0.25) was shown to reduce the capacity fading as compared to undoped sample by mitigating the spinel-to-double-hexagonal transition at 4.5 V, and due to the presence of $Ru^{4+} \leftrightarrow Ru^{5+}$ redox couple and better electronic conductivity of the Ru doped samples [92]. The other dopants studied for $LiMn_2O_4$ were Co, Cr, Al, Ni and Fe [90,91,95,96]. Sakunthala et al. obtained a notable 94% capacity retention at the end of 1000 cycles at 5C rate by the Co doped $LiMn_2O_4$ (i.e., $Li(Co_{1/6}Mn_{11/6})O_4$). Most of the aforementioned coatings and dopants are inactive materials, and in practice, these materials lower the overall capacity of the spinel $LiMn_2O_4$ and increase the cost of material preparation. An alternative strategy is to mix the $LiMn_2O_4$ with a more stable cathode material such as NMC [86].

4.3. Olivine, $LiMPO_4$ (M = Fe, Co, Mn and Their Mixtures)

Olive $LiFePO_4$ stands out among cathode materials by its superior cycle life and intrinsic safety [97]. Additionally, Fe is one of the most abundant elements on Earth's crust (i.e., even less costly than Mn [63]), making $LiFePO_4$ highly cost competitive. $LiFePO_4$ has an average operating voltage of 3.4 V vs. Li/Li^+, and a specific capacity of ~170 mAh/g corresponding to one Li per $LiFePO_4$ formula unit. The enhanced cycle life and safety of $LiFePO_4$ are associated with its lower operating voltage (i.e., 3.4 V) compared to other cathode materials such as $LiCoO_2$ (i.e., 4.2 V) and $LiMn_2O_4$ (i.e., 4 V). At such low potentials, non-aqueous electrolytes are highly stable. However, $LiFePO_4$ has one-dimensional Li-ion diffusion channels leading to poor ionic conductivity, and it has an intrinsically low electronic conductivity. Moreover, $LiFePO_4$'s density is lower (i.e., 3.6 g/cm^3) than the layered oxides and spinels (~5 g/cm^3) leading to poor volumetric energy density.

The low electronic conductivity of $LiFePO_4$ can be enhanced by carbon coating [98] and by decreasing the $LiFePO_4$ particle size [99], or both. Saranavan et al. showed that $LiFePO_4$ nanoplates (~30–40 nm along the b-axis) with a 5 nm amorphous carbon layer coating prepared by the solvothermal method significantly improved the high rate performance by effectively reducing the Li^+ diffusion length as compared to mesoporous spherical $LiFePO_4/C$ cathodes [100]. However, decreasing the particle size has a negative effect on tap density and accelerates the side reactions due to increased surface area; therefore, carbon coating is the most widely used method to improve the performance of $LiFePO_4$. It was shown that, the quantity of carbon coating in $LiFePO_4$ is not always proportional to the performance enhancement because several other parameters also play a key role in the performance of a carbon coated $LiFePO_4$ such as structure of carbon [101], especially sp^2/sp^3 and disordered/graphene (D/G) ratios [102], carbon source [103], carbon thickness [104], and processing conditions [105]. The single doping and co-doping of the carbon layer to improve the electronic conductivities are also explored. Phosphorus doped carbon layer on $LiFePO_4$ has shown to decrease the charge transfer resistance significantly which is considered to be associated with the free carriers donated by the phosphorus. Additionally, phosphorus doped carbon layer improves the capacity and capacity retention at high C rates [106]. In another study, the $LiFePO_4$ composite with boron and nitrogen co-doped carbon layers were synthesized by the ball milling process followed by calcination [107]. The sequence of doping shown to be critical (i.e., N + B is preferred), and improvements in high-rate capacity are attributed to electron and hole type carriers donated by nitrogen and boron atoms, respectively [107]. Large scale and low-cost material preparation methods are critical for the higher penetration of Li-ion batteries. LiFePO$_4$ prepared by the carbothermal reduction method, which is suitable for large scale synthesis, showed almost no capacity fade for 400 cycles [108].

Carbon nanotubes (i.e., CNT) are excellent electronic conductors and have high surface area, and their tubular structure can enhance the overall electronic and ionic conductivity of the active cathode/anode materials by connecting the inert "dead" zones in these materials [109]. The schematic of the CNT dispersed in a porous LiFePO$_4$–CNT composite is shown in Figure 10. The LiFePO$_4$–CNT composite performed significantly better than the pristine LiFePO$_4$ at both low and high C rates [109].

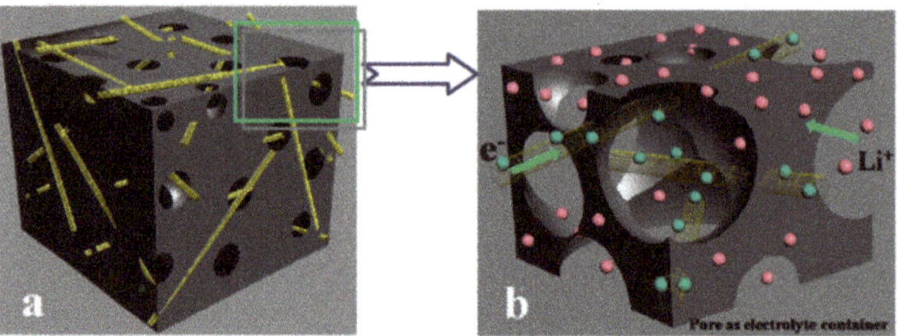

Figure 10. Schematic illustrations of the porous LiFePO$_4$–CNT composite: (**a**) showing the corresponding electron transport and ion diffusion mechanisms; (**b**) is an enlarged zone from (**a**). Adapted with permission from [109], © Royal Society of Chemistry, 2010.

Another superior electronic conductor is the graphene. Utilizing graphene in LiFePO$_4$ composites resulted in achieving near theoretical specific capacity (i.e., 170 mAh/g), exceptional rate capability and improved the cycle life due to intimate contact between non-aqueous electrolyte and the LiFePO$_4$ active materials as well as enhanced the charge and mass transfer [110,111]. The schematic of the graphene wrapped LiFePO$_4$ particles are shown in Figure 11 [111]. Hu et al. reported that electrochemically exfoliated graphene wrapped commercial LiFePO$_4$ particles can deliver 208 mAh/g capacity which is more than the theoretical capacity of LiFePO$_4$. This extra capacity is attributed to the reversible redox reaction between Li ion in the electrolyte and the electrochemically exfoliated graphene sheets [112].

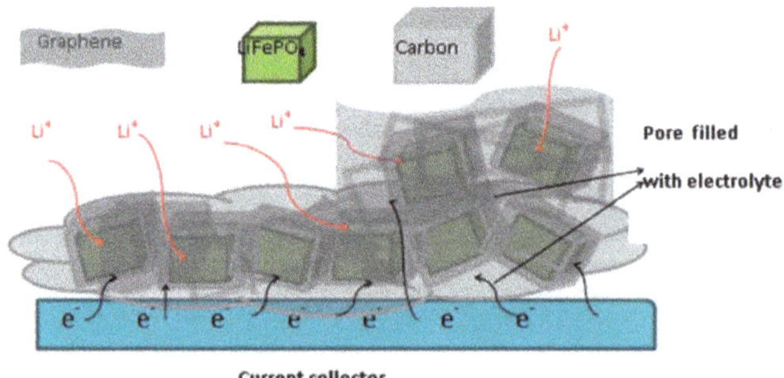

Figure 11. Schematic diagram of the graphene nanosheet/LiFePO$_4$ structure. Adapted with permission from [111], © Royal Society of Chemistry, 2012.

Further improvement in the performance of LiFePO$_4$ was obtained by encapsulating one-dimensional LiFePO$_4$ nanowires (~20–30 nm) into CNTs (i.e., LiFePO$_4$@CNT). These core–shell structures were synthesized by a sol–gel route followed by the post-heating treatment [113]. The dispersion of the CNT and its intimate contact with the LiFePO$_4$ active material are critical for the performance enhancement [114]. By utilizing nitrogen doped CNT (i.e., N–CNT), instead of pristine

CNT, it is possible to facilitate the uniform dispersion of CNT in the $LiFePO_4$ composite due to higher defect density and hydrophilic properties of N–CNTs. Additionally, nitrogen groups enhance the electronic conductivity of N–CNT as a result of increased electron density [114].

One of the limitations of $LiFePO_4$ is its low operating voltage (i.e., 3.42 V), which limits its energy density. It is possible to push the operating voltage to higher voltages (i.e., 4–4.2 V) using vanadium-based cathodes such as $LiVPO_4F$ and $LiVOPO_4$ [115–121]. Novel porous materials such as metal-organic frameworks (MOF) were also studied as electrode materials for Li-ion batteries [118,119,122–124]. Metal organic phosphate open frameworks (MOPOF) and their reduced graphene oxide nanocomposite derivatives showed promising Li storage, but further work is required to improve their rate capability and cycle life [118,119].

5. Anode Materials

5.1. Carbon Based Anodes

Carbon based anode materials are the industry standard in Li-ion batteries, and graphitic carbon is the dominant one due to its favorable electrochemical properties (i.e., low and near flat operating voltage of ~0.25 V vs. Li/Li^+), low cost, and chemical and mechanical stability [125,126]. Graphite has a theoretical capacity of 372 mAh/g with the stoichiometry of LiC_6. Non-graphitic carbons such as hard carbons have higher specific capacity and cycle life, good rate capabilities and lower cost of production; however, they suffer from low density and larger irreversible capacity and hysteresis in the voltage profile [127].

The metal-carbon composites have been extensively studied to improve the performance of carbon anode materials [128–135]. The composite anodes, in general, showed improved performance than simple mechanical mixing of the metals with the anode materials [125]. The chemically deposited Ag on graphite improved the cycle life, which was attributed to the enhanced electron conduction between graphite particles due to low electrical resistance of Ag. Additionally, the capacity of the Ag-graphite composite cathode was superior to the graphite cathode alone since Ag can form LiAg alloy upon interaction with Li [128]. Moreover, Ag deposited graphite was shown to be more resistant to humidity (i.e., ~1000 ppm) which is beneficial for easy manufacturing of Li-ion batteries [129,130]. The microencapsulation of graphite with nanosized Ni showed improvement in initial charge–discharge performance, Coulombic efficiency and cycle life. The proposed mechanism for the improvement in the initial charge–discharge behavior (i.e., decrease in irreversible capacity) was the passivation of the graphite edge surfaces by Ni deposition which in return reduces the highly active edge surface area that is exposed to the electrolyte, and makes the edge surfaces less permeable to the solvated Li ions [131,132]. Similar effects were also observed for Cu deposited natural graphite, in addition to the suppression of electrolyte decomposition and solvated Li ion co-intercalation, improvement in high rate capability was reported due to enhanced charge-transfer properties [133]. The Al deposition of natural graphite improved the cycle life and rate capability remarkably due to decreased charge-transfer resistance during cycling [134]. The other metals studied for metal-graphite composites include Au, Bi, In, Pb, Pd, Sn and Zn, and all these metal coatings, which were prepared by vacuum evaporation technique, improved the rate capability [135]. The rate capability was found to be dependent on the film thickness, too thick coatings limit the Li mobility hence reduces the rate capability.

Carbon coating is another strategy to improve the electrochemical performance of the graphite [136–139]. One of the methods to enhance the low temperature performance of Li-ion batteries is to utilize propylene carbonate (i.e., PC, melting point = $-49\,°C$) based solvents instead of ethylene carbonate (i.e., EC, melting point = $38\,°C$) based ones. However, it is well known that PC decomposes on the graphite surface; therefore, cannot form a stable solid electrolyte interphase (SEI) layer on the graphite surface which leads to diminished cycle life [140]. The studies conducted on carbon coated graphite prepared by thermal and chemical vapor decomposition techniques showed that Coulombic efficiency can be improved significantly since carbon coating acts as a buffer zone between graphite and the PC based electrolyte, and leads to compact and thinner SEI layer as shown in Figure 12 [137–139]. If carbon coated graphite anode is utilized in Li-ion battery manufacturing, high rolling and calendaring pressures should be avoided since high pressures can destroy the carbon coating [137]. The carbon coating also stores Li hence does not adversely affect the specific capacity of the graphite [139].

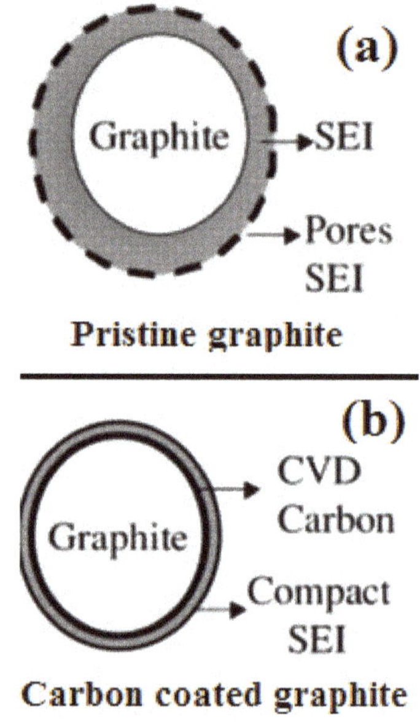

Figure 12. Schematic of the SEI layer on (**a**) pristine and (**b**) carbon coated graphite. Adapted with permission from [138], © Elsevier, 2001.

Another promising approach to protect pristine graphite from PC based solvents is rolling graphite flakes into spherical shape, and then coating with carbon [136]. The spherical shape mitigates the high orientation of the graphite flakes when coated over a current collector, hence improves the Li ion diffusion rate. Additionally, spherical graphite formed by concentric alignment of multiple graphene layers reduces the number of edge planes exposed to solvents (i.e., PC); in other words, inert basal planes are mostly exposed to PC instead of active edge planes which is favorable to improve the Coulombic efficiency [136].

5.2. Alloy Anodes

To meet the battery system targets set by the U.S. DOE given in Table 2, it is necessary to improve the gravimetric and volumetric energy density of the batteries significantly which can only be achieved by developing novel materials with enhanced Li storage properties. To achieve this goal, one of the alternatives is to replace graphite, either completely or partially, with alloy anode materials. The most promising and well-studied alloy anodes are presented in Figure 13.

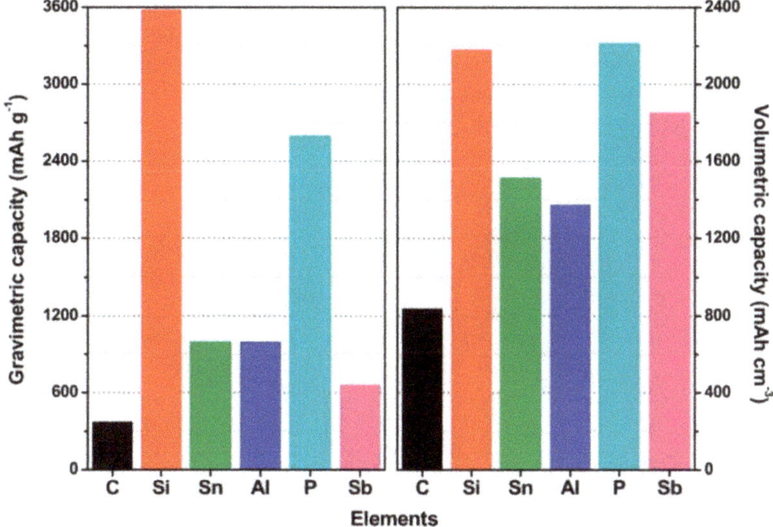

Figure 13. Theoretical gravimetric and volumetric capacities of C (LiC$_6$), Si (Li$_{3.75}$Si), Sn (Li$_{4.4}$Sn), Al (LiAl), P (Li$_3$P) and Sb (Li$_3$Sb). Adapted with permission from [141], © Royal Society of Chemistry, 2011.

One should note that, currently Li metal anode is not considered to be a viable candidate due to dendrite formation and the accompanying safety concerns [142]. Among different alloy anodes given in Figure 13, Si and Sn have attracted the most attention due their superior gravimetric and volumetric capacity. However, these alloy anodes experience large volume change (>300%) upon Li de-/intercalation which causes particle cracking and pulverization upon cycling which leads to poor cycle life. Additionally, this large volume change during cycling prevents the formation of a stable SEI layer on the alloy anodes, which continuously consumes Li ions, and results in low Coulombic efficiency. One simple but effective strategy to suppress the rapid deterioration of cycle life in Si anodes is to prepare a composite anode with low Si (i.e., Si < 33 wt %) and high binder (i.e., 33–56 wt %) content [143]. However, high inactive mass in this composite cathode severely limits the specific capacity (i.e., ~250 mAh/g after 400 cycles).

Another approach is to partially replace graphite with Si. The graphite-Si composite anodes with carbon coating showed better performance than Si alone, but quick degradation of cycle life is still a concern in these type of materials [144,145]. The carbon and metal coating techniques were also applied to improve the cycle life and Coulombic efficiency of Si anodes [146,147]. These coatings reduce the electrolyte decomposition on Si anodes by creating a buffer zone between the electrolyte and Si, but due to significant volume change of Si anode, the carbon coating loses its structural integrity by cycling, and fresh Si surfaces are exposed to electrolyte eventually. In another study, nanocrystalline Si particles of 20–80 nm that were encapsulated in a carbon aerogel showed a stable reversible capacity of 1450 mAh/g for 50 cycles [147]. The carbon aerogel works as a cushion during the expansion-contraction of the nanocrystalline Si particles, and maintains the three dimensional electrical network; this way, Si particles stay in physical contact with the carbon matrix at all times. More recent studies on Si anodes concentrated on minimizing the Si surface area exposed to electrolyte by encapsulating single Si nanoparticle in a carbon shell (i.e., yolk-shell design) which has enough void space to accommodate significant volume changes during de-/lithiation of Si [148]. This strategy minimizes the carbon shell deformation upon cycling, hence helps with obtaining a stable SEI layer over the carbon shell. This Si-carbon composite showed a reversible capacity of 1500 mAh/g at 1C rate after the initial formation cycles, and kept 74% of its initial capacity after 1000 cycles. In a similar approach, inspired by a pomegranate fruit, carbon coated Si nanoparticles are further encapsulated in a thick secondary carbon layer as shown in Figure 14 [149]. As a result, the SEI layer mostly forms on this secondary carbon layer (i.e., lower surface area) instead of the individual Si nanoparticles (i.e., higher surface area), Figure 14c. This pomegranate inspired Si-carbon composite anode showed 97% percent capacity retention (i.e., 1160 mAh/g) after 1000 cycles.

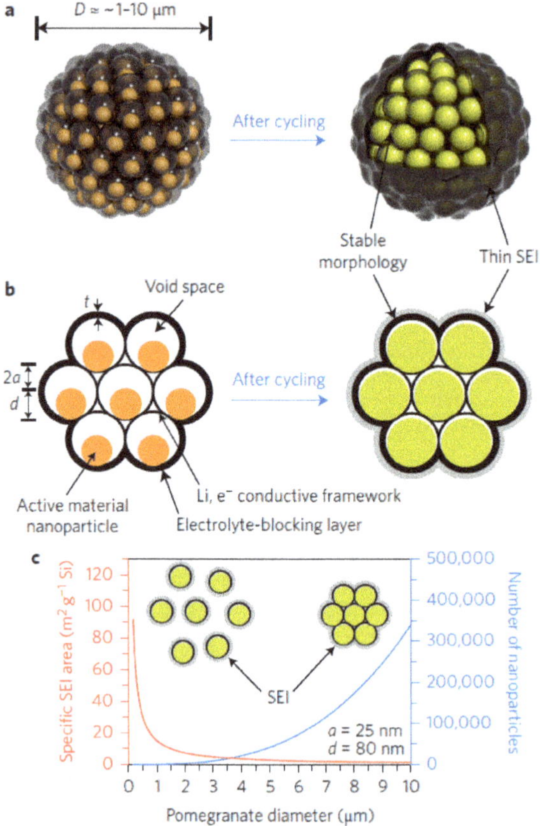

Figure 14. The schematic of the pomegranate-inspired Si anode design in (**a**) 3D and (**b**) 2D before and after cycling (in the lithiated state); (**c**) The variation of the specific SEI area and the number of primary nanoparticles in one pomegranate particle versus its diameter. Adapted with permission from [149], © Nature Publishing Group, 2014.

6. Binder

Binder is one of the inactive components in a Li-ion battery electrode, but its role is critical for proper operation of the Li-ion batteries. Binder serves as a polymeric matrix that connects active materials to each other and to the current collector in a composite Li-ion battery electrode, and accommodates volume changes due to de-/lithiation of active materials during battery operation. An ideal binder should have: (a) high binding strength, (b) favorable mechanical properties to accommodate volume changes, (c) high electronic conductivity, (d) optimal porous structure for high ionic conductivity and (e) chemical inertness [16,150]. Currently, polyvinylidene fluoride (PVDF) is the most common binder despite its shortcomings. PVDF is not conductive and does not store Li hence considered as an inactive material. Several different electroactive polymers such as polyaniline and polypyrrole were proposed to replace PVDF. These multifunctional binders improved the rate capability and cycle life of the LiFePO$_4$ which has an intrinsically poor electronic conductivity [151–154]. The polyaniline binder also improved the irreversible capacity loss during the first cycle in a graphite anode [155]. As discussed earlier, the alloy anodes show significant volume change upon de-/lithiation; therefore, an ideal binder should be highly flexible and mechanically robust to keep active materials together during cycling. Different composite elastomeric binders were developed for Si alloy anodes [156–158]. The PVDF-tetrafluoroethylene-propylene (PVDF-TFE-P) binder can be extended to 100% strain, whereas PVDF can only be extended to 10% strain. The PVDF-TFE-P binder enhanced the cycle life of a-Si$_{0.64}$Sn$_{0.36}$ alloy anode reasonably [157]. Another promising elastomeric binder is styrene-butadiene-rubber (SBR) and sodium-carboxyl-methylcellulose (SCMC) composite. As compared to PVDF, SBR + SCMC binder showed smaller moduli, a larger maximum elongation, and a stronger adhesion strength hence better cycle life in Si alloy anodes [158]. Crosslinking is an effective strategy to improve the mechanical properties of polymeric binders. By crosslinking poly(acrylic acid) with SCMC, Koo et al. obtained a capacity of 2000 mAh/g for Si anode after 100 cycles [159]. Recently, a new type of binder, hyperbranched β-cyclodextrin polymer and a dendritic gallic acid cross-linker with six adamantane units, for Si anodes based on "dynamic crosslinking" method was developed as shown in Figure 15 [160]. The dynamic crosslinking refers to supramolecular host-guest interactions. The distinctive feature of this polymer is that it can restore the broken connections between polymer chains during cycling; therefore, it can be considered as a self-healing binder. The Si anode based on this novel composite binder retained 90% of its initial capacity after 150 cycles (~1550 mAh/g).

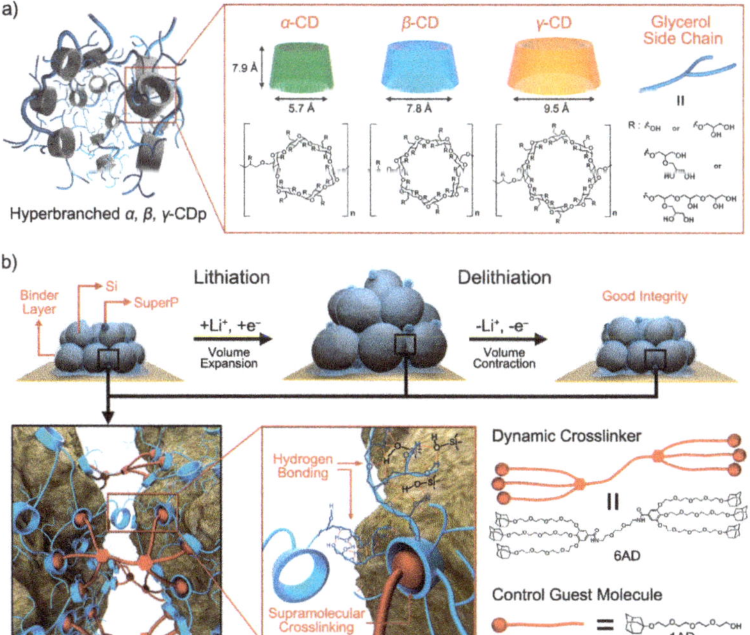

Figure 15. (a) Schematic and structures of hyperbranched R-, β-, and γ-cyclodextrin polymer (CDp); (b) Dynamic cross-linking mechanism of β-CDp and 6 adamantane (AD) in an electrode matrix along with schematic and structures of guest molecules incorporating AD moiety. Adapted with permission from [160], © American Chemical Society, 2015.

Separator

Separator is another inactive component in Li-ion batteries, but it plays a critical role in the operation and safety of such batteries. Separator physically separates the anode and cathode to prevent the electrical short circuit, while allowing Li-ion diffusion due to its porous structure in between the anode and cathode during cycling of a cell. Therefore, an ideal separator should be a good electronic insulator, mechanically robust, chemically inert under the operating conditions of the battery, and should have good wettability in non-aqueous liquid electrolytes [18]. The most common types of Li-ion battery separators are microporous and non-woven separators, former being the dominant in the Li-ion battery market [161]. The microporous separators are mainly made of polyolefins such as polyethylene (PE) and polypropylene (PP). The scanning electron microscope (SEM) images of a multilayered composite separator manufactured by Celgard™ are given in Figure 16. The multilayer structure (PP/PE/PP) is instrumental in preventing the thermal runaway of the Li-ion battery via thermal shut-down mechanism [161].

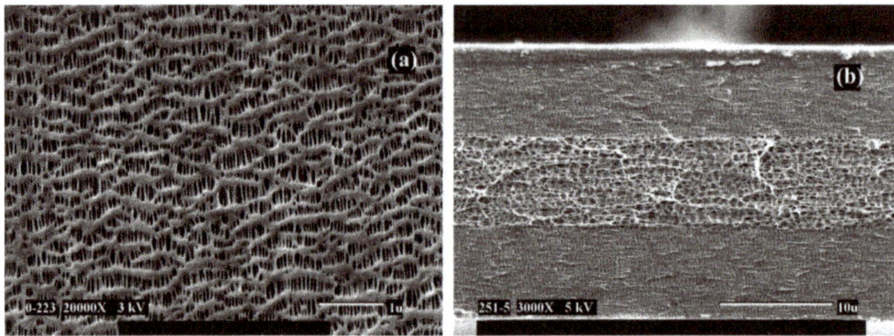

Figure 16. SEM images of Celgard 2325 (PP/PE/PP) separator used in lithium-ion batteries: (**a**) surface SEM and (**b**) cross-section SEM images. Adapted with permission from [161], © American Chemical Society, 2004.

The thermal stability, wettability and mechanical properties of the separators can be improved by surface coating, grafting and blending techniques [18]. The surface coating of the microporous separators with high temperature polymers such as polyimide, inorganic nanoparticles such as Al_2O_3 and SiO_2, or the combination of polymers and inorganic nanoparticles have shown to reduce the thermal shrinkage, and improved the mechanical properties and wettability of the separators [162–165]. It is important to note that surface coating creates an additional layer on the separator, which may limit the Li-ion diffusion rate hence the performance (i.e., cycle life and rate capability) of the cell. Although the surface coating is effective in improving the aforementioned properties of the separator, uniformity, thickness and adhesion of the coated layer are of concern in practical applications [18]. To circumvent these issues associated with the surface coating technique, surface grafting can be utilized. Unlike coting, grafting results in the permanent attachment of the desired functional groups via covalent bonding. The surface grafting can be achieved by UV-irradiation, plasma treatment and high energy radiation methods [18]. The siloxane grafted PE separator prepared by electron beam irradiation showed enhanced ionic conductivity and wettability as well as improved cycle life at high voltage operating conditions [166]. Similar results were obtained for glycidyl methacrylate grafted PE separator [167].

Traditionally, non-woven separators are not considered to be competitive with the microporous separators due to their large pore structure and difficulty in manufacturing them thin (<25 μm) with the desired mechanical properties [18,161]. One of the promising non-woven (polyethylene terephthalate) separators with ceramic coating (i.e., Al_2O_3) on both sides (i.e., Degussa-Separion™) is shown in

Figure 17. This commercially available separator have good wettability, superior chemical resistance and very low shrinkage at high temperatures [18].

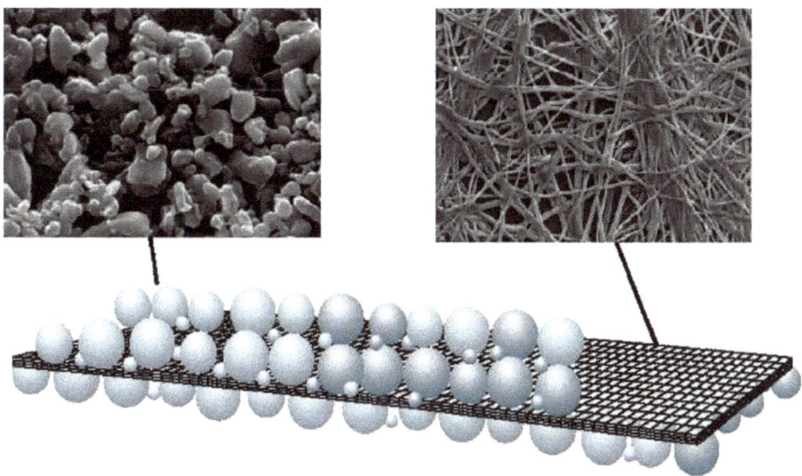

Figure 17. Schematic structure of the Degussa-Separion™ separators. Adapted with permission from [168], © Elsevier, 2007.

7. Conclusions

Rechargeable batteries play a critical role in portable electronics, transportation, back-up power and load-leveling applications. Currently, lead acid and Li-ion chemistries are the most important types of rechargeable batteries. The Li-ion chemistry is expected to play a bigger role in the future due to its superior gravimetric and volumetric densities as compared to other battery chemistries. Additionally, the cost of Li-ion batteries continues to drop due to engineering improvements and materials advancements. The nanocomposite materials are at the center of the material developments in cathode, anode, binder and separator of Li-ion batteries as discussed in this article. Particularly, nanocomposite materials can improve the safety, cycle-life, rate capability and specific capacity of Li-ion batteries. It should be noted that further improvements in the nanocomposites and their manufacturing processes are required to reach the U.S. DOE targets for energy density and cost of battery systems.

Acknowledgments: The authors greatly acknowledge the financial support from the Texas A&M University-Kingsville.

Author Contributions: Dervis Emre Demirocak conceived this work and wrote the manuscript. Sesha S. Srinivasan and Elias K. Stefanakos proofread the manuscript.

Conflicts of Interest: The authors declare no conflicts of interest.

References

1. Chen, H.; Cong, T.N.; Yang, W.; Tan, C.; Li, Y.; Ding, Y. Progress in electrical energy storage system: A critical review. *Prog. Nat. Sci.* **2009**, *19*, 291–312. [CrossRef]
2. Weinert, J.X.; Burke, A.F.; Wei, X. Lead-acid and lithium-ion batteries for the Chinese electric bike market and implications on future technology advancement. *J. Power Sources* **2007**, *172*, 938–945. [CrossRef]
3. Frost & Sullivan. Frost & Sullivan Research Service—World Lithium-ion Battery Market. 2009. Available online: http://www.frost.com/prod/servlet/report-brochure.pag?id=N76F-27-00-00-00#Further (accessed on 10 May 2017).

4. Ceraolo, M.; Huria, T.; Pede, G.; Vellucci, F. Lithium-ion starting-lighting-ignition batteries: Examining the feasibility. In Proceedings of the 2011 IEEE Vehicle Power and Propulsion Conference (VPPC), Chicago, IL, USA, 6–9 September 2011.

5. Doeff, M.M. Battery cathodes. In *Batteries for Sustainability*; Springer: New York, NY, USA, 2013; pp. 5–49.

6. Battery University. What's the Best Battery? Available online: http://batteryuniversity.com/learn/archive/whats_the_best_battery (accessed on 9 May 2017).

7. Zu, C.X.; Li, H. Thermodynamic analysis on energy densities of batteries. *Energy Environ. Sci.* **2011**, *4*, 2614–2624. [CrossRef]

8. U.S. Department of Energy. *The EV Everywhere Grand Challenge: Road to Success*; U.S. Department of Energy: Washington, DC, USA, 2014. Available online: http://energy.gov/sites/prod/files/2016/05/f31/eveverywhere_road_to_success.pdf (accessed on 11 May 2017).

9. Voelcker, J. Electric-car battery costs: Tesla $190 per kwh for pack, GM $145 for cells. *Green Car Reports*. 28 April 2016. Available online: http://www.greencarreports.com/news/1103667_electric-car-battery-costs-tesla-190-per-kwh-for-pack-gm-145-for-cells (accessed on 16 June 2017).

10. Nykvist, B.; Nilsson, M. Rapidly falling costs of battery packs for electric vehicles. *Nat. Clim. Chang.* **2015**, *5*, 329–332. [CrossRef]

11. Nitta, N.; Wu, F.; Lee, J.T.; Yushin, G. Li-ion battery materials: Present and future. *Mater. Today* **2015**, *18*, 252–264. [CrossRef]

12. Fergus, J.W. Recent developments in cathode materials for lithium ion batteries. *J. Power Sources* **2010**, *195*, 939–954. [CrossRef]

13. Tarascon, J.; Guyomard, D. The $Li_{1+x}Mn_2O_4$/C rocking-chair system: A review. *Electrochim. Acta* **1993**, *38*, 1221–1231. [CrossRef]

14. Reddy, M.; Subba Rao, G.; Chowdari, B. Metal oxides and oxysalts as anode materials for Li ion batteries. *Chem. Rev.* **2013**, *113*, 5364–5457. [CrossRef] [PubMed]

15. Goriparti, S.; Miele, E.; De Angelis, F.; Di Fabrizio, E.; Zaccaria, R.P.; Capiglia, C. Review on recent progress of nanostructured anode materials for Li-ion batteries. *J. Power Sources* **2014**, *257*, 421–443. [CrossRef]

16. Lestriez, B. Functions of polymers in composite electrodes of lithium ion batteries. *C. R. Chim.* **2010**, *13*, 1341–1350. [CrossRef]

17. Chou, S.L.; Pan, Y.; Wang, J.Z.; Liu, H.K.; Dou, S.X. Small things make a big difference: Binder effects on the performance of Li and Na batteries. *Phys. Chem. Chem. Phys.* **2014**, *16*, 20347–20359. [CrossRef] [PubMed]

18. Zhang, H.; Zhou, M.Y.; Lin, C.E.; Zhu, B.K. Progress in polymeric separators for lithium ion batteries. *RSC Adv.* **2015**, *5*, 89848–89860. [CrossRef]

19. Wakihara, M. Recent developments in lithium ion batteries. *Mater. Sci. Eng. R Rep.* **2001**, *33*, 109–134. [CrossRef]

20. Winter, M.; Brodd, R.J. What are batteries, fuel cells, and supercapacitors? *Chem. Rev.* **2004**, *104*, 4245–4270. [CrossRef] [PubMed]

21. Hu, M.; Pang, X.; Zhou, Z. Recent progress in high-voltage lithium ion batteries. *J. Power Sources* **2013**, *237*, 229–242. [CrossRef]

22. MIT Electric Vehicle Team. A Guide to Understanding Battery Specifications. Available online: http://web.mit.edu/evt/summary_battery_specifications.pdf (accessed on 7 May 2017).

23. Reddy, M.V.; Wei Wen, B.L.; Loh, K.P.; Chowdari, B.V.R. Energy storage studies on InVO$_4$ as high performance anode material for Li-ion batteries. *ACS Appl. Mater. Interfaces* **2013**, *5*, 7777–7785. [CrossRef] [PubMed]

24. Reddy, M.; Subba Rao, G.; Chowdari, B. Preparation and characterization of $LiNi_{0.5}Co_{0.5}O_2$ and $LiNi_{0.5}Co_{0.4}Al_{0.1}O_2$ by molten salt synthesis for Li ion batteries. *J. Phys. Chem. C* **2007**, *111*, 11712–11720.

25. Levi, M.; Aurbach, D. Impedance of a single intercalation particle and of non-homogeneous, multilayered porous composite electrodes for Li-ion batteries. *J. Phys. Chem. B* **2004**, *108*, 11693–11703. [CrossRef]

26. Xu, K. Nonaqueous liquid electrolytes for lithium-based rechargeable batteries. *Chem. Rev.* **2004**, *104*, 4303–4418. [CrossRef] [PubMed]

27. Nishi, Y. Lithium ion secondary batteries; past 10 years and the future. *J. Power Sources* **2001**, *100*, 101–106. [CrossRef]

28. Mizushima, K.; Jones, P.; Wiseman, P.; Goodenough, J. Li_xCoO_2 ($0 < x < 1$): A new cathode material for batteries of high energy density. *Solid State Ion.* **1981**, *3*, 171–174. [CrossRef]

29. Mizushima, K.; Jones, P.; Wiseman, P.; Goodenough, J. Li_xCoO_2 (0< x < −1): A new cathode material for batteries of high energy density. *Mater. Res. Bull.* **1980**, *15*, 783–789. [CrossRef]

30. Zou, H.; Gratz, E.; Apelian, D.; Wang, Y. A novel method to recycle mixed cathode materials for lithium ion batteries. *Green Chem.* **2013**, *15*, 1183–1191. [CrossRef]

31. Thackeray, M.M.; David, W.I.F.; Bruce, P.G.; Goodenough, J.B. Lithium insertion into manganese spinels. *Mater. Res. Bull.* **1983**, *18*, 461–472. [CrossRef]

32. Padhi, A.K.; Nanjundaswamy, K.; Goodenough, J. Phospho-olivines as positive-electrode materials for rechargeable lithium batteries. *J. Electrochem. Soc.* **1997**, *144*, 1188–1194. [CrossRef]

33. Brodd, R.J. *Batteries for Sustainability: Selected Entries from the Encyclopedia of Sustainability Science and Technology*; Springer: New York, NY, USA, 2013.

34. Cho, J.; Kim, Y.J.; Park, B. Novel $LiCoO_2$ cathode material with Al_2O_3 coating for a Li ion cell. *Chem. Mater.* **2000**, *12*, 3788–3791. [CrossRef]

35. Chen, Z.; Dahn, J. Methods to obtain excellent capacity retention in $LiCoO_2$ cycled to 4.5 V. *Electrochim. Acta* **2004**, *49*, 1079–1090. [CrossRef]

36. Chebiam, R.; Kannan, A.; Prado, F.; Manthiram, A. Comparison of the chemical stability of the high energy density cathodes of lithium-ion batteries. *Electrochem. Commun.* **2001**, *3*, 624–627. [CrossRef]

37. Cho, J.; Kim, G. Enhancement of thermal stability of $LiCoO_2$ by $LiMn_2O_4$ coating. *Electrochem. Solid State Lett.* **1999**, *2*, 253–255. [CrossRef]

38. Cho, J.; Kim, C.S.; Yoo, S.I. Improvement of structural stability of $LiCoO_2$ cathode during electrochemical cycling by sol–gel coating of SnO_2. *Electrochem. Solid State Lett.* **2000**, *3*, 362–365. [CrossRef]

39. Chen, Z.; Dahn, J. Effect of a ZrO_2 coating on the structure and electrochemistry of Li_xCoO_2 when cycled to 4.5 V. *Electrochem. Solid State Lett.* **2002**, *5*, A213–A216. [CrossRef]

40 Wang, Z.; Wu, C.; Liu, L.; Wu, F.; Chen, L.; Huang, X. Electrochemical evaluation and structural characterization of commercial $LiCoO_2$ surfaces modified with MgO for lithium-ion batteries. *J. Electrochem. Soc.* **2002**, *149*, A466–A471. [CrossRef]

41. Hong, W.; Ming-Cai, C. Modification of $LiCoO_2$ by surface coating with $MgO/TiO_2/SiO_2$ for high-performance lithium-ion battery. *Electrochem. Solid-State Lett.* **2006**, *9*, A82–A85. [CrossRef]

42. Chen, Z.; Qin, Y.; Amine, K.; Sun, Y.K. Role of surface coating on cathode materials for lithium-ion batteries. *J. Mater. Chem.* **2010**, *20*, 7606–7612. [CrossRef]

43. Cho, J.; Lee, J.G.; Kim, B.; Park, B. Effect of P_2O_5 and $AlPO_4$ coating on $LiCoO_2$ cathode material. *Chem. Mater.* **2003**, *15*, 3190–3193. [CrossRef]

44. Lee, J.G.; Kim, T.G.; Park, B. Metal-phosphate coating on $LiCoO_2$ cathodes with high cutoff voltages. *Mater. Res. Bull.* **2007**, *42*, 1201–1211. [CrossRef]

45. Kim, J.; Noh, M.; Cho, J.; Kim, H.; Kim, K.B. Controlled nanoparticle metal phosphates (Metal = Al, Fe, Ce, and Sr) coatings on $LiCoO_2$ cathode materials. *J. Electrochem. Soc.* **2005**, *152*, A1142–A1148. [CrossRef]

46. Tan, K.; Reddy, M.; Rao, G.S.; Chowdari, B. Effect of $AlPO_4$-coating on cathodic behaviour of $Li(Ni_{0.8}Co_{0.2})O_2$. *J. Power Sources* **2005**, *141*, 129–142. [CrossRef]

47. Sun, Y.K.; Han, J.M.; Myung, S.T.; Lee, S.W.; Amine, K. Significant improvement of high voltage cycling behavior AlF_3-coated $LiCoO_2$ cathode. *Electrochem. Commun.* **2006**, *8*, 821–826. [CrossRef]

48. Yang, Z.; Qiao, Q.; Yang, W. Improvement of structural and electrochemical properties of commercial $LiCoO_2$ by coating with LaF_3. *Electrochim. Acta* **2011**, *56*, 4791–4796. [CrossRef]

49. Park, J.S.; Mane, A.U.; Elam, J.W.; Croy, J.R. Amorphous metal fluoride passivation coatings prepared by atomic layer deposition on $LiCoO_2$ for Li-ion batteries. *Chem. Mater.* **2015**, *27*, 1917–1920. [CrossRef]

50. Bai, Y.; Liu, N.; Liu, J.; Wang, Z.; Chen, L. Coating material-induced acidic electrolyte improves $LiCoO_2$ performances. *Electrochem. Solid State Lett.* **2006**, *9*, A552–A556. [CrossRef]

51. Li, C.; Zhang, H.; Fu, L.; Liu, H.; Wu, Y.; Rahm, E.; Holze, R.; Wu, H. Cathode materials modified by surface coating for lithium ion batteries. *Electrochim. Acta* **2006**, *51*, 3872–3883. [CrossRef]

52. Cho, J.; Kim, Y.J.; Park, B. $LiCoO_2$ cathode material that does not show a phase transition from hexagonal to monoclinic phase. *J. Electrochem. Soc.* **2001**, *148*, A1110–A1115. [CrossRef]

53. Kim, J.S.; Johnson, C.; Vaughey, J.; Hackney, S.; Walz, K.; Zeltner, W.; Anderson, M.; Thackeray, M. The electrochemical sability of spinel electrodes coated with ZrO_2, Al_2O_3, and SiO_2 from colloidal suspensions. *J. Electrochem. Soc.* **2004**, *151*, A1755–A1761. [CrossRef]

54. Dahéron, L.; Dedryvere, R.; Martinez, H.; Flahaut, D.; Ménétrier, M.; Delmas, C.; Gonbeau, D. Possible explanation for the efficiency of Al-based coatings on $LiCoO_2$: Surface properties of $LiCo_{1-x}Al_xO_2$ solid solution. *Chem. Mater.* **2009**, *21*, 5607–5616. [CrossRef]

55. Kweon, H.J.; Park, J.; Seo, J.; Kim, G.; Jung, B.; Lim, H.S. Effects of metal oxide coatings on the thermal stability and electrical performance of $LiCoO_2$ in a Li-ion cell. *J. Power Sources* **2004**, *126*, 156–162. [CrossRef]

56. Tan, K.; Reddy, M.; Rao, G.S.; Chowdari, B. High-performance $LiCoO_2$ by molten salt ($LiNO_3$:LiCl) synthesis for Li-ion batteries. *J. Power Sources* **2005**, *147*, 241–248. [CrossRef]

57. Xu, J.; Thomas, H.; Francis, R.W.; Lum, K.R.; Wang, J.; Liang, B. A review of processes and technologies for the recycling of lithium-ion secondary batteries. *J. Power Sources* **2008**, *177*, 512–527. [CrossRef]

58. Gaines, L. The future of automotive lithium-ion battery recycling: Charting a sustainable course. *Sustain. Mater. Technol.* **2014**, *1*, 2–7. [CrossRef]

59. Hirano, A.; Kanno, R.; Kawamoto, Y.; Takeda, Y.; Yamaura, K.; Takano, M.; Ohyama, K.; Ohashi, M.; Yamaguchi, Y. Relationship between non-stoichiometry and physical properties in $LiNiO_2$. *Solid State Ion.* **1995**, *78*, 123–131. [CrossRef]

60. Cho, J.; Kim, T.-J.; Kim, Y.J.; Park, B. High-performance ZrO_2-coated $LiNiO_2$ cathode material. *Electrochem. Solid State Lett.* **2001**, *4*, A159–A161. [CrossRef]

61. Dahn, J.; Fuller, E.; Obrovac, M.; Von Sacken, U. Thermal stability of Li_xCoO_2, $LixNiO_2$ and λ-MnO_2 and consequences for the safety of Li-ion cells. *Solid State Ion.* **1994**, *69*, 265–270. [CrossRef]

62. Tey, S.L.; Reddy, M.; Subba Rao, G.; Chowdari, B.; Yi, J.; Ding, J.; Vittal, J.J. Synthesis, Structure, and Magnetic Properties of $[Li(H_2O)M(N_2H_3CO_2)_3]\cdot 0.5H_2O$ (M = Co, Ni) as Single Precursors to $LiMO_2$ Battery Materials. *Chem. Mater.* **2006**, *18*, 1587–1594.

63. InvestmentMine. Commodity and Metal Prices. Available online: http://www.infomine.com/investment/metal-prices/ (accessed on 10 May 2017).

64. Armstrong, A.R.; Bruce, P.G. Synthesis of layered $LiMnO_2$ as an electrode for rechargeable lithium batteries. *Nature* **1996**, *381*, 499–500. [CrossRef]

65. Vitins, G.; West, K. Lithium intercalation into layered $LiMnO_2$. *J. Electrochem. Soc.* **1997**, *144*, 2587–2592. [CrossRef]

66. Cho, J.; Kim, T.J.; Park, B. The effect of a metal-oxide coating on the cycling behavior at 55 °C in orthorhombic $LiMnO_2$ cathode materials. *J. Electrochem. Soc.* **2002**, *149*, A288–A292. [CrossRef]

67. Cho, J.; Kim, Y.J.; Kim, T.J.; Park, B. Enhanced structural stability of o-$LiMnO_2$ by sol–gel coating of Al_2O_3. *Chem. Mater.* **2001**, *13*, 18–20. [CrossRef]

68. Chikkannanavar, S.B.; Bernardi, D.M.; Liu, L. A review of blended cathode materials for use in Li-ion batteries. *J. Power Sources* **2014**, *248*, 91–100. [CrossRef]

69. Reddy, M.; Tung, B.D.; Yang, L.; Minh, N.D.Q.; Loh, K.; Chowdari, B. Molten salt method of preparation and cathodic studies on layered-cathode materials $Li(Co_{0.7}Ni_{0.3})O_2$ and $Li(Ni_{0.7}Co_{0.3})O_2$ for Li-ion batteries. *J. Power Sources* **2013**, *225*, 374–381.

70. Reddy, M.; Rao, G.S.; Chowdari, B. Synthesis by molten salt and cathodic properties of $Li(Ni_{1/3}Co_{1/3}Mn_{1/3})O_2$. *J. Power Sources* **2006**, *159*, 263–267. [CrossRef]

71. Reddy, M.; Rao, G.S.; Chowdari, B. Synthesis and electrochemical studies of the 4 V cathode, $Li(Ni_{2/3}Mn_{1/3})O_2$. *J. Power Sources* **2006**, *160*, 1369–1374. [CrossRef]

72. Zhao, X.; Reddy, M.; Liu, H.; Rao, G.S.; Chowdari, B. Layered $Li(Ni_{0.2}Mn_{0.2}Co_{0.6})O_2$ synthesized by a molten salt method for lithium-ion batteries. *RSC Adv.* **2014**, *4*, 24538–24543.

73. Tan, T.Q.; Idris, M.S.; Osman, R.A.M.; Reddy, M.; Chowdari, B. Structure and electrochemical behaviour of $LiNi_{0.4}Mn_{0.4}Co_{0.2}O_2$ as cathode material for lithium ion batteries. *Solid State Ion.* **2015**, *278*, 43–48.

74. Doughty, D.; Roth, E.P. A general discussion of Li ion battery safety. *Electrochem. Soc. Interface* **2012**, *21*, 37–44. [CrossRef]

75. Jang, D.H.; Shin, Y.J.; Oh, S.M. Dissolution of spinel oxides and capacity losses in 4 V $Li/Li_xMn_2O_4$ cells. *J. Electrochem. Soc.* **1996**, *143*, 2204–2211. [CrossRef]

76. Amatucci, G.; Blyr, A.; Sigala, C.; Alfonse, P.; Tarascon, J. Surface treatments of $Li_{1+x}Mn_{2-x}O_4$ spinels for improved elevated temperature performance. *Solid State Ion.* **1997**, *104*, 13–25. [CrossRef]

77. Tsunekawa, H.; Tanimoto, S.; Marubayashi, R.; Fujita, M.; Kifune, K.; Sano, M. Capacity fading of graphite electrodes due to the deposition of manganese ions on them in Li-ion batteries. *J. Electrochem. Soc.* **2002**, *149*, A1326–A1331. [CrossRef]

78. Arumugam, D.; Kalaignan, G.P. Synthesis and electrochemical characterizations of Nano-SiO$_2$-coated LiMn$_2$O$_4$ cathode materials for rechargeable lithium batteries. *J. Electroanal. Chem.* **2008**, *624*, 197–204. [CrossRef]

79. Wang, L.; Zhao, J.; Guo, S.; He, X.; Jiang, C.; Wan, C. Investigation of SnO$_2$-modified LiMn$_2$O$_4$ composite as cathode material for lithium-ion batteries. *Int. J. Electrochem. Sci.* **2010**, *5*, 1113–1126.

80. Gnanaraj, J.; Pol, V.; Gedanken, A.; Aurbach, D. Improving the high-temperature performance of LiMn$_2$O$_4$ spinel electrodes by coating the active mass with MgO via a sonochemical method. *Electrochem. Commun.* **2003**, *5*, 940–945. [CrossRef]

81. Zhou, W.J.; He, B.L.; Li, H.L. Synthesis, structure and electrochemistry of Ag-modified LiMn$_2$O$_4$ cathode materials for lithium-ion batteries. *Mater. Res. Bull.* **2008**, *43*, 2285–2294. [CrossRef]

82. Huang, S.; Wen, Z.; Yang, X.; Zhu, X.; Lin, B. Synthesis and the Improved High-Rate Performance of LiMn$_2$O$_4$/Ag Composite Cathode for Lithium-Ion Batteries. *Electrochem. Solid State Lett.* **2006**, *9*, A443–A447. [CrossRef]

83. Ding, Y.; Xie, J.; Cao, G.; Zhu, T.; Yu, H.; Zhao, X. Enhanced elevated-temperature performance of Al-doped single-crystalline LiMn$_2$O$_4$ nanotubes as cathodes for lithium ion batteries. *J. Phys. Chem. C* **2011**, *115*, 9821–9825. [CrossRef]

84. Jia, X.; Yan, C.; Chen, Z.; Wang, R.; Zhang, Q.; Guo, L.; Wei, F.; Lu, Y. Direct growth of flexible LiMn$_2$O$_4$/CNT lithium-ion cathodes. *Chem. Commun.* **2011**, *47*, 9669–9671. [CrossRef] [PubMed]

85. Cho, J.; Kim, G.B.; Lim, H.S.; Kim, C.S.; Yoo, S.I. Improvement of structural stability of LiMn$_2$O$_4$ cathode material on 55 °C cycling by sol–gel coating of LiCoO$_2$. *Electrochem. Solid State Lett.* **1999**, *2*, 607–609. [CrossRef]

86. Kitao, H.; Fujihara, T.; Takeda, K.; Nakanishi, N.; Nohma, T. High-temperature storage performance of Li-ion batteries using a mixture of Li-Mn spinel and Li-Ni-Co-Mn oxide as a positive electrode material. *Electrochem. Solid State Lett.* **2005**, *8*, A87–A90. [CrossRef]

87. Yi, T.F.; Zhu, Y.R.; Zhu, X.D.; Shu, J.; Yue, C.B.; Zhou, A.N. A review of recent developments in the surface modification of LiMn$_2$O$_4$ as cathode material of power lithium-ion battery. *Ionics* **2009**, *15*, 779–784. [CrossRef]

88. Prabu, M.; Reddy, M.; Selvasekarapandian, S.; Rao, G.S.; Chowdari, B. (Li, Al)-co-doped spinel, Li(Li$_{0.1}$Al$_{0.1}$Mn$_{1.8}$)O$_4$ as high performance cathode for lithium ion batteries. *Electrochim. Acta* **2013**, *88*, 745–755.

89. Zhao, X.; Reddy, M.; Liu, H.; Ramakrishna, S.; Rao, G.S.; Chowdari, B.V. Nano LiMn$_2$O$_4$ with spherical morphology synthesized by a molten salt method as cathodes for lithium ion batteries. *RSC Adv.* **2012**, *2*, 7462–7469. [CrossRef]

90. Sakunthala, A.; Reddy, M.; Selvasekarapandian, S.; Chowdari, B.; Selvin, P.C. Synthesis of compounds, Li(MMn$_{11/6}$)O$_4$ (M = Mn$_{1/6}$, Co$_{1/6}$,(Co$_{1/12}$Cr$_{1/12}$),(Co$_{1/12}$Al$_{1/12}$),(Cr$_{1/12}$Al$_{1/12}$)) by polymer precursor method and its electrochemical performance for lithium-ion batteries. *Electrochim. Acta* **2010**, *55*, 4441–4450. [CrossRef]

91. Reddy, M.; Sakunthala, A.; SelvashekaraPandian, S.; Chowdari, B. Preparation, Comparative Energy Storage Properties, and Impedance Spectroscopy Studies of Environmentally Friendly Cathode, Li(MMn$_{11/6}$) O4 (M = Mn$_{1/6}$,Co$_{1/6}$, (Co$_{1/12}$Cr$_{1/12}$)). *J. Phys. Chem. C* **2013**, *117*, 9056–9064. [CrossRef]

92. Reddy, M.; Manoharan, S.S.; John, J.; Singh, B.; Rao, G.S.; Chowdari, B. Synthesis, Characterization, and Electrochemical Cycling Behavior of the Ru-Doped Spinel, Li [Mn$_{2-x}$Ru$_x$]O$_4$ (x = 0, 0.1, and 0.25). *J. Electrochem. Soc.* **2009**, *156*, A652–A660. [CrossRef]

93. Reddy, M.; Cheng, H.; Tham, J.; Yuan, C.; Goh, H.; Chowdari, B. Preparation of Li(Ni$_{0.5}$Mn$_{1.5}$)O$_4$ by polymer precursor method and its electrochemical properties. *Electrochim. Acta* **2012**, *62*, 269–275.

94. Reddy, M.; Raju, M.S.; Sharma, N.; Quan, P.; Nowshad, S.H.; Emmanuel, H.C.; Peterson, V.; Chowdari, B. Preparation of Li$_{1.03}$Mn$_{1.97}$O$_4$ and Li$_{1.06}$Mn$_{1.94}$O$_4$ by the polymer precursor method and X-ray, neutron diffraction and electrochemical studies. *J. Electrochem. Soc.* **2011**, *158*, A1231–A1236.

95. Wei, Y.; Yan, L.; Wang, C.; Xu, X.; Wu, F.; Chen, G. Effects of Ni doping on [MnO$_6$] octahedron in LiMn$_2$O$_4$. *J. Phys. Chem. B* **2004**, *108*, 18547–18551. [CrossRef]

96. Liu, Q.; Wang, S.; Tan, H.; Yang, Z.; Zeng, J. Preparation and doping mode of doped LiMn$_2$O$_4$ for Li-ion batteries. *Energies* **2013**, *6*, 1718–1730. [CrossRef]

97. Masquelier, C.; Croguennec, L. Polyanionic (phosphates, silicates, sulfates) frameworks as electrode materials for rechargeable Li (or Na) batteries. *Chem. Rev.* **2013**, *113*, 6552–6591. [CrossRef] [PubMed]

98. Ravet, N.; Goodenough, J.; Besner, S.; Simoneau, M.; Hovington, P.; Armand, M. Improved iron based cathode material. In Proceedings of the 196th ECS meeting, Honolulu, HI, USA, 17–22 October 1999.

99. Yamada, A.; Chung, S.C.; Hinokuma, K. Optimized LiFePO$_4$ for lithium battery cathodes. *J. Electrochem. Soc.* **2001**, *148*, A224–A229. [CrossRef]

100. Saravanan, K.; Reddy, M.; Balaya, P.; Gong, H.; Chowdari, B.; Vittal, J.J. Storage performance of LiFePO$_4$ nanoplates. *J. Mater. Chem.* **2009**, *19*, 605–610. [CrossRef]

101. Doeff, M.M.; Hu, Y.; McLarnon, F.; Kostecki, R. Effect of surface carbon structure on the electrochemical performance of LiFePO$_4$. *Electrochem. Solid State Lett.* **2003**, *6*, A207–A209. [CrossRef]

102. Wilcox, J.D.; Doeff, M.M.; Marcinek, M.; Kostecki, R. Factors influencing the quality of carbon coatings on LiFePO$_4$. *J. Electrochem. Soc.* **2007**, *154*, A389–A395. [CrossRef]

103. Zaghib, K.; Shim, J.; Guerfi, A.; Charest, P.; Striebel, K. Effect of carbon source as additives in LiFePO$_4$ as positive electrode for lithium-ion batteries. *Electrochem. Solid State Lett.* **2005**, *8*, A207–A210. [CrossRef]

104. Dominko, R.; Bele, M.; Gaberscek, M.; Remskar, M.; Hanzel, D.; Pejovnik, S.; Jamnik, J. Impact of the carbon coating thickness on the electrochemical performance of LiFePO$_4$/C composites. *J. Electrochem. Soc.* **2005**, *152*, A607–A610. [CrossRef]

105. Hu, Y.; Doeff, M.M.; Kostecki, R.; Finones, R. Electrochemical performance of sol–gel synthesized LiFePO$_4$ in lithium batteries. *J. Electrochem. Soc.* **2004**, *151*, A1279–A1285. [CrossRef]

106. Jinli, Z.; Jiao, W.; Yuanyuan, L.; Ning, N.; Junjie, G.; Feng, Y.; Wei, L. High-performance lithium iron phosphate with phosphorus-doped carbon layers for lithium ion batteries. *J. Mater. Chem. A* **2015**, *3*, 2043–2049. [CrossRef]

107. Zhang, J.; Nie, N.; Liu, Y.; Wang, J.; Yu, F.; Gu, J.; Li, W. Boron and nitrogen codoped carbon layers of LiFePO$_4$ improve the high-rate electrochemical performance for lithium ion batteries. *ACS Appl. Mater. Interfaces* **2015**, *7*, 20134–20143. [CrossRef] [PubMed]

108. Rao, R.P.; Reddy, M.; Adams, S.; Chowdari, B. Preparation, temperature dependent structural, molecular dynamics simulations studies and electrochemical properties of LiFePO$_4$. *Mater. Res. Bull.* **2015**, *66*, 71–75. [CrossRef]

109. Zhou, Y.; Wang, J.; Hu, Y.; O'Hayre, R.; Shao, Z. A porous LiFePO$_4$ and carbon nanotube composite. *Chem. Commun.* **2010**, *46*, 7151–7153. [CrossRef] [PubMed]

110. Yang, J.; Wang, J.; Wang, D.; Li, X.; Geng, D.; Liang, G.; Gauthier, M.; Li, R.; Sun, X. 3D porous LiFePO$_4$/graphene hybrid cathodes with enhanced performance for Li-ion batteries. *J. Power Sources* **2012**, *208*, 340–344. [CrossRef]

111. Shi, Y.; Chou, S.L.; Wang, J.Z.; Wexler, D.; Li, H.J.; Liu, H.K.; Wu, Y. Graphene wrapped LiFePO$_4$/C composites as cathode materials for Li-ion batteries with enhanced rate capability. *J. Mater. Chem.* **2012**, *22*, 16465–16470. [CrossRef]

112. Hu, L.H.; Wu, F.Y.; Lin, C.T.; Khlobystov, A.N.; Li, L.J. Graphene-modified LiFePO$_4$ cathode for lithium ion battery beyond theoretical capacity. *Nat. Commun.* **2013**, *4*, 1687. [CrossRef] [PubMed]

113. Yang, J.; Wang, J.; Tang, Y.; Wang, D.; Xiao, B.; Li, X.; Li, R.; Liang, G.; Sham, T.K.; Sun, X. In situ self-catalyzed formation of core–shell LiFePO$_4$@CNT nanowires for high rate performance lithium-ion batteries. *J. Mater. Chem. A* **2013**, *1*, 7306–7311. [CrossRef]

114. Yang, J.; Wang, J.; Li, X.; Wang, D.; Liu, J.; Liang, G.; Gauthier, M.; Li, Y.; Geng, D.; Li, R. Hierarchically porous LiFePO$_4$/nitrogen-doped carbon nanotubes composite as a cathode for lithium ion batteries. *J. Mater. Chem.* **2012**, *22*, 7537–7543. [CrossRef]

115. Reddy, M.; Rao, G.S.; Chowdari, B. Long-term cycling studies on 4 V-cathode, lithium vanadium fluorophosphate. *J. Power Sources* **2010**, *195*, 5768–5774. [CrossRef]

116. Hameed, A.S.; Nagarathinam, M.; Reddy, M.; Chowdari, B.; Vittal, J.J. Synthesis and electrochemical studies of layer-structured metastable α i-LiVOPO$_4$. *J. Mater. Chem.* **2012**, *22*, 7206–7213. [CrossRef]

117. Hameed, A.S.; Reddy, M.; Chowdari, B.; Vittal, J.J. Carbon coated Li$_3$V$_2$(PO$_4$)$_3$ from the single-source precursor, Li$_2$(VO)$_2$(HPO$_4$)$_2$(C$_2$O$_4$)·6H$_2$O as cathode and anode materials for Lithium ion batteries. *Electrochim. Acta* **2014**, *128*, 184–191. [CrossRef]

118. Nagarathinam, M.; Saravanan, K.; Phua, E.J.H.; Reddy, M.; Chowdari, B.; Vittal, J.J. Redox-Active Metal-Centered Oxalato Phosphate Open Framework Cathode Materials for Lithium Ion Batteries. *Angew. Chem. Int. Ed.* **2012**, *51*, 5866–5870. [CrossRef] [PubMed]
119. Hameed, A.S.; Reddy, M.; Nagarathinam, M.; Runčevski, T.; Dinnebier, R.E.; Adams, S.; Chowdari, B.; Vittal, J.J. Room temperature large-scale synthesis of layered frameworks as low-cost 4 V cathode materials for lithium ion batteries. *Sci. Rep.* **2015**, *5*, 16270. [CrossRef] [PubMed]
120. Hameed, A.S.; Reddy, M.; Sarkar, N.; Chowdari, B.; Vittal, J.J. Synthesis and electrochemical investigation of novel phosphite based layered cathodes for Li-ion batteries. *RSC Adv.* **2015**, *5*, 60630–60637. [CrossRef]
121. Hameed, A.S.; Nagarathinam, M.; Schreyer, M.; Reddy, M.; Chowdari, B.; Vittal, J.J. A layered oxalatophosphate framework as a cathode material for Li-ion batteries. *J. Mater. Chem. A* **2013**, *1*, 5721–5726. [CrossRef]
122. Ke, F.S.; Wu, Y.S.; Deng, H. Metal-organic frameworks for lithium ion batteries and supercapacitors. *J. Solid State Chem.* **2015**, *223*, 109–121. [CrossRef]
123. Bai, L.; Tu, B.; Qi, Y.; Gao, Q.; Liu, D.; Liu, Z.; Zhao, L.; Li, Q.; Zhao, Y. Enhanced performance in gas adsorption and Li ion batteries by docking Li$^+$ in a crown ether-based metal-organic framework. *Chem. Commun.* **2016**, *52*, 3003–3006. [CrossRef] [PubMed]
124. Banerjee, A.; Singh, U.; Aravindan, V.; Srinivasan, M.; Ogale, S. Synthesis of CuO nanostructures from Cu-based metal organic framework (MOF-199) for application as anode for Li-ion batteries. *Nano Energy* **2013**, *2*, 1158–1163. [CrossRef]
125. Wu, Y.P.; Rahm, E.; Holze, R. Carbon anode materials for lithium ion batteries. *J. Power Sources* **2003**, *114*, 228–236. [CrossRef]
126. Roberts, A.D.; Li, X.; Zhang, H. Porous carbon spheres and monoliths: Morphology control, pore size tuning and their applications as Li-ion battery anode materials. *Chem. Soc. Rev.* **2014**, *43*, 4341–4356. [CrossRef] [PubMed]
127. Buiel, E.; Dahn, J. Li-insertion in hard carbon anode materials for Li-ion batteries. *Electrochim. Acta* **1999**, *45*, 121–130. [CrossRef]
128. Nishimura, K.; Honbo, H.; Takeuchi, S.; Horiba, T.; Oda, M.; Koseki, M.; Muranaka, Y.; Kozono, Y.; Miyadera, H. Design and performance of 10 Wh rechargeable lithium batteries. *J. Power Sources* **1997**, *68*, 436–439. [CrossRef]
129. Wu, Y.; Jiang, C.; Wan, C.; Holze, R. Composite materials of silver and natural graphite as anode with low sensibility to humidity. *J. Power Sources* **2002**, *112*, 255–260. [CrossRef]
130. Wu, Y.; Jiang, C.; Wan, C.; Tsuchida, E. Composite anode material for lithium ion battery with low sensitivity to water. *Electrochem. Commun.* **2000**, *2*, 626–629. [CrossRef]
131. Yu, P.; Ritter, J.A.; White, R.E.; Popov, B.N. Ni-composite microencapsulated graphite as the negative electrode in lithium-ion batteries I. Initial irreversible capacity study. *J. Electrochem. Soc.* **2000**, *147*, 1280–1285. [CrossRef]
132. Yu, P.; Ritter, J.A.; White, R.E.; Popov, B.N. Ni-composite microencapsulated graphite as the negative electrode in lithium-ion batteries II: Electrochemical impedance and self-discharge studies. *J. Electrochem. Soc.* **2000**, *147*, 2081–2085. [CrossRef]
133. Guo, K.; Pan, Q.; Wang, L.; Fang, S. Nano-scale copper-coated graphite as anode material for lithium-ion batteries. *J. Appl. Electrochem.* **2002**, *32*, 679–685. [CrossRef]
134. Kim, S.-S.; Kadoma, Y.; Ikuta, H.; Uchimoto, Y.; Wakihara, M. Electrochemical performance of natural graphite by surface modification using aluminum. *Electrochem. Solid State Lett.* **2001**, *4*, A109–A112. [CrossRef]
135. Takamura, T.; Sumiya, K.; Suzuki, J.; Yamada, C.; Sekine, K. Enhancement of Li doping/undoping reaction rate of carbonaceous materials by coating with an evaporated metal film. *J. Power Sources* **1999**, *81*, 368–372. [CrossRef]
136. Yoshio, M.; Wang, H.; Fukuda, K.; Umeno, T.; Abe, T.; Ogumi, Z. Improvement of natural graphite as a lithium-ion battery anode material, from raw flake to carbon-coated sphere. *J. Mater. Chem.* **2004**, *14*, 1754–1758. [CrossRef]
137. Wang, H.; Yoshio, M. Carbon-coated natural graphite prepared by thermal vapor decomposition process, a candidate anode material for lithium-ion battery. *J. Power Sources* **2001**, *93*, 123–129. [CrossRef]
138. Natarajan, C.; Fujimoto, H.; Tokumitsu, K.; Mabuchi, A.; Kasuh, T. Reduction of the irreversible capacity of a graphite anode by the CVD process. *Carbon* **2001**, *39*, 1409–1413. [CrossRef]

139. Yoshio, M.; Wang, H.; Fukuda, K.; Hara, Y.; Adachi, Y. Effect of carbon coating on electrochemical performance of treated natural graphite as lithium-ion battery anode material. *J. Electrochem. Soc.* **2000**, *147*, 1245–1250. [CrossRef]

140. Dey, A.; Sullivan, B. The electrochemical decomposition of propylene carbonate on graphite. *J. Electrochem. Soc.* **1970**, *117*, 222–224. [CrossRef]

141. Jeong, G.; Kim, Y.U.; Kim, H.; Kim, Y.J.; Sohn, H.J. Prospective materials and applications for Li secondary batteries. *Energy Environ. Sci.* **2011**, *4*, 1986–2002. [CrossRef]

142. Aurbach, D.; Zinigrad, E.; Cohen, Y.; Teller, H. A short review of failure mechanisms of lithium metal and lithiated graphite anodes in liquid electrolyte solutions. *Solid State Ion.* **2002**, *148*, 405–416. [CrossRef]

143. Beattie, S.D.; Larcher, D.; Morcrette, M.; Simon, B.; Tarascon, J.M. Si electrodes for Li-ion batteries-a new way to look at an old problem. *J. Electrochem. Soc.* **2008**, *155*, A158–A163. [CrossRef]

144. Lee, H.Y.; Lee, S.M. Carbon-coated nano-Si dispersed oxides/graphite composites as anode material for lithium ion batteries. *Electrochem. Commun.* **2004**, *6*, 465–469. [CrossRef]

145. Wen, Z.; Yang, J.; Wang, B.; Wang, K.; Liu, Y. High capacity silicon/carbon composite anode materials for lithium ion batteries. *Electrochem. Commun.* **2003**, *5*, 165–168. [CrossRef]

146. Yoshio, M.; Wang, H.; Fukuda, K.; Umeno, T.; Dimov, N.; Ogumi, Z. Carbon-coated Si as a lithium-ion battery anode material. *J. Electrochem. Soc.* **2002**, *149*, A1598–A1603. [CrossRef]

147. Wang, G.; Ahn, J.; Yao, J.; Bewlay, S.; Liu, H. Nanostructured Si–C composite anodes for lithium-ion batteries. *Electrochem. Commun.* **2004**, *6*, 689–692. [CrossRef]

148. Liu, N.; Wu, H.; McDowell, M.T.; Yao, Y.; Wang, C.; Cui, Y. A yolk-shell design for stabilized and scalable Li-ion battery alloy anodes. *Nano Lett.* **2012**, *12*, 3315–3321. [CrossRef] [PubMed]

149. Liu, N.; Lu, Z.; Zhao, J.; McDowell, M.T.; Lee, H.W.; Zhao, W.; Cui, Y. A pomegranate-inspired nanoscale design for large-volume-change lithium battery anodes. *Nat. Nanotechnol.* **2014**, *9*, 187–192. [CrossRef] [PubMed]

150. Mazouzi, D.; Karkar, Z.; Hernandez, C.R.; Manero, P.J.; Guyomard, D.; Roué, L.; Lestriez, B. Critical roles of binders and formulation at multiscales of silicon-based composite electrodes. *J. Power Sources* **2015**, *280*, 533–549. [CrossRef]

151. Huang, Y.H.; Goodenough, J.B. High-rate LiFePO$_4$ lithium rechargeable battery promoted by electrochemically active polymers. *Chem. Mater.* **2008**, *20*, 7237–7241. [CrossRef]

152. Chen, W.M.; Qie, L.; Yuan, L.X.; Xia, S.A.; Hu, X.L.; Zhang, W.X.; Huang, Y.H. Insight into the improvement of rate capability and cyclability in LiFePO$_4$/polyaniline composite cathode. *Electrochim. Acta* **2011**, *56*, 2689–2695. [CrossRef]

153. Tamura, T.; Aoki, Y.; Ohsawa, T.; Dokko, K. Polyaniline as a functional binder for LiFePO$_4$ cathodes in lithium batteries. *Chem. Lett.* **2011**, *40*, 828–830. [CrossRef]

154. Huang, Y.H.; Park, K.S.; Goodenough, J.B. Improving lithium batteries by tethering carbon-coated LiFePO$_4$ to polypyrrole. *J. Electrochem. Soc.* **2006**, *153*, A2282–A2286. [CrossRef]

155. Dominko, R.; Gaberšček, M.; Drofenik, J.; Bele, M.; Pejovnik, S. A novel coating technology for preparation of cathodes in Li-ion batteries. *Electrochem. Solid State Lett.* **2001**, *4*, A187–A190. [CrossRef]

156. Chen, Z.; Christensen, L.; Dahn, J. Comparison of PVDF and PVDF-TFE-P as binders for electrode materials showing large volume changes in lithium-ion batteries. *J. Electrochem. Soc.* **2003**, *150*, A1073–A1078. [CrossRef]

157. Chen, Z.; Christensen, L.; Dahn, J. Large-volume-change electrodes for Li-ion batteries of amorphous alloy particles held by elastomeric tethers. *Electrochem. Commun.* **2003**, *5*, 919–923. [CrossRef]

158. Liu, W.R.; Yang, M.H.; Wu, H.C.; Chiao, S.; Wu, N.L. Enhanced cycle life of Si anode for Li-ion batteries by using modified elastomeric binder. *Electrochem. Solid State Lett.* **2005**, *8*, A100–A103. [CrossRef]

159. Koo, B.; Kim, H.; Cho, Y.; Lee, K.T.; Choi, N.S.; Cho, J. A Highly cross-linked polymeric binder for high-performance silicon negative electrodes in lithium ion batteries. *Angew. Chem. Int. Ed.* **2012**, *51*, 8762–8767. [CrossRef] [PubMed]

160. Kwon, T.W.; Jeong, Y.K.; Deniz, E.; AlQaradawi, S.Y.; Choi, J.W.; Coskun, A. Dynamic cross-linking of polymeric binders based on host–guest interactions for silicon anodes in lithium ion batteries. *ACS Nano* **2015**, *9*, 11317–11324. [CrossRef] [PubMed]

161. Arora, P.; Zhang, Z. Battery separators. *Chem. Rev.* **2004**, *104*, 4419–4462. [CrossRef] [PubMed]

162. Song, J.; Ryou, M.H.; Son, B.; Lee, J.N.; Lee, D.J.; Lee, Y.M.; Choi, J.W.; Park, J.K. Co-polyimide-coated polyethylene separators for enhanced thermal stability of lithium ion batteries. *Electrochim. Acta* **2012**, *85*, 524–530. [CrossRef]

163. Liu, H.; Xu, J.; Guo, B.; He, X. Effect of Al_2O_3/SiO_2 composite ceramic layers on performance of polypropylene separator for lithium-ion batteries. *Ceram. Int.* **2014**, *40*, 14105–14110. [CrossRef]

164. Choi, J.A.; Kim, S.H.; Kim, D.W. Enhancement of thermal stability and cycling performance in lithium-ion cells through the use of ceramic-coated separators. *J. Power Sources* **2010**, *195*, 6192–6196. [CrossRef]

165. Ryou, M.H.; Lee, Y.M.; Park, J.K.; Choi, J.W. Mussel-inspired polydopamine-treated polyethylene separators for high-power Li-ion batteries. *Adv. Mater.* **2011**, *23*, 3066–3070. [CrossRef] [PubMed]

166. Lee, J.Y.; Lee, Y.M.; Bhattacharya, B.; Nho, Y.C.; Park, J.K. Separator grafted with siloxane by electron beam irradiation for lithium secondary batteries. *Electrochim. Acta* **2009**, *54*, 4312–4315. [CrossRef]

167. Ko, J.; Min, B.; Kim, D.W.; Ryu, K.; Kim, K.; Lee, Y.; Chang, S. Thin-film type Li-ion battery, using a polyethylene separator grafted with glycidyl methacrylate. *Electrochim. Acta* **2004**, *50*, 367–370. [CrossRef]

168. Zhang, S.S. A review on the separators of liquid electrolyte Li-ion batteries. *J. Power Sources* **2007**, *164*, 351–364. [CrossRef]

Article

A State of Charge Estimator Based Extended Kalman Filter Using an Electrochemistry-Based Equivalent Circuit Model for Lithium-Ion Batteries

Xin Lai *, Chao Qin, Wenkai Gao, Yuejiu Zheng * and Wei Yi

School of Mechanical Engineering, University of Shanghai for Science and Technology, Shanghai 200093, China; Qin_Chao1995@126.com (C.Q.); wenkai_gao@163.com (W.G.); 13072182082@163.com (W.Y.)
* Correspondence: laixin@usst.edu.cn (X.L); yuejiu_zheng@163.com (Y.Z.); Tel.: +86-21-55275287 (Y.Z.)

Received: 22 August 2018; Accepted: 6 September 2018; Published: 8 September 2018

Abstract: In this paper, an improved equivalent circuit model (ECM) considering partial electrochemical properties is developed for accurate state-of-charge (SOC). In the proposed model, the solid-phase diffusion process is calculated by a simple equation about particle surface SOC, and the double layer is simulated by two resistance-capacitance (RC) networks. To improve the global accuracy of the model, a subarea parameter-identification method based on particle swarm optimization is proposed, in order to determine the optimal model parameters in the entire SOC area. Then, an SOC estimator is developed based on extended kalman filter. The comparative study shows that a model considering solid-phase diffusion with two RC networks is the best choice. Finally, experimental results show that the accuracy of the proposed model is one times higher than that of the traditional ECM in the low SOC area, and is able to estimate SOC with errors less than 1% in the entire SOC area. Furthermore, estimation results of two types of batteries under two working conditions indicate that the developed model and SOC estimator have satisfactory global accuracy and guaranteed robustness with low computational complexity, which can be applied in real-time situations.

Keywords: lithium-ion batteries; simplified electrochemical model; state of charge estimator; extended kalman filter

---

1. Introduction

Electrochemical energy storage systems (EESS) are power sources for many devices, e.g., cell phones, laptops, medical devices, electric vehicles (EVs), smart grid systems, etc. [1]. With a high demand of superior EESS, lithium-ion batteries (LIBs) have gained increasing popularity over other existing typical electrochemical batteries due to their favorable performances in high energy density, lightweight, long cyclic lifetime, low self-discharge rate, and almost zero memory effect [2,3]. However, LIB is a nonlinear dynamic system with a very narrow operating range, and some incorrect operations could lead to irreversible damage and shortened life [4–6]. Therefore, a reliable and effective battery management system (BMS) is required to optimize performance, improve safety and prolong life of LIBs. The critical state estimation, such as state of charge (SOC), state of health, and state of power, is fundamental in a BMS. Specifically, the SOC is an essential indicator used to regulate the operating decisions and to avoid overcharge or overdischarge [7]. However, the SOC cannot be directly measured by sensors, and the battery itself is highly nonlinear, which makes accurate SOC estimation very difficult [8].

1.1. Literature Review

Since the SOC estimation is essentially based on the battery model for most SOC estimation methods, an accurate battery model and matched model parameters are the prerequisite for accurate SOC estimation. Obviously, model accuracy is closely related to model structure and model parameter identification algorithm. The equivalent circuit models (ECMs) are the most common battery models adopted in the actual BMS because of few computations and acceptable precision [9]. The popular and widely reported ECMs include the Rint model, Thevenin model, partnership for a new generation of vehicles (PNGV) model, and general nonlinear (GNL) model [10–12], which are all based on resistance-capacitance (RC) networks with different orders. The model structure and electronic component values of the ECM directly affect the features and accuracy of the model. Ref. [13] investigated eleven ECMs and stated that the second-order RC models are the best choice owing to their balance of accuracy and reliability. However, ECMs are not empirical models, which means the model parameters have no clear electrochemical meaning [14]. Therefore, the model errors may be large, especially in low SOC area (SOC lower than 20%) [15]. Generally, the empirical model needs large computations and memories due to its complex equations, and currently it is still hard to use in a BMS for on-line estimation and real-time control. Based on the structure of ECM, Ref. [15] proposed an extended equivalent circuit model (EECM) considering partial electrochemical properties. However, the reliability of parameter identification is doubtful because nine parameters need to be identified in proposed EECM.

The most popular existing approaches for parameter identification of ECMs include the genetic algorithm (GA) [16], particle swarm optimization (PSO) algorithm [17], and the least-squares method [18]. The appropriate identification algorithm should match the battery model to pursue the balance between accuracy and computation. Moreover, the local optimization problem should also be avoided in the process of model parameter identification.

We could conclude that an improved ECM considering partial electrochemical properties is a better choice to balance accuracy and computational burden for online application. Moreover, appropriate model parameter identification and SOC estimation algorithm are essential for improving the accuracy and robustness of SOC estimation.

1.2. Main Contributions

This paper aims at developing an onboard battery model considering partial electrochemical properties and an accurate and robust SOC estimator in the entire SOC area. The unique contributions brought about in this paper are the following.

(1) Four typical ECMs and four improved ECMs considering partial electrochemical properties are compared under the new European Driving Cycle (NEDC) and the dynamic stress test (DST) working conditions to obtain a more suitable battery model for the entire SOC area.
(2) A subarea parameter-identification method based on PSO is proposed to improve the global model accuracy in the entire SOC area.
(3) A SOC estimator based on extended kalman filter (EKF) for our proposed model is developed, and its accuracy and robustness are verified by experiments.

1.3. Organization of the Paper

The rest of this paper is organized as follows: Section 2 describes the developed model. In Section 3, a model-parameter identification method in the entire SOC area is proposed, and model errors are compared for various models to obtain a more appropriate model. Section 4 describes an EKF-based SOC estimator for the developed model, and its advantages are verified by experiments. Finally, conclusions drawn and the closing remarks are presented in Section 5.

2. Electrochemistry-Based Equivalent Circuit Model

2.1. Single-Particle Model

The pseudo-two-dimensional model (P2D) are widely used to describe the electrochemical behavior of lithium-ion batteries [14,19,20]. However, P2D is hard to use in onboard cases because of its complexity. Therefore, a series of model simplification attempts were made to reduce the computational complexity. The single-particle model (SPM) is a simplified model that is derived by approximating the electrode by a single spherical particle, and is becoming a popular model in recent years for SOC estimation [21,22]. The schematic of the SPM, which is illustrated in Figure 1.

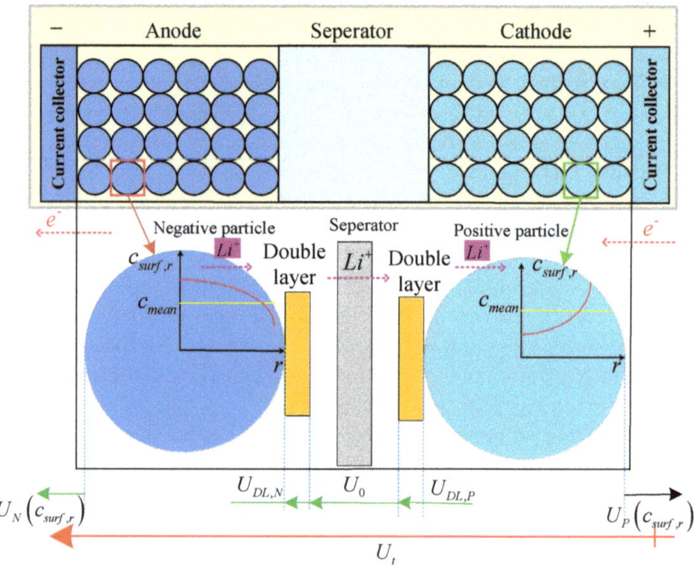

Figure 1. Schematic of the single-particle model (SPM) (modified from Reference [23]).

Based on Reference [15], the terminal voltage (U_t) can be expressed as:

$$U_t = \left(U_P\left(c_{surf,r}\right) - U_N\left(c_{surf,r}\right)\right) - U_0 - U_{DL} \tag{1}$$

where $c_{surf,r}$ is lithium concentration at electrode particle surface, $U_P\left(c_{surf,r}\right)$ and $U_N\left(c_{surf,r}\right)$ are the surface potential of positive electrode and negative electrode, respectively. U_0 is the sum of the liquid phase voltage drop caused by the separator and electrolyte, and the voltage drop of the collector. U_{DL} is the voltage drop of the double layer, and it can be determined as:

$$U_{DL} = IR_{CT}\left(1 - e^{\left(\frac{-t}{\tau_{DL}}\right)}\right) \tag{2}$$

where R_{CT} is the resistance of the double layer, and τ_{DL} is time constant.

Moreover, Equation (1) can be rewritten as:

$$U_t = \left(U_P\left(SOC_{surf,r}\right) - U_N\left(SOC_{surf,r}\right)\right) - U_0 - U_{DL} \tag{3}$$

where $SOC_{surf,r}$ is the SOC at particle surface. However, $SOC_{surf,r}$ could not be directly determined, and it can be expressed as:

$$SOC_{surf,r} = SOC_{avg} + \Delta SOC \tag{4}$$

where SOC_{avg} is the average SOC, which is represented by the average concentration of Li^+ in the electrode particle (C_{mean}). ΔSOC is the difference between SOC_{avg} and $SOC_{surf,r}$, which is related to the difference of $c_{surf,r}$ and C_{mean} in solid-phase diffusion process. According to Reference [15], ΔSOC follows the Equation (5):

$$\Delta SOC = K_{SD}I_F\left(1 - e^{\left(\frac{-t}{\tau_{SD}}\right)}\right) \tag{5}$$

where K_{SD} is concentration difference parameter in solid diffusion, I_F is the faradaic current, and τ_{SD} is time constant.

When the traditional ECM is used for SOC estimation, the open circuit voltage (OCV) of the battery U_{OCV} is looked up by the SOC_{avg} with the OCV-SOC curve. However, SOC_{avg} could not reflect the solid-phase diffusion. In this study, we used $U_{OCV}\left(SOC_{surf}\right)$ instead of $U_{OCV}\left(SOC_{avg}\right)$, and Equation (3) is hence rewritten as:

$$U_t = U_{OCV}\left(SOC_{surf}\right) - U_0 - U_{DL} \tag{6}$$

It is noted that the difference between the surface concentration c_{surf} and the average concentration C_{mean} indicates the solid-phase diffusion results. c_{surf} reflects the dynamic process of lithium-ion, which indirectly reflects the process of solid-phase diffusion. SOC_{surf} is closely related to c_{surf}. Therefore, our model includes the solid phase diffusion process and is a simplified electrochemical model.

2.2. ECM Considering Electrochemical Properties

A typical ECM generally uses the RC network comprising resistors and capacitors to simulate the dynamic characteristics of the battery, and these ECMs with n RC-networks is called the nRC model hereafter. The terminal voltage of the battery determined by the Kirchhoff voltage law can be expressed as [13,24]:

$$U_t = U_{OCV}(SOC) - IR_0 - \sum_{i=1}^{n} R_i\left(1 - e^{-t/R_iC_i}\right) \tag{7}$$

where R_0 is ohmic resistance, I is charge or discharge current, R_i and C_i are the i-th polarization resistance and i-th polarization capacitance, respectively.

Comparing Equation (7) with Equation (6), we can see that the model equations are very similar, the small difference is that the Equation (7) uses the SOC_{avg}, while the Equation (6) uses the SOC_{surf}. Combining traditional ECM and considering the solid-phase diffusion process inside the battery, an electrochemistry-based ECM is developed, and the schematic of this model is shown in Figure 2. In this model, the RC network is used to simulate the influence of the double electric layer, and a simple equation is used to calculate the solid-phase diffusion process. The number of RC networks corresponds to the number of double electric layers. To clarify the effect of the double layer and solid-phase diffusion on the model accuracy, eight models are chosen for comparison, and their mathematical equations are shown in Table 1. In Table 1, nRC represents ECMs with n RC-networks, EnRC (n = 0, 1, 2, 3) represents ECMs considering electrochemical properties. Obviously, n is the number of electric double layers in the model. In Section 3, model parameters will be identified, and errors of eight models will be compared to choose the right model.

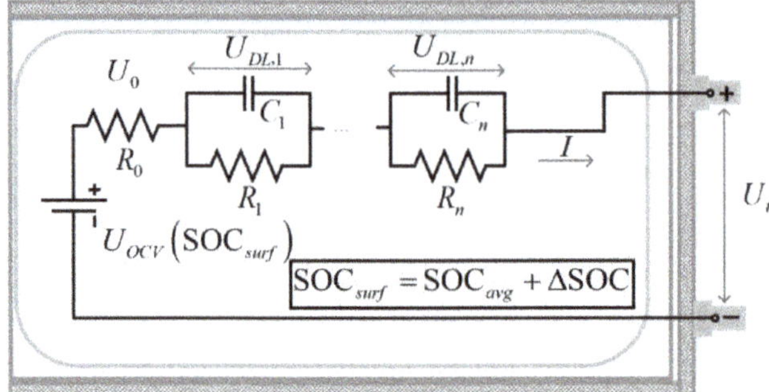

Figure 2. Schematic of the proposed EnRC model.

Table 1. Discretization equations of various resistance-capacitance (RC) models.

Model Name	Discretization Equations (Discharge Is Negative)	Number of Parameters
0RC	$U_t(k) = U_{OCV}(\text{SOC}_{avg}(k)) + IR_0$	2
E0RC	$\begin{cases} U_t(k) = U_{OCV}(\text{SOC}_{surf}(k)) + IR_0 \\ \text{SOC}_{surf}(k) = \text{SOC}_{avg} + \Delta\text{SOC}(k) \\ \Delta\text{SOC}(k) = \Delta\text{SOC}(k-1)e^{-\frac{T_s}{\tau_{SD}}} + K_{SD}I\left(1 - e^{-\frac{T_s}{\tau_{SD}}}\right) \end{cases}$	4
nRC: 1RC 2RC 3RC	$\begin{cases} U_t(k) = U_{OCV}(\text{SOC}_{avg}) + I_k R_0 + \sum\limits_{i=1}^{n} U_i(k) \quad (n=1,2,3) \\ U_i(k) = U_i(k-1)e^{-\frac{T_s}{\tau_i}} + I_k R_i\left(1 - e^{-\frac{T_s}{\tau_i}}\right) \end{cases}$	1RC: 4 2RC: 6 3RC: 8
EnRC: E1RC E2RC E3RC	$\begin{cases} U_t(k) = U_{OCV}(\text{SOC}_{avg}(k)) + I_k R_0 + \sum\limits_{i=1}^{n} U_i(k) \quad (n=1,2,3) \\ U_i(k) = U_i(k-1)e^{-\frac{T_s}{\tau_i}} + I_k R_i\left(1 - e^{-\frac{T_s}{\tau_i}}\right) \\ \text{SOC}_{surf}(k) = \text{SOC}_{avg} + \Delta\text{SOC}(k) \\ \Delta\text{SOC}(k) = \Delta\text{SOC}(k-1)e^{-\frac{T_s}{\tau_{SD}}} + K_{SD}I\left(1 - e^{-\frac{T_s}{\tau_{SD}}}\right) \end{cases}$	E1RC: 6 E2RC: 8 E3RC: 10

3. Model Parameter Identification and Comparison

3.1. Experiments

The experiments were performed using a commercial LIB with cathode of $\text{LiNi}_x\text{Co}_y\text{Mn}_{1-x-y}$ (NCM). To fully verify the effectiveness of the proposed model, two types of LIBs were selected for experiments. The basic parameters of two LIBs are listed in Table 2. As shown in Figure 3a, the experiments were conducted in a battery tester made by DIGATRON which has a current range of -100 A to $+100$ A and a voltage range of 0 V to 20 V. The voltage accuracy is 1 mV and the current accuracy is $\pm0.1\%$ full scale. A software (BTS-600, Digatron Power Electronics, Aachen, Germany) installed on PC is used to control the charging and discharging of the battery according to the given operating conditions, and record the terminal voltage and current of the battery at a frequency of 1 Hz. The acquired data was used for model parameter identification and SOC estimation.

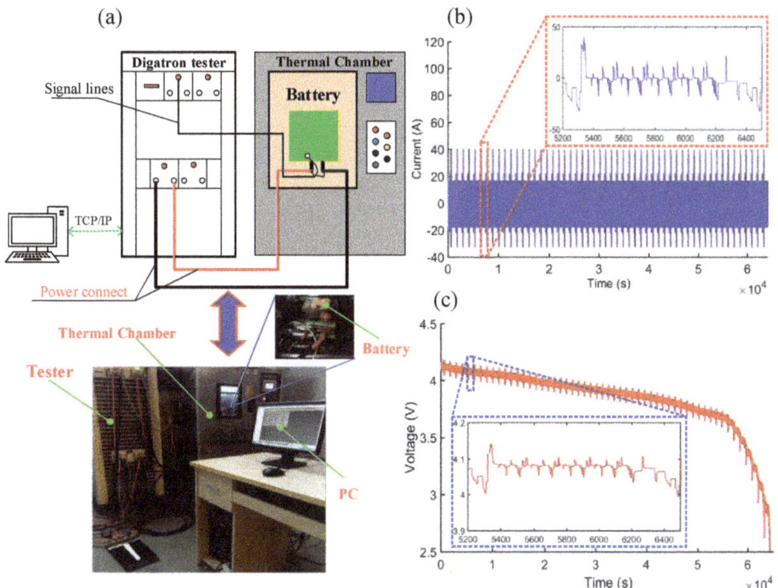

Figure 3. Experimental equipment and results. (**a**) Schematic of the cell test system; (**b**) Measured current under European Driving Cycle (NEDC) working condition; (**c**) Measured voltage under NEDC working condition.

Capacity and hybrid pulse power characterization (HPPC) experiments [25] were first performed to determine the capacity and OCV of LIBs. The capacity test process is as follows: Place the test LIB in the temperature chamber at 25 °C for 3 h. Then, discharge the LIB at a constant discharge current 1/3 C to 2.5 V. After waiting for 1 h, fully charge the LIB using the constant current-constant voltage (CC-CV) method. In this method, the LIB is charged at a constant current (1/3 C) until the voltage reaches 4.15 V, and then, the LIB is charged at a constant voltage until the charging current falls to 1.6 A; then, charging is paused for 1 h. This process is repeated three times, and the mean value of the test capacity is chosen as the battery capacity. The HPPC test is designed to determine the open circuit voltage (OCV). In this test, a series of pulse power sequences are provided to the fully charged battery. Following one pulse power sequence, the battery is discharged to a SOC of 97.5% at 1/3 C and rested for 3 h before the next pulse power sequence is provided. The battery is tested at decrements of 2.5% SOC (10% when the SOC is less than 90%) until the cutoff voltage of 2.5 V is reached.

Then, the two test LIBs were subsequently fully charged at 1/3 C. The discharge experiment was then performed, until the batteries were fully discharged at 1/3 C under the NEDC and DST working cycles until the cutoff voltage of 2.5 V is reached, respectively. It is noted that the charge and discharge currents are controlled by the program according to the cyclic working curve. The current and voltage curves under NEDC cycles on Cell #1 obtained from the experimental results are displayed in Figure 3b,c. The experimental data of Cell #2 under two cycle conditions are not listed here for brevity.

Table 2. Main parameters of experimental lithium-ion batteries (LIBs).

Type	Nominal Capacity (Ah)	Lower Cut-Off Voltage (V)	Upper Cut-Off Voltage (V)	Maximum Charge Current (A)
Cell #1	32.5	2.5	4.15	65
Cell #2	40	2.8	4.2	100

3.2. Model Parameter Identification Using PSO

For the nRC and EnRC, the model parameters that need to be identified and optimized can be expresses as:

$$
\begin{cases}
V_j = \begin{bmatrix} R_0^+ & R_0^- \end{bmatrix} \text{(if the model is 0RC)} \\[4pt]
V_j = \begin{bmatrix} R_0^+ & R_0^- & \underbrace{\tau_1 \ R_1}_{1st\ RC} & \underbrace{\tau_2 \ R_2}_{2nd\ RC} & \cdots & \underbrace{\tau_n \ R_n}_{n-th\ RC} \end{bmatrix} \text{(if the model is } n\text{RC, } n = 1, 2, 3\text{)} \\[4pt]
V_j = \begin{bmatrix} R_0^+ & R_0^- & \underbrace{\tau_1 \ R_1}_{1st\ RC} & \underbrace{\tau_2 \ R_2}_{2nd\ RC} & \cdots & \underbrace{\tau_n \ R_n}_{n-th\ RC} & K_{SD} & \tau_{SD} \end{bmatrix} \text{(if the model is E}n\text{RC)}
\end{cases}
\tag{8}
$$

From Equation (8), it can be easily found that the identification parameters of the EnRC model are only two more than that of the nRC model. In the process of identification and optimization, the closer the model terminal voltage to the measured terminal voltage, the more accurate are the model parameters. Therefore, the root-mean-square error (RMSE) between the model terminal voltage and the measured terminal voltage can be employed as the fitness value to assess the model parameters and acquire the optimal model parameters that make the model voltage closest to the measured voltage. The objective function for the optimization can be expressed as:

$$
\min : g(V) = \sqrt{\frac{1}{n} \sum_{k=1}^{n} \left(u_{i,k}(V) - \hat{u}_{i,k} \right)^2},
\tag{9}
$$

where $g(V)$ is the objective function, $u_{i,k}$ is the model voltage, and $\hat{u}_{i,k}$ is the measured voltage.

During parameter optimization, the upper and lower bounds of the parameters can be obtained by experimental results. The bounds of the same parameters for different models are maintained the same for fair comparison.

In this study, the PSO algorithm is used for global optimization for parameter identification of the above model. The PSO is a typical swarm intelligence algorithm, inspired by flocks of birds in search for food; it has been successfully applied for artificial neural-network training, function optimization, and pattern classification [26]. Because of the emergence of several variants over time, certain researchers have attempted to define a standard PSO version, with updates to incorporate the latest advances. The most recent standard PSO version was defined in 2011 and is referred to as the Standard PSO2011 (SPSO2011). The following is a brief description of the basic principles of PSO algorithm, and detailed descriptions of the PSO can be found in Reference [27].

Each particle in the algorithm represents a potential solution to the problem. The state of each particle is represented by its position x, velocity v, and fitness. In the iteration process, particle states are continuously updated until the termination criteria are met. Assume that in a D-dimensional search space, there is a swarm consisting of n particles, $\mathbf{X} = (X_1, X_2, \ldots, X_n)$, where the velocity of the ith particle is expressed as a D-dimensional vector $\mathbf{X}_i = (X_{i1}, X_{i2}, \ldots, X_{nD})^\mathrm{T}$, the individual extremum is expressed as $\mathbf{P}_i = (P_{i1}, P_{i2}, \ldots, P_{nD})^\mathrm{T}$, and the swarm extremum is expressed as $\mathbf{P}_{best} = (P_{g1}, P_{g2}, \ldots, P_{gD})^\mathrm{T}$. Then, the following relationship exists between particle velocity and position update during the iteration process:

$$
\mathbf{V}_{id}^{k+1} = \omega \mathbf{V}_{id}^{k} + c_1 r_1 \left(\mathbf{P}_{id}^{k} - \mathbf{X}_{id}^{k} \right) + c_2 r_2 \left(\mathbf{P}_{gd}^{k} - \mathbf{X}_{id}^{k} \right),
\tag{10}
$$

$$
\mathbf{X}_{id}^{k+1} = \mathbf{X}_{id}^{k} + \mathbf{V}_{id}^{k+1},
\tag{11}
$$

where ω is the inertia weight; $d = 1, 2, \ldots, D$; $i = 1, 2, \ldots, n$; k is the current iteration number; c_1 and c_2 are acceleration factors, and r_1 and r_2 are random numbers subject to uniform distribution within [0,1].

The calculation flow of PSO is shown in Table 3. Similar to other evolution-based algorithms, PSO is a random search algorithm. However, PSO preliminarily conducts a search based on its own velocity in order to avoid complex genetic operations; hence, it is a very efficient optimization algorithm.

Table 3. Calculation flow of particle swarm optimization (PSO).

Step 1: Initialize the variables, randomly generate a particle swarm, and calculate the particle fitness values;
Step 2: Repeat the following steps until the termination criterion is satisfied (the error is sufficiently small, or the maximum loop count is reached):   Implement the following operations on each individual: Update the velocity and position state (according to Equations (10) and (11));   Update the variable representing the individual's best position;
Step 3: Output the optimization results.

To improve the accuracy of the ECM in the entire SOC range, a subarea parameter-identification method is adopted in this study. The basic principles of this method are as follows: The entire SOC range (0–100%) is divided into 10 areas, and the PSO algorithm is then used to identify the model parameters in each area. Ten sets of model parameters are thereby obtained to form model parameters for the entire SOC area. For SOC estimation, the model parameters are selected based on the corresponding area of the SOC.

3.3. Results and Discussion

Model parameters of eight models are identified by the above method in the whole SOC area, and results are shown in Figures 4 and 5. Figure 4 shows identification results under NEDC working condition. As shown in Figure 4a, the accuracy of the E*n*RC model is obviously higher than that of the *n*RC model with the same order in the low SOC area, indicating that the proposed model provide satisfactory accuracy in the whole SOC area due to the consideration of the solid-phase diffusion process. Moreover, the model accuracy increases with the increase of the order of RC network, but it will not increase continuously. The accuracy of E2RC and E3RC models is almost the same. Therefore, the E2RC model is the best choice owing to it balance of accuracy and complexity. Figure 4b shows the estimated and measured terminal voltages of E2RC and 2RC models under NEDC working condition. Obviously, the model error of E2RC model is only half of that of 2RC model in the low SOC area. Table 4 list the identification time of eight battery models, we can see that the computation time of our proposed model is slightly larger than that of the ECM with the same order. Through the above comparative analysis, we can conclude that E2RC model is the best choice with the best accuracy and low computation.

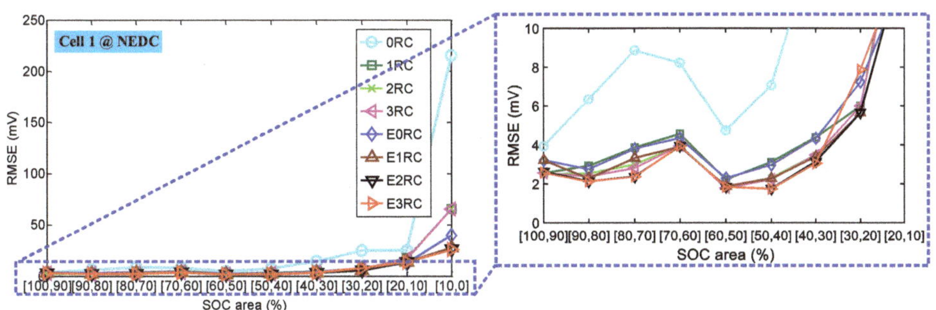

(a)

Figure 4. *Cont.*

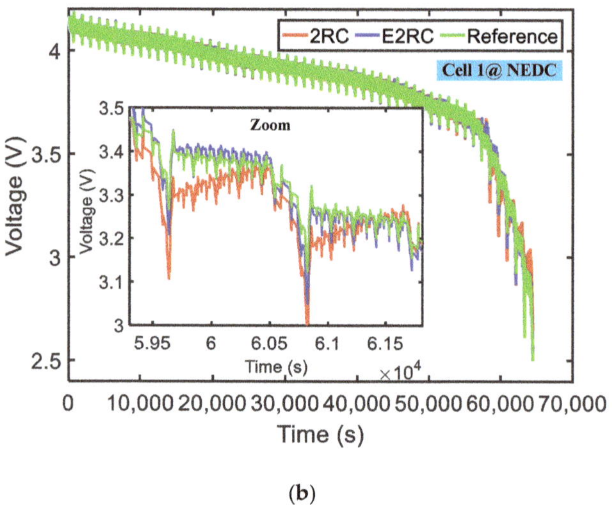

(**b**)

Figure 4. Parameter identification results under NEDC working condition. (**a**) Root-mean-square error (RMSE) comparison for eight models; (**b**) terminal voltage profiles for 2RC and E2RC.

Table 4. Identification time of eight battery models (Cell 1@ European Driving Cycle (NEDC)) in the entire state-of-charge (SOC) area.

Model Name	Identification Time (s)
0RC	1.7796
E0RC	4.0983
1RC	5.2645
E1RC	5.9482
2RC	5.9897
E2RC	6.1520
3RC	8.5727
E3RC	11.3445

Figure 5 shows RMSE and terminal voltage comparisons of various models under DST working condition. We can see that the E2RC model is the best choice under DST. Based on the comparative study of the above eight models under two working conditions, we can conclude that the model error in the low SOC are is only half of that for traditional ECM, and the E2RC model is the most suitable model. Therefore, the battery model used in this study for SOC estimation is the E2RC model.

It is noted that the n in EnRC model represents the number of double layers in the SPM. It can be seen that the models with two double layers have the highest accuracy, which is consistent with the model shown in Figure 1. Moreover, the proposed model uses only one solid-phase diffusion equation in the SPM to achieve satisfactory model accuracy, which is different with the model proposed in Reference [14].

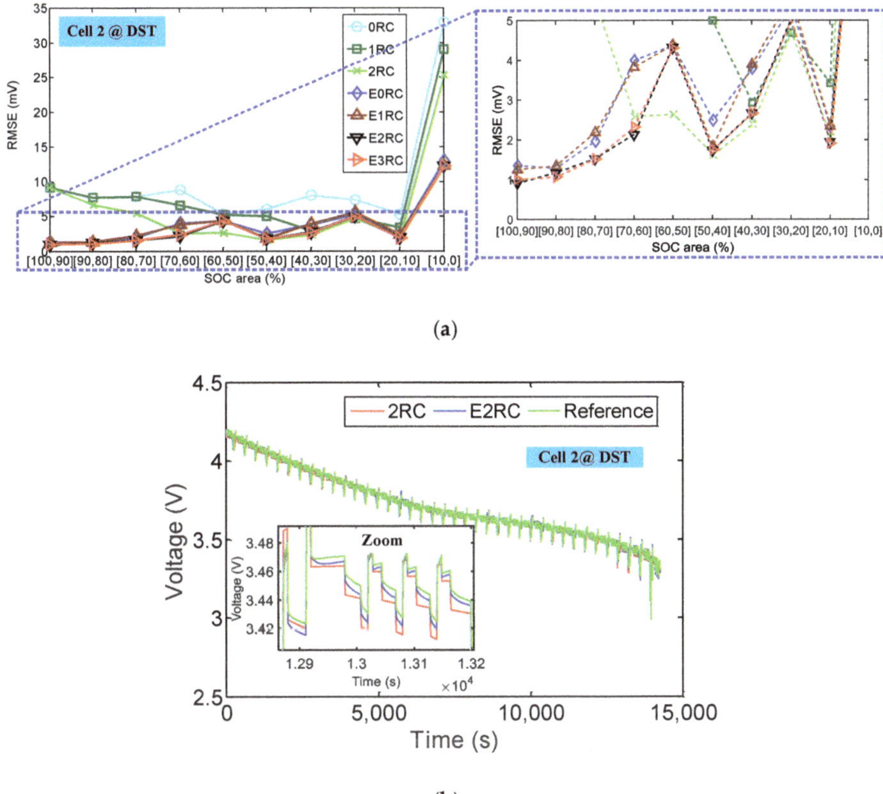

(a)

(b)

Figure 5. Parameter identification results under dynamic stress test (DST) working condition. (a) RMSE comparison; (b) Terminal voltage profiles for 2RC and E2RC.

4. SOC Estimation in the Entire SOC Area

4.1. EKF-Based SOC Estimator

The EKF algorithm is used to estimate the SOC in this paper. The EKF considers the noise characteristics of the current and voltage sensors, and effectively overcomes the problem of random errors [28]. The schematic of EKF is illustrated in Fig. 6. The EKF is based on dynamic equations. Assuming that k is the discrete-time index, $\mathbf{x_k}$ is the state vector to be estimated, $\mathbf{z_k}$ is the output vector, and the system input vector is $\mathbf{u_k}$. The battery model can be expressed by the following state equations:

$$\mathbf{x_{k+1}} = \mathbf{f}(\mathbf{x_k}, \mathbf{u_k}) + w_k \tag{12}$$

$$\mathbf{z_k} = \mathbf{h}(\mathbf{x_k}, \mathbf{u_k}) + v_k \tag{13}$$

where w_k denotes random process noise, v_k denotes measurement error, $\mathbf{f}(\mathbf{x_k}, \mathbf{u_k})$ is a nonlinear state transition function, and $\mathbf{h}(\mathbf{x_k}, \mathbf{u_k})$ is a nonlinear measurement function.

The state equations of the SOC estimator can be expressed as:

$$\mathbf{x_k} = (\mathrm{SOC}_k, u_{1,k}, u_{2,k}, \Delta\mathrm{SOC}) \tag{14}$$

$$\mathbf{h_k} = OCV(\mathrm{SOC}_{surf,k}) - u_{1,k} - u_{2,k} - R_0 I_k + v_k \tag{15}$$

$$\mathbf{u_k} = I_k \tag{16}$$

In Figure 6, the superscript "−" and "+" indicate a priori estimate and a posteriori estimate at time step k, respectively. $\mathbf{P}$ is the covariance matrix of uncertainty in state estimation; $\mathbf{Q}$ is the covariance matrix of process noise; $\mathbf{R}$ is covariance matrix for measuring uncertainty; $\mathbf{\overline{K}_k}$ is kalman gain matrix. The detailed EKF algorithm can be found in Reference [29].

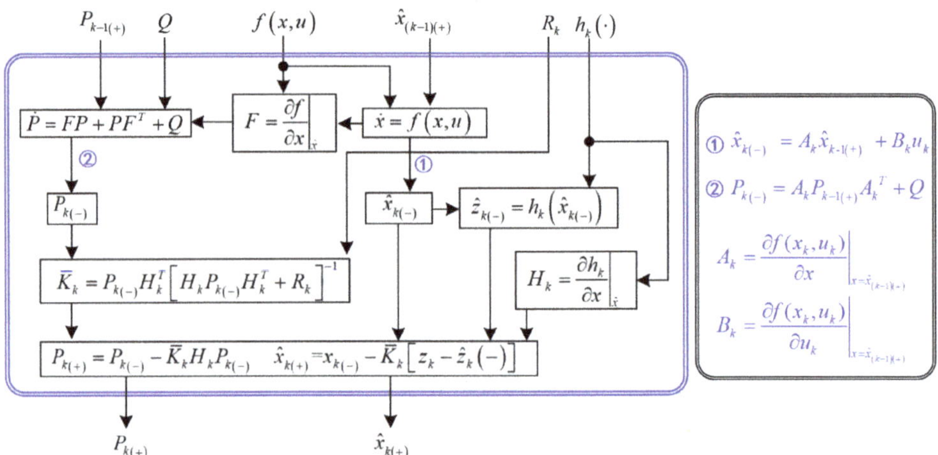

Figure 6. Schematic of extended kalman filter (EKF).

4.2. Results and Discussion

The EKF-based SOC estimator described in Section 4.1 is used to estimate the SOC under NEDC and DST working conditions in the entire SOC area (0–100%), respectively. The results are shown in Figures 7 and 8. Results imply that the SOC estimation error based on 2RC and E2RC model is similar in high SOC area, however, the SOC estimation error based on E2RC model is less 50% of that based on 2RC model in low SOC area. Moreover, the SOC estimation error based on E2RC model is always less than 1% in the entire SOC area.

Furthermore, in order to evaluate the robustness of our proposed model, we set various sensor errors and model errors to calculate the SOC estimation errors in the entire SOC area. The sensor error mainly originates from the voltage sensor and current sensor, whereas the model error has two types, namely, voltage drift or voltage noise. Reference [12] indicated that the voltage noise has no effect on the SOC error for a large time scale. Hence, the effect of the voltage drift (U_{drift}), model error (M_{drift}), and current sensor error (I_{drift}) on the SOC is considered in our study.

Figure 9 describes the influence of various model and sensor errors on the SOC estimated by EKF estimator under NEDC and DST working conditions. Figure 9a shows the relationship between the model error and the RMSE of SOC (R_{SOC}) based on E2RC model. We can see that as long as the M_{drift} is within $\pm 20 mV$, R_{SOC} can be kept within 5%. Figure 9b shows the relationship between the U_{drift} and R_{SOC}. The U_{drift} is generally within 10 mV according to Reference [30]. In this case, R_{SOC} can be maintained at less than 3%. Figure 9b shows the relationship between the I_{drift} and R_{SOC}, indicating that I_{drift} has little effect on the R_{SOC} obtained by EKF estimator. From the above analysis, we can see that the proposed battery model can achieve satisfactory accuracy in a wide range of model and sensor errors, which implies that our proposed model has good robustness in the SOC entire area.

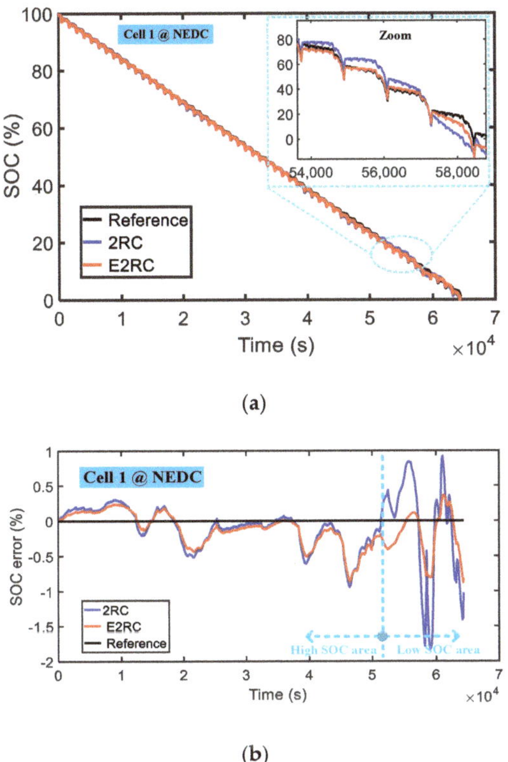

(**a**)

(**b**)

Figure 7. State-of-charge (SOC) estimation results under NEDC working condition. (**a**) SOC estimated value; (**b**) SOC error.

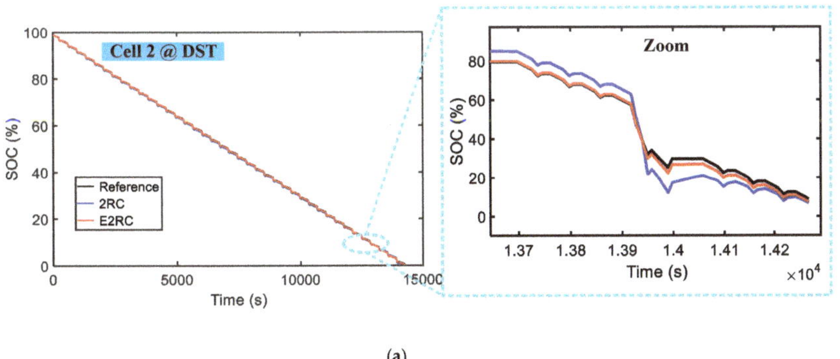

(**a**)

Figure 8. *Cont.*

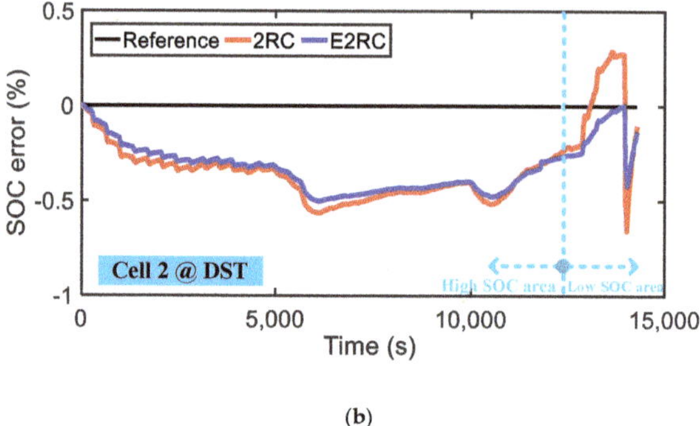

(**b**)

Figure 8. SOC estimation results under DST working condition. (**a**) SOC estimated value; (**b**) SOC error.

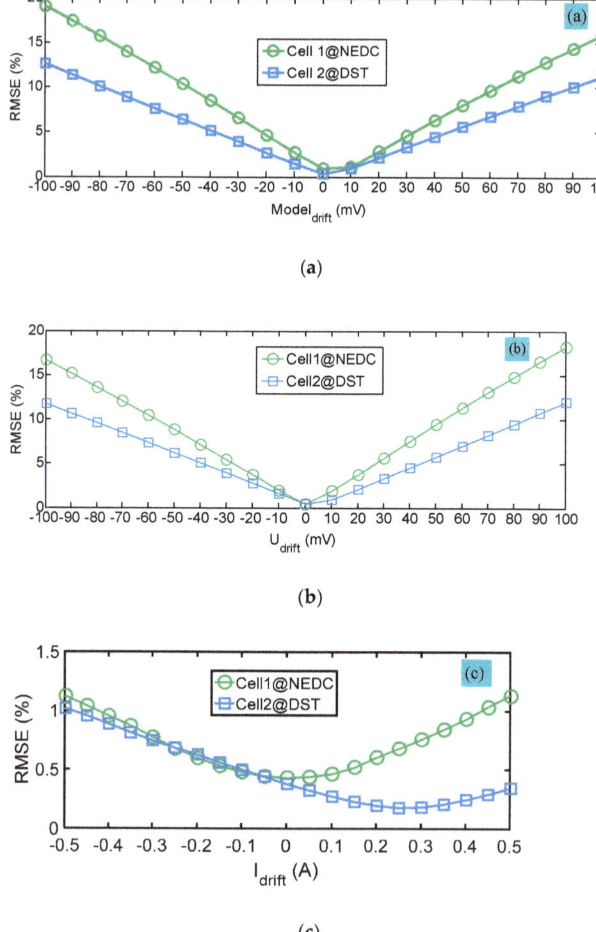

(**a**)

(**b**)

(**c**)

Figure 9. Influence of model and sensor errors on SOC estimated by EKF estimator under NEDC and DST working conditions. (**a**) Relationship between model error and SOC error. (**b**) Relationship between U_{drift} and SOC error. (**c**) Relationship between I_{drift} and SOC error.

5. Conclusions

This paper proposes a SOC estimator based on a SPM model in the entire SOC area. In this study, the PSO algorithm is employed to identify the global model parameters in the subarea, and the SOC estimation algorithm is estimated by EKF. The experimental results of two types of batteries under NEDC and DST conditions can be concluded as follows:

(1) Comparative studies show that E2RC model is the best choice. The accuracy of the proposed model is one times higher than that of the traditional ECM in the low SOC area, and slightly better than that of the ECM in the high SOC area.

(2) An EKF-based SOC estimator using our proposed model has higher SOC estimation accuracy than the ECM, especially in low SOC area. The SOC estimation error is less than 1% in the entire SOC area.

(3) The proposed battery model and SOC estimation algorithm have satisfactory accuracy and robustness with low computational complexity.

Author Contributions: X.L. wrote the manuscript; X.L. and C.Q. conceived, designed and performed the experiments; W.Y. analyzed the data; W.G. contributed to the analysis of results; Y.Z. contributed to the analysis of results, provided feedback to the content and participated in writing the manuscript.

Funding: This research was funded by the National Natural Science Foundation of China (NSFC) under the grant number of 51505290 and 51507102.

Conflicts of Interest: The authors declare no conflict of interest.

References

1. Dong, G.Z.; Wei, J.W.; Chen, Z.H. Constrained bayesian dual-filtering for state of charge estimation of lithium-ion batteries. *Int. J. Electr. Power Energy Syst.* **2018**, *99*, 516–524. [CrossRef]

2. Zheng, Y.J.; Ouyang, M.G.; Han, X.B.; Lu, L.G.; Li, J.Q. Investigating the error sources of the online state of charge estimation methods for lithium-ion batteries in electric vehicles. *J. Power Sources* **2018**, *377*, 161–188. [CrossRef]

3. Chen, L.; Wang, Z.Z.; Lu, Z.Q.; Li, J.Z.; Ji, B.; Wei, H.Y.; Pan, H.H. A novel state-of-charge estimation method of lithium-ion batteries combining the grey model and genetic algorithms. *IEEE Trans. Power Electron.* **2018**, *33*, 8797–8807. [CrossRef]

4. Sahinoglu, G.O.; Pajovic, M.; Sahinoglu, Z.; Wang, Y.B.; Orlik, P.V.; Wada, T. Battery state-of-charge estimation based on regular/recurrent gaussian process regression. *IEEE Trans. Ind. Electron.* **2018**, *65*, 4311–4321. [CrossRef]

5. Ladpli, P.; Kopsaftopoulos, F.; Chang, F.K. Estimating state of charge and health of lithium-ion batteries with guided waves using built-in piezoelectric sensors/actuators. *J. Power Sources* **2018**, *384*, 342–354. [CrossRef]

6. Lai, X.; Zheng, Y.J.; Zhou, L.; Gao, W.K. Electrical behavior of overdischarge-induced internal short circuit in lithium-ion cells. *Electrochim. Acta* **2018**, *278*, 245–254. [CrossRef]

7. Wei, Z.B.; Zou, C.F.; Leng, F.; Soong, B.H.; Tseng, K.J. Online model identification and state-of-charge estimate for lithium-ion battery with a recursive total least squares-based observer. *IEEE Trans. Ind. Electron.* **2018**, *65*, 1336–1346. [CrossRef]

8. Chemali, E.; Kollmeyer, P.J.; Preindl, M.; Ahmed, R.; Emadi, A. Long short-term memory networks for accurate state-of-charge estimation of li-ion batteries. *IEEE Trans. Ind. Electron.* **2018**, *65*, 6730–6739. [CrossRef]

9. Panchal, S.; Mcgrory, J.; Kong, J.; Fraser, R.; Fowler, M.; Dincer, I.; Agelin-Chaab, M. Cycling degradation testing and analysis of a lifepo4 battery at actual conditions. *Int. J. Energy Res.* **2017**, *41*, 2565–2575. [CrossRef]

10. Liu, C.Z.; Liu, W.Q.; Wang, L.Y.; Hu, G.D.; Ma, L.P.; Ren, B.Y. A new method of modeling and state of charge estimation of the battery. *J. Power Sources* **2016**, *320*, 1–12. [CrossRef]

11. He, H.W.; Xiong, R.; Guo, H.Q.; Li, S.C. Comparison study on the battery models used for the energy management of batteries in electric vehicles. *Energy Convers. Manag.* **2012**, *64*, 113–121. [CrossRef]

12. Ma, Y.; Duan, P.; Sun, Y.S.; Chen, H. Equalization of lithium-ion battery pack based on fuzzy logic control in electric vehicle. *IEEE Trans. Ind. Electron.* **2018**, *65*, 6762–6771. [CrossRef]

13. Lai, X.; Zheng, Y.J.; Sun, T. A comparative study of different equivalent circuit models for estimating state-of-charge of lithium-ion batteries. *Electrochim. Acta* **2018**, *259*, 566–577. [CrossRef]

14. Han, X.B.; Ouyang, M.G.; Lu, L.G.; Li, J.Q. Simplification of physics-based electrochemical model for lithium ion battery on electric vehicle. Part i: Diffusion simplification and single particle model. *J. Power Sources* **2015**, *278*, 802–813. [CrossRef]

15. Ouyang, M.G.; Liu, G.M.; Lu, L.G.; Li, J.Q.; Han, X.B. Enhancing the estimation accuracy in low state-of-charge area: A novel onboard battery model through surface state of charge determination. *J. Power Sources* **2014**, *270*, 221–237. [CrossRef]

16. Zhang, C.P.; Jiang, J.C.; Gao, Y.; Zhang, W.G.; Liu, Q.J.; Hu, X.S. Charging optimization in lithium-ion batteries based on temperature rise and charge time. *Appl. Energy* **2017**, *194*, 569–577. [CrossRef]

17. Tian, Y.; Li, D.; Tian, J.D.; Xia, B.Z. State of charge estimation of lithium-ion batteries using an optimal adaptive gain nonlinear observer. *Electrochim. Acta* **2017**, *225*, 225–234. [CrossRef]

18. Lim, K.; Bastawrous, H.A.; Duong, V.H.; See, K.W.; Zhang, P.; Dou, S.X. Fading kalman filter-based real-time state of charge estimation in lifepo4 battery-powered electric vehicles. *Appl. Energy* **2016**, *169*, 40–48. [CrossRef]

19. Sturm, J.; Ennifar, H.; Erhard, S.V.; Rheinfeld, A.; Kosch, S.; Jossen, A. State estimation of lithium-ion cells using a physicochemical model based extended kalman filter. *Appl. Energy* **2018**, *223*, 103–123. [CrossRef]

20. Deng, Z.W.; Yang, L.; Deng, H.; Cai, Y.S.; Li, D.D. Polynomial approximation pseudo-two-dimensional battery model for online application in embedded battery management system. *Energy* **2018**, *142*, 838–850. [CrossRef]

21. Allam, A.; Onori, S. An interconnected observer for concurrent estimation of bulk and surface concentration in the cathode and anode of a lithium-ion battery. *IEEE Trans. Ind. Electron.* **2018**, *65*, 7311–7321. [CrossRef]

22. Meng, J.H.; Luo, G.Z.; Ricco, M.; Swierczynski, M.; Stroe, D.I.; Teodorescu, R. Overview of lithium-ion battery modeling methods for state-of-charge estimation in electrical vehicles. *Appl. Sci.* **2018**, *8*, 659. [CrossRef]

23. Liu, G.M.; Ouyang, M.G.; Lu, L.G.; Li, J.Q.; Han, X.B. Online estimation of lithium-ion battery remaining discharge capacity through differential voltage analysis. *J. Power Sources* **2015**, *274*, 971–989. [CrossRef]

24. Farmann, A.; Sauer, D.U. Comparative study of reduced order equivalent circuit models for on-board state-of-available-power prediction of lithium-ion batteries in electric vehicles. *Appl. Energy* **2018**, *225*, 1102–1122. [CrossRef]

25. Panchal, S.; Rashid, M.; Long, F.; Mathew, M.; Fraser, R.; Fowler, M. Degradation testing and modeling of 200 ah LiFePO$_4$ battery. *SAE Tech. Pap.* **2018**. [CrossRef]

26. Ma, Z.Y.; Wang, Z.P.; Xiong, R.; Jiang, J.C. A mechanism identification model based state-of-health diagnosis of lithium-ion batteries for energy storage applications. *J. Clean Prod.* **2018**, *193*, 379–390. [CrossRef]

27. Altinoz, O.T.; Yilmaz, A.E.; Duca, A.; Ciuprina, G. Incorporating the avoidance behavior to the standard particle swarm optimization 2011. *Adv. Electr. Comput. Eng.* **2015**, *15*, 51–58. [CrossRef]

28. Lin, X.F. Theoretical analysis of battery soc estimation errors under sensor bias and variance. *IEEE Trans. Ind. Electron.* **2018**, *65*, 7138–7148. [CrossRef]

29. Wei, J.W.; Dong, G.Z.; Chen, Z.H.; Kang, Y. System state estimation and optimal energy control framework for multicell lithium-ion battery system. *Appl. Energy* **2017**, *187*, 37–49. [CrossRef]

30. Lu, L.G.; Han, X.B.; Li, J.Q.; Hua, J.F.; Ouyang, M.G. A review on the key issues for lithium-ion battery management in electric vehicles. *J. Power Sources* **2013**, *226*, 272–288. [CrossRef]

 *applied sciences*

Article

Optimization of Battery Energy Storage System Capacity for Wind Farm with Considering Auxiliary Services Compensation

Xin Jiang [1], Guoliang Nan [2], Hao Liu [2], Zhimin Guo [3], Qingshan Zeng [1] and Yang Jin [1,*]

[1] School of Electrical Engineering, Zhengzhou University, Zhengzhou 450001, China; jiangxin@zzu.edu.cn (X.J.); qszeng@zzu.edu.cn (Q.Z.)
[2] State Grid Henan Comprehensive Energy Service Company Limited, Zhengzhou 450052, China; smxngl@126.com (G.N.); 13838252779@163.com (H.L.)
[3] Henan EPRI Hitech Group Company Limited, Zhengzhou 450052, China; gzm514@163.com
* Correspondence: yangjin@zzu.edu.cn; Tel.: +86-037-1167783113

Received: 27 September 2018; Accepted: 15 October 2018; Published: 17 October 2018

Abstract: An optimal sizing model of the battery energy storage system (BESS) for large-scale wind farm adapting to the scheduling plan is proposed in this paper. Based on the analysis of the variability and uncertainty of wind output, the cost of auxiliary services of systems that are eased by BESS is quantized and the constraints of BESS accounting for the effect of wind power on system dispatching are proposed. Aiming to maximum the benefits of wind-storage union system, an optimal capacity model considering BESS investment costs, wind curtailment saving, and auxiliary services compensation is established. What's more, the effect of irregular charge/discharge process on the life cycle of BESS is considered into the optimal model by introducing an equivalent loss of the cycle life. Finally, based on the typical data of a systems, results show that auxiliary services compensation can encourage wind farm configuration BESS effectively. Various sensitivity analyses are performed to assess the effect of the auxiliary services compensation, on-grid price of wind power, investment cost of BESS, cycle life of BESS, and wind uncertainty reserve level of BESS on this optimal capacity.

Keywords: large-scale wind farm; auxiliary services compensation; battery energy storage system; optimal capacity; equivalent loss of cycle life

---

1. Introduction

As a flexible and adjustable power supply, the energy storage system provides a new idea to cope with the intermittent power integration [1]. In various types of large-scale energy storage systems (such as pumped storage, compressed air storage, etc.), battery energy storage system (BESS) has the most promising broad in power applications benefiting from its high energy efficiency and weak requirement of geographical conditions [2]. Wind farm with BESS configuration will become a common model for large-scale wind power development in the future. However, in addition to the high investment cost of BESS, how to optimal BESS size to balance the investment cost and the effect of levelling wind power fluctuation and uncertainty has been a research hotspot in recent years.

Current research of the optimal storage capacity with adapting to the scheduling plan are mainly focused on the two parts: smoothing the fluctuation of wind output to deal with the wind peaking demand [3–5]; compensating the uncertainty of wind output to make up the wind forecast error [6,7]. In literature [8], an optimization model of BESS aiming to adapt the scheduling plan based on the reference output of wind power during each designated period is proposed. Taking an hour as time-scale, the literature [9] built an optimization capacity model of BESS that is based on the unit commitment. However, the above literatures only consider the fluctuation of wind power at each time window, and not account for the influence of wind power integration on the system peaking and sparing demand. In reference [10], an optimization method of BESS capacity was proposed with taking hourly and inner-hour fluctuation of wind output. By providing some climbing ability, BESS can effectively reduce the addition peaking and sparing demand that is caused by wind power integration. A multi-objective optimization method for BESS configuration and capacity optimization is built in literature [11], and the uncertainty of wind output is added to the model in the form of probability. In [12], the benefit of electricity price on the difference between peak and valley in BESS is added to the optimization model based on the electricity market. In [13], when considering the duration time of BESS reserving the wind power uncertainty, the cost-benefit analysis model of BESS is established based on the optimal power flow. In [14], BESS is used to provide sparing reserve capacity for wind power integration, and it is of great significance to optimize the storage capacity with taking the sparing auxiliary service of BESS into account. All of these research focus on minimizing the cost of BESS investment based on the cost-benefit analysis. However, the above literatures have not analyzed the economy from the point view of wind-energy union system.

Actually, the most direct benefit of wind-energy union system is the additional electricity of wind power integration though transferring the wind power during the hard peaking periods. What is more, BESS can also mitigate the peaking and sparing auxiliary services costs of systems that are caused by the fluctuation and uncertainty of wind power integration, which can be regarded as a certain degree of compensation to the wind energy union system. But, there is little literature to consider such auxiliary service compensation into the optimization storage capacity.

Herein, from the point view of wind-energy storage, this paper puts forward a method to optimize the storage capacity with considering auxiliary service compensation. First of all, the fluctuation and uncertainty of the hourly wind output are analyzed. Based on description of BESS participating in the scheduling plan, the auxiliary service cost of BESS mitigation is quantified. Secondly, the equivalent life loss is introduced by considering the BESS irregular charge and discharge on the impact of cycle life. The BESS constraints adapting to the scheduling plan is put forward. Taking the wind power curtailment as one of decision variable, an optimal capacity model of BESS with considering BESS investment cost, wind curtailment saving, and auxiliary service compensation is built based on the cost benefit analysis. Finally, an example is given to validate the effectiveness of the proposed model, and the influence of the auxiliary services compensation, on-grid price of wind power, BESS investment cost, BESS cycle life, and BESS reserve level on the optimization result is analyzed. Results show that the auxiliary service compensation can effectively encourage the wind farm configuration BESS.

The reminder of this paper is organized, as follows: Section 2 describes the optimization problem of BESS adapting to the scheduling with involved the auxiliary service compensation as one of the benefit of wind-energy union system. Section 3 provides the mathematical formulation of the optimization storage capacity problem. Section 4 presents the case results and Section 5 outlines the conclusions.

2. Problem Formulation

2.1. Wind Output Characteristics

2.1.1. Variability

Considering the wind power as a "negative" load, the net load of systems can be described as Formula (1). With the wind power participating in the scheduling plan, the climbing constraint of conventional units that responds to the peaking requirement of the net load fluctuation can be expressed as Formula (2).

$$P_{net,t} = P_{load,t} - P_{wind,t} \tag{1}$$

$$R^{dn}_{amp,t} \leq P_{net,t} - P_{net,t-1} \leq R^{up}_{amp,t} \tag{2}$$

Depending on the fluctuation magnitude and direction of the wind output and the load demand, the variability of wind output can be divided into positive peaking characteristics and anti-peaking characteristics [15]. Figure 1 shows the typical daily load demand and wind output of an actual wind farm in central China.

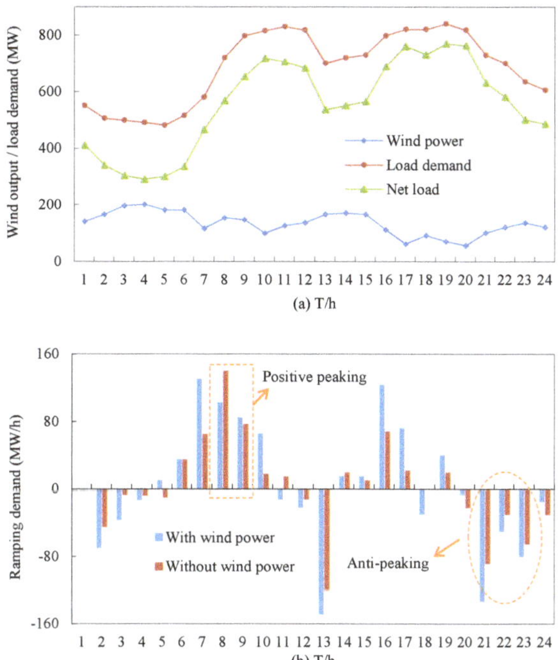

Figure 1. Load demand/wind output under four seasons; (**a**) net load fluctuation with and without wind power integration; and, (**b**) ramping demand of systems with and without wind power.

As we can see from Figure 1, the output of the wind farm exhibits obviously anti-peaking characteristics at most moments, especially at night, when the wind power output is high and the load is small. In which situation, the usual peaking strategy of systems is to on-off peaking units or curtail the wind power to meet the additional peaking demand that is caused by wind power integration, which aggravate the operation cost of conventional units and limit the benefit of wind farm owners at the same time.

2.1.2. Uncertainty

Plenty of statistical analyses of the wind forecast error show that the hourly forecast error of wind output tends to be normal distribution and different wind speed segments appear in different deviation normal distributions [16]. Assuming that the hourly maximum deviation from the mean value is three times to the standard deviation, the prediction interval of wind power can be illustrated in Figure 2 based on the probability density function (PDF) of wind forecast error in literature [17].

Figure 2 shows that the forecast error interval of wind power corrected by the deviation normal distribution is not symmetric with the forecast value, which provides a more accurate spinning reserve information for the wind power participation in the scheduling plan. Unlike the highly repetitive of load demand [18], the wind output has a great range of uncertainty. In order to cope with the uncertainty of wind power, conventional units need to reserve additional climbing ability as the spinning reserve, which is also the main limitation reason for "wind power difficult integration".

The spinning reserve constraint of conventional units with wind power integration can be expressed as Equation (3).

$$\begin{cases} \sum_{i=1}^{N_g} u_{i,t} P_{gi}^{\max} - \sum_{i=1}^{N_g} u_{i,t} P_{gi,t} \geq \Delta P_{load,t} + \Delta P_{wind,t}^{up} \\ \sum_{i=1}^{N_g} u_{i,t} P_{gi,t} - \sum_{i=1}^{N_g} u_{i,t} P_{gi}^{\min} \geq \Delta P_{load,t} + \Delta P_{wind,t}^{dn} \end{cases} \tag{3}$$

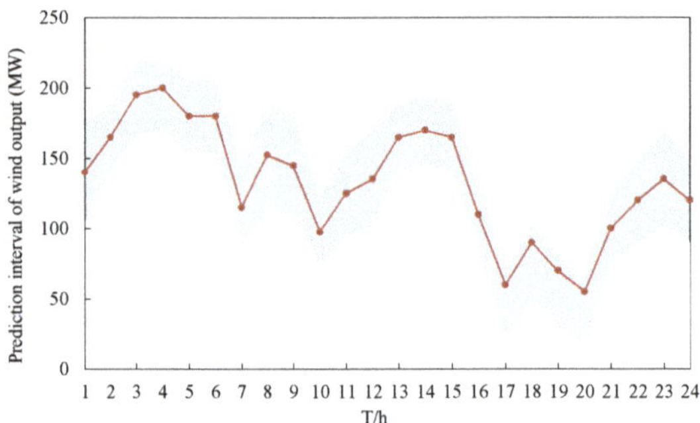

Figure 2. Prediction interval of wind output though a day.

2.2. Auxiliary Services Eased by BESS

2.2.1. BESS Participation in the Scheduling Plan

As can be seen in Figure 3, BESS enables wind power controllable by transferring wind power in space and time. Storing wind energy during anti-peaking periods and releasing wind energy during positive-peaking periods are helpful for easing the additional peaking demand of conventional units in tracking wind power fluctuation.

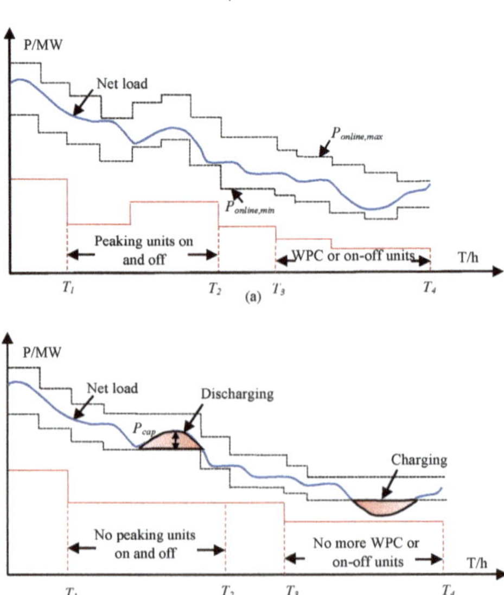

Figure 3. Schematic diagram of battery energy storage system (BESS) participating in the scheduling; (**a**) wind power integration without BESS; and, (**b**) wind power integration with BESS. BEES: battery energy storage system; WPC: wind power curtailment.

As shown in Figure 4, the forecast value of net load is between P_{net} and P'_{net} with considering a certain confidence interval. BESS provides a good choose for wind farm to handle with the forecast error by leaving some reserve capacity. By comparing Figure 4a,b, it is corresponding to the lower peaking capacity of conventional units by increasing BESS reserve capacity P_{cap}, which can effectively improve the economics operation of conventional units.

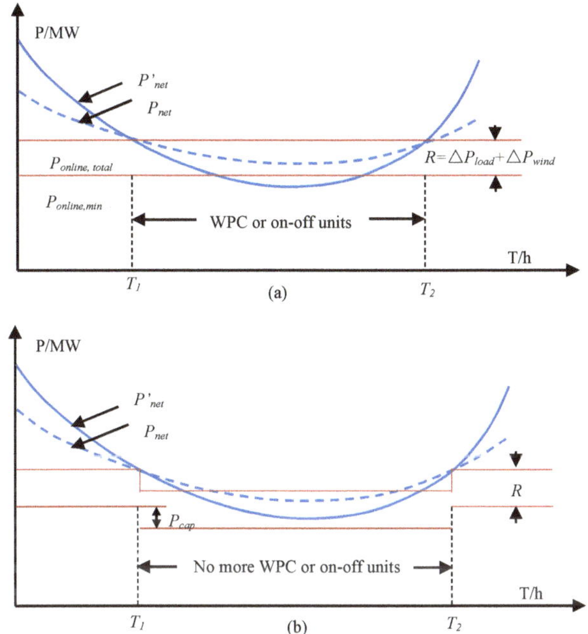

Figure 4. Schematic diagram of BESS making up the sparing reserve of the wind power uncertainty; (**a**) wind power integration without BESS; and, (**b**) wind power integration with BESS.

2.2.2. Quantification the Ancillary Services Cost

The variability and uncertainty of wind power generation require conventional units to provide corresponding auxiliary service support [19]. Above analyses show that the auxiliary services that are provided by conventional units include peaking and spinning reserve two parts. The variability of wind power mainly affects the peaking auxiliary service and the uncertainty of wind power mainly affects the spinning reserve auxiliary service.

This paper defines that the auxiliary service cost of BESS mitigation for wind power integration is the difference between the ancillary services cost provided by conventional units with and without configuration BESS, which can be generally divided into two categories: fixed cost and variable cost. Among them, the fixed cost is mainly the investment cost of the conventional units, the variable cost is the fuel cost [20].

$$C_{serve} = C_{fixed} + C_{vary} \tag{4}$$

$$\begin{cases} C_{fixed} = C_{AI} \cdot \left(\sum_{i=1}^{M} P_{gi}^{N} - \sum_{i=1}^{M_{BESS}} P_{gi}^{N} \right) \\ C_{vary} = \left(c_{g}^{BESS} - c_{g}^{Wind} \right) \cdot \sum_{t=1}^{T} \sum_{i=1}^{M_{BESS}} P_{gi,t}^{B} \end{cases} \tag{5}$$

2.3. Mathematical Description of BESS

There are two key parameters influencing the optimization capacity of BESS: the investment cost and the cycle life. Researches have shown that BESS will be widely used in electricity market with the investment cost being less than 250 \$/kW·h and the cycle life being more than 4000 times [1,2]. At present, the lithium-ion battery is considered to be the most promising energy storage technology for its high rate in the MW-level electrochemical energy storage project [3]. Therefore, this paper chooses the lithium-ion battery as an example to carry out the following research, and the unit investment cost of BESS set as 250 \$/kW·h and the cycle life takes 4500 times.

2.3.1. Equivalent Loss of Cycle Life

The cycle life of BESS is mainly effect by the charge/discharge depth [21]. According to the test results of the Lithium-ion battery system in literature [22], BESS has a corresponding cycle life at each depth of discharge, as listed in Table 1.

Table 1. Relationship between the charge/discharge depth and cycle life of Lithium-ion battery system.

Discharge Depth (%)	0.2	0.4	0.6	0.8	1.0
Cycle (time)	9000	7200	5700	5200	4500

Table 1 shows that shallow charge/discharge depth is conducive to extend the cycle life of BESS. Based on the power function method in Formula (6) [23], the fitting curve of charge/discharge depth with cycle life is illustrated in Figure 5.

$$L_{cyc,D} = 4500D^{-0.795} \tag{6}$$

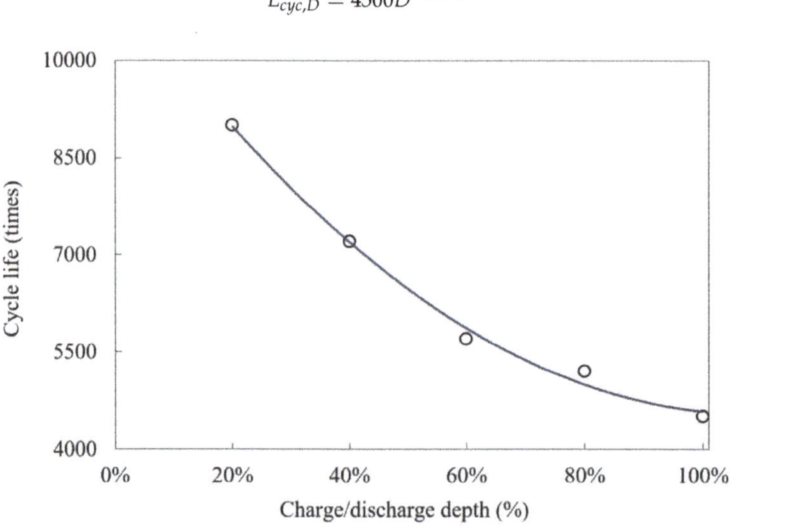

Figure 5. Depth of charge/discharge versus the cycle life of the Lithium-ion battery.

Based on the irreversible electrochemical loss for each charge/discharge process on the cycle life of BESS [24], the equivalent lifetime loss rate is used to obtain the equivalent service year of BESS, calculated as Equation (7):

$$T_{life} = 1 / \sum_{i=1}^{N_B} \frac{1}{L_{cyc,Di}} \tag{7}$$

2.3.2. Constraints of BESS Accounting to the Scheduling

The wind-energy union system participating in the scheduling aims to achieve the coordination with conventional units, that is, give full play to BESS on conventional units of the complementary role, while trying to avoid BESS frequent charge and discharge. Generally, the rated power and energy capacity are two key indicators in describing the sizing of BESS [25]. Under the condition of allowing wind power curtailment, the charge/discharge power of BESS can be illustrated as Formula (8):

$$P_{s,t} = P_{wind,t} - P_{union,t} - P_{wloss,t} \tag{8}$$

where, $P_{s,t} > 0$ refers to charging, and $P_{s,t} < 0$ refers to discharging.

The net load of systems with wind-energy union system integration can be rewritten, as follows:

$$P'_{net,t} = P_{load,t} - P_{wind,t} - P_{s,t} \tag{9}$$

The state of charge (SOC) of BESS at hour t can be expressed, as follows:

$$\begin{cases} S_{soc,t} = S_{soc,t-1} + \eta_s P_{s,t} \Delta t, & P_{s,t} > 0 \\ S_{soc,t} = S_{soc,t-1} + \dfrac{P_{s,t}}{\eta_s \Delta t}, & P_{s,t} < 0 \end{cases} \tag{10}$$

Set λ_t as the charge/discharge status of BESS at time t, and λ_t can only have one state in each time period, that is:

$$\lambda_1, \lambda_2, \cdots \lambda_t \in \{-1, 0, 1\} \tag{11}$$

where, $\lambda_t = 0$ means the BESS being the idle float status; $\lambda_t = 1$ means the BESS being the discharging status; and, $\lambda_t = 1$ means the charging status.

Subject to the limitations of the rated charge/discharge power and rated energy storage capacity of BESS, the constraints of BESS at time t are shown, as follows:

$$\begin{cases} |P_{s,t}| \leq \eta_s P_{cap} \\ 0 \leq S_{soc,t} \leq S_{cap} \end{cases} \tag{12}$$

Equation (12) limit the fluctuation range of wind power in adjacent scheduling intervals, which alleviates the additional peaking demand that is caused by wind power integration. If BESS can keep a certain reserve capacity for the uncertainty of wind output, it will further alleviate the additional spinning reserve requirement. Thus, constraints of BESS that adapting to the scheduling can be expressed as:

$$\begin{cases} -\eta_s P_{cap} + \Delta P_{wind,t}^{dn} \leq P_{s,t} \leq \eta_s P_{cap} - \Delta P_{wind,t}^{up} \\ \Delta P_{wind,t}^{dn} \Delta t \leq S_{soc,t} \leq S_{cap} + \Delta P_{wind,t}^{up} \Delta t \end{cases} \tag{13}$$

It should be noted that when BESS provides spinning reserve for wind power forecast error, conventional units only need to reserve the load forecast error, that is, only $\Delta P_{load,t}$ need to be retained in the Equation (3) given in Section 2.1.2.

As shown in Formula (13), the BESS capacity for wind farm will be increased while considering BESS as reserve capacity for the wind uncertainty, which further increases the investment cost of the wind-energy union system. However, the equivalent cycle life of BESS can be expanded under this situation of shallow charge/discharge (detailed analysis shown in Section 2.3.1). It is equivalent to decrease the unit investment cost of BESS among the full cycle life to some extent. In addition, BESS providing the spinning reserve capacity will also receive more additional auxiliary services compensation benefit. Thus, how to reasonably consider the BESS reserve degree for the wind uncertainty is also the problem to be discussed in the following research.

3. Optimal Model

3.1. Objective Function

Based on the cost-benefit analysis, an optimization capacity model of BESS aiming to maximize the net income of wind-energy union system is proposed in this part.

$$\max f = S_{ave} + C_{serve} - C_{ost} \tag{14}$$

(1) Investment cost of BESS C_{ost}

The amortized capital cost model of BESS in [26] is adopted and modified as the cost function to be minimized, where the influences of the depth and times of the charge/discharge on the equivalent loss of cycle life are taken into account. The cost function can be written as:

$$\begin{cases} C_{ost} = \frac{C_c(p,n)}{365}(\alpha^s \cdot P_{cap} + \beta^s \cdot S_{cap}) \\ C_c(p,n) = \frac{p(1+p)^{T_{life}}}{(1+p)^{T_{life}}-1} \\ \beta^s = \frac{C_E}{T_{life}} + C_{OM} \\ \alpha^s = r \cdot \beta^s \end{cases} \tag{15}$$

(2) Directly benefit of saving wind curtailed energy S_{ave}

Wind power curtailment will be mitigated by BESS shifting this part of energy during wind anti-peaking periods. So that the reduction amount of wind curtailed energy with and without considering BESS can be regarded as the direct benefit of the wind-energy union system. During a scheduling period, it can be expressed as Formula (15):

$$S_{ave} = \rho^w \left(\int_{t=1}^{T} (P_{wloss,t} - P^B_{wloss,t}) \Delta t \right) \tag{16}$$

(3) Additional benefit of auxiliary service compensation C_{serve}

This paper proposes that the auxiliary service cost of wind power integration eased by BESS should be as a part of the revenue of wind-energy union system. Based on the analysis of Section 2.2, the ancillary service costs of the "net load" fluctuation curve with and without BESS can be calculated, as follows:

$$C_{serve} = C_{serve}^{Wind} - C_{serve}^{BESS} \tag{17}$$

3.2. Constraints

Other constraints including the wind curtailment constraint, unit output constraint, climbing constraint, and on/off time constraint, are still traditional constraints and not discussed in this paper.

$$s.t \begin{cases} P_{gi}^{min} \leq P_{gi,t} \leq P_{gi}^{max} \\ 0 \leq P_{wloss,t} \leq P_{wind,t} \\ 0 \leq |P_{gi,t} - P_{gi,(t-1)}| \leq R_i \\ \begin{cases} (u_{i,t} - u_{i,(t-1)})[T^{on}_{i,(t-1)} - T^{on}_{i,min}] \leq 0 \\ (u_{i,(t-1)} - u_{i,t})[T^{off}_{i,(t-1)} - T^{off}_{i,min}] \leq 0 \end{cases} \end{cases} \tag{18}$$

3.3. System Performance Indices

The following performance indices related to the economics of conventional units and wind farm are used to compare different cases in the model.

(1) Unit coal cost of conventional units ($/MW·h) c_g

$$c_g = C_{Gen} / \sum_{t=1}^{T} \sum_{i=1}^{N_g} P_{gi,t}$$

$$\begin{cases} C_{Gen} = \sum_{t=1}^{T} \sum_{i=1}^{N_g} [u_{i,t} f(P_{gi,t}) + u_{i,t}(1 - u_{i,(t-1)})S_i] \\ f(P_{gi,t}) = a_i + b_i P_{gi,t} + c_i P_{gi,t}^2 \end{cases} \tag{19}$$

(2) Wind energy curtailment rate (%) q

$$q = \int_{t=1}^{T} P_{wloss,t} \Delta t / \int_{t=1}^{T} P_{wind,t} \Delta t \times 100\% \tag{20}$$

4. Case Study

4.1. Basic Data

Taking the 10 units system as example, shown in Table 2. The installed capacity of wind farm selected from the North China is 250 MW, and the forecast value of wind power and load during a scheduling period are shown in Figure 1. The forecast error interval of wind power is shown in Figure 2. Since the charge/discharge efficiency of the lithium-ion battery is high, η_s is regarded as 1.0. r set as 1.172, and the C_{OM} is taken as 26 $/kW·h/year [24]. The on grid price of wind power ρ^w is 0.084 $/kW·h (0.54 ¥/kW·h). The computation for all cases are carried out using YALMP toolbox and CPLEX solver [27].

Table 2. Parameters of 10 units system.

Units	1	2	3	4	5	6	7	8	9	10
P_{max} (MW)	455	455	130	130	162	80	85	55	55	55
P_{min} (MW)	150	150	20	20	25	20	25	10	10	10
c ($/h)	1000	970	700	680	450	370	480	660	665	670
b ($/MW·h^2)	16.19	17.26	16.60	16.50	19.70	22.26	26.74	25.92	27.27	27.29
a (10^{-3} $/MW·h^2)	0. 48	0.31	2.0	2.1	3.98	7.12	7.9	4.13	2.22	1.73
R_i (MW/h)	130	130	60	60	90	40	40	20	20	20
S_i ($)	4500	5000	550	560	900	170	260	30	30	30
Initial status (h)	8	−8	−5	−5	−5	−3	−3	−1	−1	−1

4.2. Operation Results Without BESS

In order to provide a basis for the subsequent calculation of the wind power curtailment and the auxiliary service cost mitigation by BESS, scheduling results of systems without BESS are given in Table 3. The dispatch output of each conventional unit and the wind curtailed energy during a scheduling period are described in Figure 6.

- Case 1: The optimal scheduling results of systems without wind power integration is calculated.
- Case 2: With allowing wind power curtailment, the scheduling results with wind power integration is considered.

While comparing case 1 with case 2 in Table 3, due to its anti-peaking characteristics and uncertainty of wind output, the unit coal cost of conventional units is significantly increased from 21.221 $/MW·h to 21.534 $/ MW·h, and the total installed capacity of conventional units participating in auxiliary service is also increased by 55 MW. Figure 6 shows more intuitively that due to the limitation of the minimum output of conventional units, a large amount of wind power curtailment occurs at 2:00 to 6:00 during the difficult peaking period with high wind output and low load demand.

Table 3. Scheduling results of systems with and without wind power integration.

Case	c_g ($/MW·h)	q (%)	$\sum P_{gi}^N$ (MW)
1	21.201	-	382
2	21.579	6.09%	437

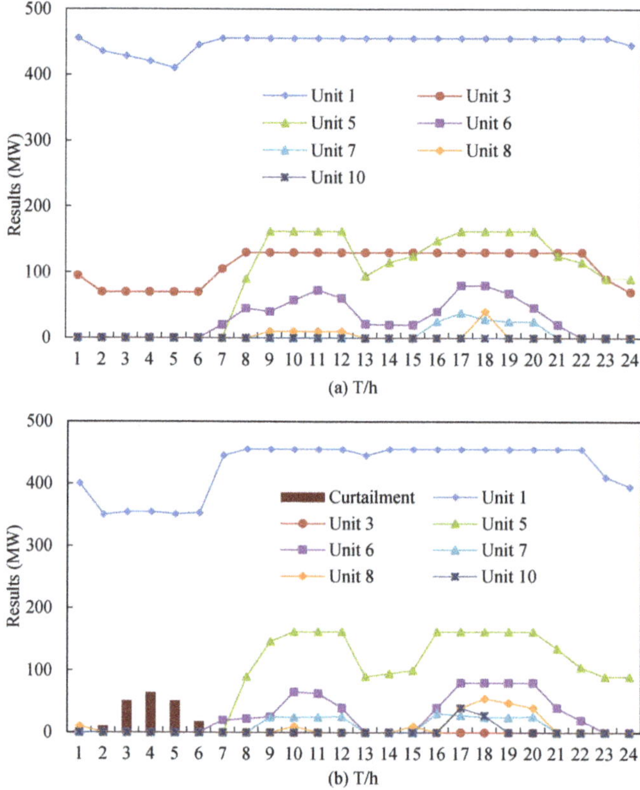

Figure 6. Scheduling results of systems without BESS; (**a**) Case1 results without wind power integration; and, (**b**) Case 2 results with wind power integration.

4.3. Operation Results with BESS

Based on the proposed optimization model (Case 3), the effect of auxiliary service compensation benefit and BESS reserve wind uncertainty on optimal capacity are analyzed, as below in Table 4. The charge/discharge power and SOC of BESS under different cases over a scheduling period are described in Figure 7.

- Case 3: both the auxiliary service compensation and BESS reserve wind forecast error are considered, that is, the proposed model;
- Case 4: without auxiliary service compensation and with BESS reserve wind uncertainty;
- Case 5: with auxiliary service compensation, and without BESS reserve wind uncertainty; and,
- Case 6: neither auxiliary service compensation nor BESS reserve wind uncertainty is considered.

It should be noted that, the objective functions of Case 4 and Case 6 without considering the auxiliary service compensation will be rewritten as:

$$\max f = S_{ave} - C_{cap} \tag{21}$$

Table 4. Optimal results with BESS for different cases.

Case	c_g ($/MW·h)	f ($)	q (%)	P_{cap} (MW)	S_{cap} (MW·h)	N_{cyc} (time)
Case 3	21.238	2990	0.0%	58	122.4	1.73
Case 4	21.305	−1540	0.64%	55	88.9	1.71
Case 5	21.290	1790	3.61%	27.75	60.25	1.85
Case 6	21.544	1050	4.32%	7.15	35.75	1.56

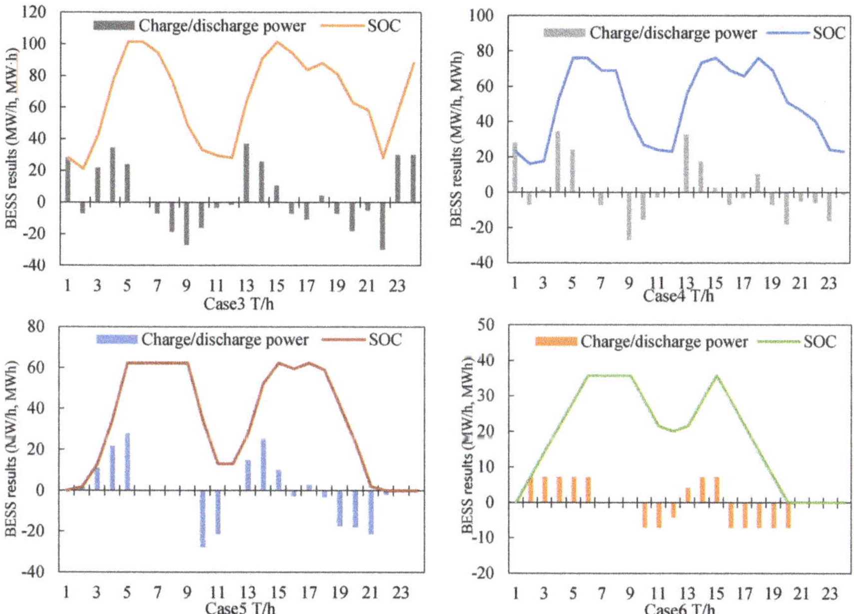

Figure 7. Charge/discharge process of BESS under different cases. SOC: state of charge.

(1) Table 4 shows that with configuring BESS, c_g of Case 3~Case 6 with BESS is significantly smaller than that of case2 without BESS (21.534 $/MW·h). It is reasonable that BESS mitigate the operational cost of conventional units caused by the wind anti-peaking characteristics through transferring the wind power in the time and space. c_g of the proposed model Case 3 (c_g = 21.236 $/MW·h) is close to the results of Case 1 without wind power integration. It means that with taking both the ancillary services compensation and BESS reserve wind uncertainty into account, wind-energy union system can achieve the "wind power friendly integration" and the economy operation of systems.

(2) Comprising c_g of Case 3 with Case 4 in Table 4, it can be seen that the economical operation of conventional units is further improved with BESS keeping reserve capacity for the wind uncertainty. From the benefits f of the wind energy union system in the Case 4, it can be seen that even if the unit investment cost of BESS is assumed as 250 $/kWh, the positive income will not be realized without considering the ancillary services compensation for BESS. That is, ignoring this part benefit of the BESS adapting to scheduling will seriously hinder the enthusiasm of wind farm configuring BESS and further harmful to the large-scale wind power integration.

(3) As the optimal results in Case 5 shows, the total benefit of wind-energy union system is less than the proposed model of Case 3. It is mainly due to the equivalent cycle number over a scheduling period is increased without considering BESS reserve wind uncertainty, which shortens the equivalent cycle life of BESS and equivalent increases the unit investment cost of BESS to some extent. That is, ignore the reduced capacity cost investment of BESS reserve wind uncertainty is not enough to make up for the loss of benefits caused by shorter cycle life in Case 5, which can be more intuitively seen from comparing the charge/discharge process of Case 5 with Case 3 in Figure 7.

(4) Comparison Case 5 with Case 6 in Table 4, it can be seen that a smaller storage capacity is configured in Case 6 without considering the auxiliary service compensation. It is because that the investment cost of BESS is too expense and not enough to be covered by the benefit of additional wind power integration. The wind farm chose to configure less storage capacity for the pursuit of the maximum benefit, which is more clearly pointed in the charge/discharge process during a scheduling period of Case 6 in Figure 7. Thus, the net load fluctuation of systems is not significantly improved and the operational costs c_g is also failed to be improved in Case 6.

4.4. Sensitivity Analysis

4.4.1. On-grid Price of Wind Power

There are differences in on-grid price of wind power for different wind sources. Therefore, the effects of different on-grid price from 0.084 \$/kW·h to 0.064 \$/kW·h on the optimal storage capacity of BESS are studied, as shown in Table 5.

Table 5. Optimal results under different on-grid prices of wind power.

On-Grid Price (\$/kW·h)	f (\$)	P_{cap} (MW·h)	S_{cap} (MW·h)
0.084	2990	58	122.4
0.080	2321	54	102.4
0.076	1845	35.25	63.25
0.072	740	27.75	60.25
0.068	322	7.15	14.25
0.064	0	0	0

From Table 5, it can be seen that the capacity of BESS has changed slowly and only reflects on the net benefit of the wind-energy union system before the on-grid price down to 0.076 \$/kW·h. However, when the on-grid price of wind power falls to 0.064 \$/kW·h, the benefit that is contributed by wind energy union operation cannot balance the investment cost of BESS, and it is no longer suitable for configuring BESS in this wind farm.

4.4.2. Investment Cost and Cycle Life of BESS

As shown in the above analysis of Case 3, wind-energy union system can achieve positive returns under the condition of C_E being 250 \$/kW·h and $L_{cyc,N}$ being 4000 times. In other words, the additional benefits for wind farm configuration BESS is greater than the additional investment costs. To find the balance point of the additional benefits and costs with C_E and $L_{cyc,N}$ varying, the impact of C_E and $L_{cyc,N}$ on the optimization results is shown in Figures 8 and 9, respectively.

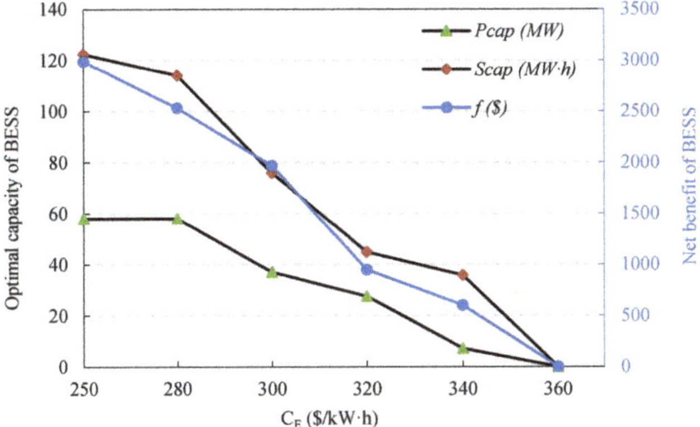

Figure 8. Optimal results under different investment costs of BESS. C_E: unit capacity cost of BESS.

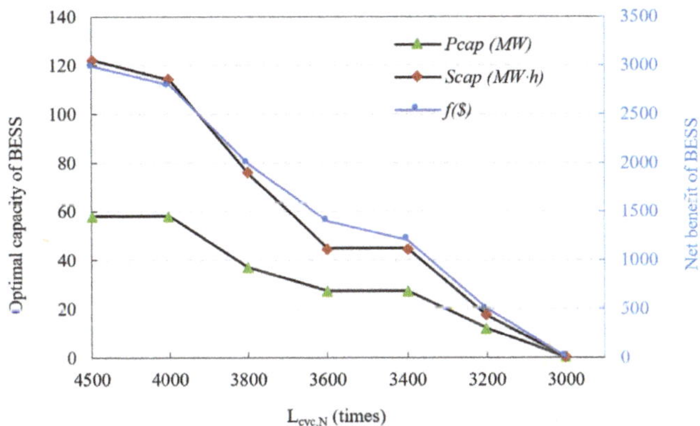

Figure 9. Optimal results under different cycle life of BESS. $L_{cyc,N}$: cycle life of BESS with fully charge/discharge.

(1) It can be seen from Figure 8 that under the condition of $L_{cyc,N}$ being 4000 times, if BESS can be provided a certain compensation for participating in the scheduling, the wind-energy union system will reach the payment balance with C_E being 360 \$/kW·h. That is to say, C_E < 360 \$/kW·h can guarantee the positive benefits of this wind farm. The auxiliary service compensation for BESS is more effective to stimulate the wind farm configuring BESS and promote the early arrival of the "BESS generation".

(2) Similarly, from Figure 9, we can see that the capacity of BESS is *0* when $L_{cyc,N}$ reducing to 2800 times under the condition of C_E = 250 \$/kW·h. In other words, compared to the above mentioned in Section 2.3, the cycle life should being more than 4000 times in large-scale BESS application, it will be more effective to incentive wind farm configuring BESS with taking the auxiliary service compensation of BESS into account.

4.4.3. Reserve Level of BESS

As mentioned earlier, the forecast error of wind power can be described by a normal distribution with a certain mean and standard deviation. That is, the forecast error interval concentrates in the larger probability of the inter-range. If considering the full reserve by the costly BESS for the 100% confidence interval of wind forecast error, there may be some reserve idle with less economical. Therefore, it is necessary to analyze the influence of the reserve level of BESS on the optimal capacity. The reserve level of BESS in Case 3 and Case 5 can be regarded as 100% and 0%. Based on Case 3 with the 100% reserve level, the reserve level is shortened successively. Thus, the effect of different reserve levels on the optimization results is analyzed, as shown in Table 6.

It is worth mentioning that, in order to ensure that the reserve capacity that is provided by conventional units and BESS meet the uncertainty of wind output, the BESS constraints and spinning reserve constraint of units can be rewritten as:

$$\begin{cases} -\eta_s P_{cap} + \Delta P_{wind,t}^{dn} \leq S_t \leq \eta_s P_{cap} - \varepsilon_s \Delta P_{wind,t}^{up} \\ \Delta P_{wind,t}^{dn} \Delta t \leq S_{soc,t} \leq S_{cap} + \varepsilon_s \Delta P_{wind,t}^{up} \Delta t \end{cases} \tag{22}$$

$$\begin{cases} \sum_{i=1}^{N_g} u_{i,t} P_{gi,t}^{max} - \sum_{i=1}^{N_g} u_{i,t} P_{gi,t} \geq \Delta P_{load,t} + (1 - \varepsilon_s) \Delta P_{wind,t}^{up} \\ \sum_{i=1}^{N_g} u_{i,t} P_{gi,t} - \sum_{i=1}^{N_g} u_{i,t} P_{gi,t}^{min} \geq \Delta P_{load,t} + (1 - \varepsilon_s) \Delta P_{wind,t}^{dn} \end{cases} \tag{23}$$

Table 6. Optimal results under different reserve level of BESS.

ε_s	c_g ($/MW·h)	f ($)	q (%)	P_{cap} (MW·h)	S_{cap} (MW·h)
100%	21.238	2990	0.0%	58	122.4
80%	21.324	3160	1.10%	44.8	95.2
60%	21.323	3930	1.84%	34.8	79.6
40%	21.312	3560	2.54%	31.2	70.35
20%	21.307	2910	3.23%	29.5	61.25
0%	21.290	1790	3.61%	27.75	60.25

As can be seen from Table 6, with decreasing the BESS reserve level for wind uncertainty, the net benefit of wind-energy union system returns to increase first and then decrease. This is mainly because that the investment cost of BESS is still expensive when compared to the benefit of auxiliary service compensation. If BESS provides full reserve capacity for the wind uncertainty, there will be a lot of waste for the wind uncertainty being mostly concentrated in the small confidence interval with high probability. However, with the reserve level decreasing, the net benefit of wind-energy union system decreases again limited by the equivalent cycle life and the auxiliary service compensation income.

In addition, it can be seen from Table 6 that, when the reserve level being 60%, both the net benefit of wind-energy union system and the operation efficiency of conventional units are the maximum. The variety rates of the performance indices (f, c_g, and q) are small with the reserve level between 40% and 60%. It means that the installed power of BESS being 68.75–76.8 $/MW·h and the installed capacity of BESS between 30.2 and 32.8MW can realize the operation efficiency of BESS and conventional units in this wind farm.

5. Conclusions

In this paper, a novel method that determines the optimal BESS capacity with considering auxiliary services compensation is proposed from the point view of wind-energy union system. By quantifying the auxiliary services cost that is caused by the variability and uncertainty of wind output and analyzing the effect of irregular charge/discharge process on the life cycle of BESS, both the auxiliary services compensation and the equivalent loss of the cycle life are introduced in this model, which is more reasonable and precise in economic and electrochemical sense. Simulation results shows that the auxiliary services compensation can encourage wind farm configuration BESS effectively.

Moreover, effect of the on-gird price of wind power, investment cost of BESS, cycle life of BESS and BESS reserve level are assessed though sensitivity analyses. Results show that BESS can be early applied in large-scale wind farm with investment cost being less than 360 $/kW·h or the cycle life being more than 2800 times with taking the auxiliary services compensation into account. It is noteworthy that the net income of wind-storage system reach maximum with the reserve interval of BESS being about 40%~60% of the wind power forecasting error.

Author Contributions: Conceptualization, Y.J.; Methodology and Software, X.J.; Validation, G.N.; Investigation, H.L.; Data Curation, Z.G.; Writing—Original Draft Preparation, X.J.; Writing—Review and Editing, Q.Z.

Funding: This research was funded by National Natural Science Foundation of China (grant number 51807180).

Abbreviations

$P_{net,t}$	net load of systems with wind power integration in hour t
$P_{load,t}$	forecast value of load demand in hour t
$P_{wind,t}$	forecast output of wind farm in hour t
$R_{amp,t}^{up} / R_{amp,t}^{dn}$	up/down ramp demand of the net load in hour t
$P_{gi}^{max} / P_{gi}^{min}$	maximum/minimum output of unit i
$u_{i,t}$	on-off state of unit i in hour t
$P_{gi,t}$	output of unit i in hour t
$\triangle P_{load,t}$	spinning reserve demand of the load demand in hour t
$\Delta P_{wind,t}^{up} / \Delta P_{wind,t}^{dn}$	upper and down limitation of wind prediction interval
N_g	number of conventional units
$P_{online,max}/P_{online,min}$	upper/lower limitation of the online units
$P_{online,total}$	total output of the online conventional units
P'_{net}	net load of systems with wind farm configuration BESS
R	total spinning reserve capacity required by the system
P_{cap}	rated power of BESS
S_{cap}	rate capacity of BESS
C_{serve}	difference auxiliary service cost of systems with and without BESS
C_{fixed}	fixed cost item of auxiliary service
C_{vary}	variable cost item of auxiliary service
C_{AI}	daily investment cost per capacity of conventional units
P_{gi}^{N}	rated power of the conventional unit i
$P_{gi,t}^{B}$	output of unit i with configuration BESS in hour t
M^{BESS}/M	number of units participating the auxiliary service with/without BESS
c_g^{BESS} / c_g^{Wind}	unit coal cost of conventional units with/without BESS
T	one scheduling period
D	charge/discharge depth of BESS
$L_{cyc,D}$	cycle life of BESS under the charge/discharge depth of D
$L_{cyc,N}$	cycle life of BESS with fully charge/discharge
N_B	charge/discharge number of BESS though the life cycle
T_{life}	equivalent operation years of BESS
$P_{s,t}$	output of BESS in hour t
$P_{union,t}$	output of wind energy union system in hour t
$S_{soc,t}$	state of charge (SOC) of BESS in hour t
$\triangle t$	scheduled interval
λ_t	charge/discharge status of BESS at time t
η_s	charge/discharge effectiveness of BESS
C_{ost}	total investment cost of BESS
S_{ave}	directly benefit from saving wind curtailed energy
$C_c(p,n)$	capital recovery factor with annual interest rate p
C_E	unit capacity cost of BESS
C_{OM}	unit operation and maintenance cost of BESS
α^s	amortized power cost per year
β^s	amortized capacity cost per year
r	kW·h/kW cost ratio of BESS

ρ^w	on-grid price of wind power
$P_{wloss,t}/P^B{}_{wloss,t}$	curtailed wind energy with and without BESS in hour t
$C^{Wind}_{serve}/C^{BESS}_{serve}$	auxiliary service cost caused by wind power with/without BESS
R_i	ramping ability of unit i
$T^{on}_{i,max}/T^{on}_{i,min}$	maximum/minimum online time of unit i
C_{Gen}	operating cost function of conventional units
$f(P_{gi,t})$	quadratic fuel cost function with coefficients a_i, b_i, c_i
S_i	on-off cost of unit i
q	curtailed rate of wind power
N_{cyc}	equivalent cycle numbers of BESS though a scheduling period
ε_s	reserve level provided by BESS for the uncertainty of wind power

References

1. Yang, Z.G.; Zhang, J.L.; Kintner-Meyer, M.C.W.; Liu, X.C.; Choi, D.; Lemmon, J.P.; Liu, J. Electrochemical energy storage for green grid. *Chem. Rev.* **2011**, *111*, 3577–3613. [CrossRef] [PubMed]
2. Dunn, B.; Kamath, H.; Tarascon, J.M. Electrical energy storage for the grid: A battery of choices. *Science* **2011**, *334*, 928–934. [CrossRef] [PubMed]
3. Qiu, J.; Zhao, J.H.; Yang, H.M.; Wang, D.X.; Dong, Z.Y. Planning of solar photovoltaics, battery energy storage system and gas micro turbine for coupled micro energy grids. *Appl. Energy* **2018**, *219*, 361–369. [CrossRef]
4. Kou, P.; Gao, F.; Guan, X.H. Stochastic predictive control of battery energy storage for wind farm dispatching: Using probabilistic wind power forecasts. *Rene. Energy* **2015**, *80*, 286–300. [CrossRef]
5. Wen, S.L.; Lan, H.; Fu, Q.; Yu, D.; Yu, D.C.; Zhang, L.J. Economic allocation for energy storage system considering wind power distribution. *IEEE Trans. Power Syst.* **2015**, *30*, 644–652. [CrossRef]
6. Ghofrani, M.; Arabali, A.; Etezadi-Amoli, M.; Fadali, M.S. Energy storage application for performance enhancement of wind integration. *IEEE Trans. Power Syst.* **2013**, *28*, 4803–4811. [CrossRef]
7. Yang, Y.; Li, H.; Aichhorn, A.; Zheng, J.P.; Greenleaf, M. Sizing strategy of distributed battery storage system with high penetration of photovoltaic for voltage regulation and peak load shaving. *IEEE Trans. Smart Grid.* **2014**, *5*, 982–991. [CrossRef]
8. Dong, J.J.; Gao, F.; Guan, X.H.; Zhai, Q.Z.; Wu, J. Storage-reserve sizing with qualified reliability for connected high renewable penetration micro-grid. *IEEE Trans. Sustain. Energy* **2016**, *7*, 732–743. [CrossRef]
9. Qin, M.W.; Chan, K.W.; Chuang, C.Y.; Luo, X.; Wu, T. Optimal planning and operation of energy storage systems in radial networks for wind power integration with reserve support. *IEEE Gener. Transm. Distrib.* **2016**, *10*, 2019–2025. [CrossRef]
10. Kargarian, A.; Hug, G. Optimal sizing of energy storage systems: A combination of hourly and intra-hour time perspectives. *IEEE Gener. Transm. Distrib.* **2016**, *10*, 594–600. [CrossRef]
11. Luo, F.; Meng, K.; Dong, Z.Y.; Zheng, Y.; Chen, Y.Y.; Wong, K.P. Coordinated operational planning for wind farm with battery energy storage system. *IEEE Trans. Sustain. Energy* **2015**, *6*, 253–262. [CrossRef]
12. Akhavan-Hejazi, H.; Mohsenian-Rad, H. Optimal operation of independent storage systems in energy and reserve markets with high wind penetration. *IEEE Trans. Smart Grid.* **2014**, *5*, 1088–1097. [CrossRef]
13. Halamay, D.A.; Brekken, T.K.; Simmons, A.; McArthur, S. Reserve requirement impacts of large-scale integration of wind, solar, and ocean wave power generation. *IEEE Trans. Sustain. Energy* **2011**, *2*, 321–328. [CrossRef]
14. Ding, H.J.; Hu, Z.H.; Song, Y.H. Rolling optimization of wind farm and energy storage system in electricity markets. *IEEE Trans. Power Syst.* **2015**, *30*, 2676–2684. [CrossRef]
15. Jiang, X.; Chen, H.K.; Xiang, T.Y. Assessing the effect of wind power peaking characteristics on the maximum penetration level of wind power. *IET Gener. Trans. Distrib.* **2015**, *9*, 2466–2473. [CrossRef]
16. Zhang, N.; Kang, C.Q.; Xia, Q.; Liang, J. Modeling conditional forecast error for wind power in generation scheduling. *IEEE Trans. Power Syst.* **2014**, *29*, 1316–1324. [CrossRef]
17. Wu, D.L.; Wang, Y.; Guo, C.X.; Liu, Y.; Gao, Z.X. An economic dispatching model considering wind power forecast error in electricity market environment. *Autom. Electr. Power Syst.* **2012**, *36*, 23–28.
18. Su, C.C.; Hsu, Y.Y. Fuzzy dynamic programming: Application to unit commitment. *IEEE Trans. Power Syst.* **1991**, *6*, 1231–1237. [CrossRef]

19. Ela, E.; Kirby, B.; Navid, N. Effective ancillary services market designs on high wind power penetration systems. In Proceedings of the 2012 IEEE Power and Energy Society General Meeting (PESGM), San Diego, CA, USA, 22–26 July 2012.
20. Østergaard, P.A. Ancillary services and the integration of substantial quantities of wind power. *Appl. Energy* **2006**, *83*, 451–463. [CrossRef]
21. Wang, J.; Liu, P.; Hicks-Garner, J.; Sherman, E.; Soukiazian, S.; Verbrugge, M.; Tataria, H.; Musser, J.; Finamore, P. Cycle-life model for graphite-LiFePO$_4$ cells. *J. Power Sources* **2011**, *196*, 3942–3948. [CrossRef]
22. Gao, F.; Yang, K.; Hui, D.; Li, D. Cycle-life energy analysis of LiFePO$_4$ batteries for energy storage. *Proc. CSEE* **2013**, *33*, 41–45.
23. Zhou, C.K.; Qin, K.J.; Allan, M.; Zhou, W.J. Modeling of the cost of EV battery wear due to V2G application in power systems. *IEEE Trans. Energy Conver.* **2011**, *26*, 1041–1049. [CrossRef]
24. Chen, W.Z.; Li, Q.B.; Shi, L.; Luo, Y.; Zhan, D.D.; Shi, N.; Liu, K. Energy storage sizing for dispatch ability of wind farm. In Proceedings of the 2012 11th International Conference on Environment and Electrical Engineering, Venice, Italy, 18–25 May 2012; IEEE: Piscataway, NJ, USA, 2012. [CrossRef]
25. Wang, X.Y.; Vilathgamuwa, D.M.; Choi, S.S. Determination of battery storage capacity in energy buffer for wind farm. *IEEE Trans. Energy Convers.* **2008**, *23*, 868–878. [CrossRef]
26. Tek, K.A.; Brekken, A.Y.; Jouanne, A.V.; Yen, Z.Z.; Hapke, H.M.; Halamay, D.A. Optimal energy storage sizing and control for wind power applications. *IEEE Trans. Sustain. Energy* **2011**, *2*, 69–77.
27. Lofberg, J. YALMIP: A toolbox for modeling and optimization in MATLAB. In Proceedings of the 2004 IEEE International Conference on Robotics and Automation (IEEE Cat. No.04CH37508), New Orleans, LA, USA, 2–4 September 2004; IEEE: Piscataway, NJ, USA, 2004; Volume 3, pp. 284–289.

 *applied sciences*

Article

Structure and Capacitance of Electrical Double Layers at the Graphene–Ionic Liquid Interface

Pengfei Lu [1], Qiaobo Dai [2], Liangyu Wu [1] and Xiangdong Liu [1,2,*

[1]   School of Hydraulic, Energy and Power Engineering, Yangzhou University, Yangzhou 225127, China; pfabc0826@126.com (P.L.); lywu@yzu.edu.cn (L.W.)
[2]   School of Energy and Environment, Southeast University, Nanjing 210096, China; qbdai@microflows.net
*   Correspondence: liuxd@yzu.edu.cn; Tel.: +86-25-8797-1315

Received: 25 August 2017; Accepted: 6 September 2017; Published: 12 September 2017

Abstract: Molecular dynamics simulations are carried out to investigate the structure and capacitance of the electrical double layers (EDLs) at the interface of vertically oriented graphene and ionic liquids $[EMIM]^+/[BF_4]^-$. The distribution and migration of the ions in the EDL on the rough and non-rough electrode surfaces with different charge densities are compared and analyzed, and the effect of the electrode surface morphology on the capacitance of the EDL is clarified. The results suggest that alternate distributions of anions and cations in several consecutive layers are formed in the EDL on the electrode surface. When the electrode is charged, the layers of $[BF_4]^-$ anions experience more significant migration than those of $[EMIM]^+$ cations. These ion layers can be extended deeper into the bulk electrolyte solution by the stronger interaction of the rough electrode, compared to those on the non-rough electrode surface. The potential energy valley of ions on the neutral electrode surface establishes a potential energy difference to compensate the energy cost of the ion accumulation, and is capable of producing a potential drop across the EDL on the uncharged electrode surface. Due to the greater effective contact area between the ions and electrode, the rough electrode possesses a larger capacitance than the non-rough one. In addition, it is harder for the larger-sized $[EMIM]^+$ cations to accumulate in the narrow grooves on the rough electrode, when compared with the smaller $[BF_4]^-$. Consequently, the double-hump-shaped C–V curve (which demonstrates the relationship between differential capacitance and potential drop across the EDL) for the rough electrode is asymmetric, where the capacitance increases more significantly when the electrode is positively charged.

Keywords: ionic liquid; vertically oriented graphene; electrical double layers; charge density; capacitance

---

1. Introduction

Due to their excellent performance in the charge/discharge rate, power delivery and cycle life, supercapacitors (or electrochemical capacitors) have already been used in parts of hybrid vehicles and emergency systems, and are also reported to be a promising energy source for future energy storage systems [1,2]. Compared with traditional batteries [3–6], lower energy density is a critical problem hindering the future successful application of supercapacitors. To overcome this drawback, superior electrode and electrolyte materials, which are two main components of supercapacitors, are urgently needed. Graphene materials [7,8] have a higher conductivity and larger electrochemical available surface, and thus the ion transport in their 2D plane structure is more efficient. Therefore, compared with other carbon-based materials (such as active carbon [9], carbon nanotubes [10], onion-like carbon [11] and porous carbon materials [12,13]), graphene materials present an immense potential to improve the energy density of supercapacitors. In addition, ionic liquids [14] could remain stable under a wider range of voltages when used as the electrolyte of supercapacitors. Therefore, based on a combination of these two components, graphene/ionic liquid supercapacitors have become an attractive object in the research field of energy storage [15–17].

As is well known, for graphene/ionic liquid supercapacitors, the processes of energy storage and release are closely related to the structure of an electrical double layer (EDL), that is, ion distributions on the surface of the graphene electrode. For example, the counterions aggregate in the inner layer of the EDL in the energy storage process and desorb from the electrode surface in the energy release process [18]. For this reason, a thorough understanding of the structure of the EDL and its effects on the capacitance of the ionic liquid EDL is crucial for further improving the efficiency of graphene/ionic liquid supercapacitor applications.

An EDL commonly exists in electrolytes at the solid/liquid interface, which usually plays an important role in fluid flow in microchannels or porous materials [19,20]. An EDL can be represented by a Helmholtz model and by the Gouy–Chapman theory [21,22]. Note that these models are usually suitable for inorganic electrolytes, where simple inorganic ions are usually regarded as particles. However, the ionic liquid only contains organic cations and organic (or inorganic) anions with a complex structure, which is free of solvent and cannot be simply regarded as a system of 'particles'. This make the ionic liquid EDL more complicated than that of traditional inorganic electrolytes. Consequently, the above theories do not seem to be tenable any more. In order to solve this problem, many researchers have turned to further study of the ionic liquid EDL on the solid surface [23,24], and they suggest that the EDL of the ionic liquid contains two components: the inner compact layer and the outer diffusive layer. In the former layer, the counterions are packed closely. Outside the compact layer, anions and cations are distributed alternately in several consecutive layers, which form the diffusive layer. From this viewpoint, the complex structures of ions are taken into consideration. Based on this viewpoint, several groups have carried out studies on the ionic liquid EDL on the graphene surface and its structures, as well as the capacitance, in an effort to support the design and optimization of graphene/ionic liquid supercapacitors. Feng et al. [25,26] examined the EDL structures of different ionic liquids near planar electrodes and found that the size and charge of ions could significantly affect the ions' distribution on the surface of the graphene electrode and the EDL capacitance. Vatamanu et al. [27] showed that an enhancement in the capacitance can be obtained when the length scale of the surface roughness of the graphene electrode is comparable to the ion size. In addition, the relationship between differential capacitance and the electrode potential (i.e., the C–V curve or capacitance-voltage curve), another important property of supercapacitors, has also been extensively investigated. Bell-shaped and double-hump-shaped C–V curves were both observed for different electrodes and electrolytes, as reported by Kornyshev, Fedorov, Vatamanu and others [28–30].

In summary, up until now, several investigations have been carried out to study the ionic liquid EDL on the graphene electrode surface, including its structures and capacitance characteristics, as well as the relative influence factors. However, the detailed distribution and migration of the ions during electrode charging, as well as the underlying molecular-level mechanisms, are not fully understood. In addition, the influence of electrode morphology on the detailed distribution and migration of the ions needs to be further explored. As a result, in order to gain a molecular-level insight into the ionic liquid EDL on the graphene electrode surface, an electrode–electrolyte model is constructed here by combining the vertically oriented graphene electrode [31] and [EMIM]$^+$/[BF$_4$]$^-$ ionic liquid, which is numerically analyzed by the molecular dynamics (MD) method in this work. Accordingly, the distribution and migration of the ions in the ionic liquid EDL on the vertically oriented graphene electrode with different charge densities and roughnesses are compared and analyzed to clarify the effect of electrode surface morphology on the capacitance of the ionic liquid EDL.

2. Simulation System and Methods

The molecular-level structures of [EMIM]$^+$/[BF$_4$]$^-$ ion pairs, vertically oriented graphene and the graphene/ionic liquid interface are illustrated in Figure 1, where the atoms on the ions are labeled by the numerical and element symbols in Figure 1a. The non-rough vertically oriented graphene electrode is modeled by 12 successive graphene nano-ribbon layers which are stacked in A–A sequence [30]. All the graphene layers orient toward the ions with their edge planes, forming the electrode surface. According to previous studies [30], as shown in Figure 1c, the layer space of the multi-graphene layers in a non-rough electrode is set as 0.34 nm. Compared with the multi-graphene layers in the non-rough electrode, the rough electrode has the same layer space and different rough surface, possessing grooves with a width of 1.01 nm and depth of 0.85 nm, as depicted in Figure 1b. The non-rough and rough electrodes both have lateral sizes of 4.26×3.94 nm^2. In the electrolyte near the electrode surface, 300 [EMIM]$^+$/[BF$_4$]$^-$ ion pairs are randomly packed with a thickness of about 4.0 nm. Outside the electrolyte region, another 4.0 nm vacuum space is added to eliminate the interaction between the particles and their images outside of the unit cell in the z-direction.

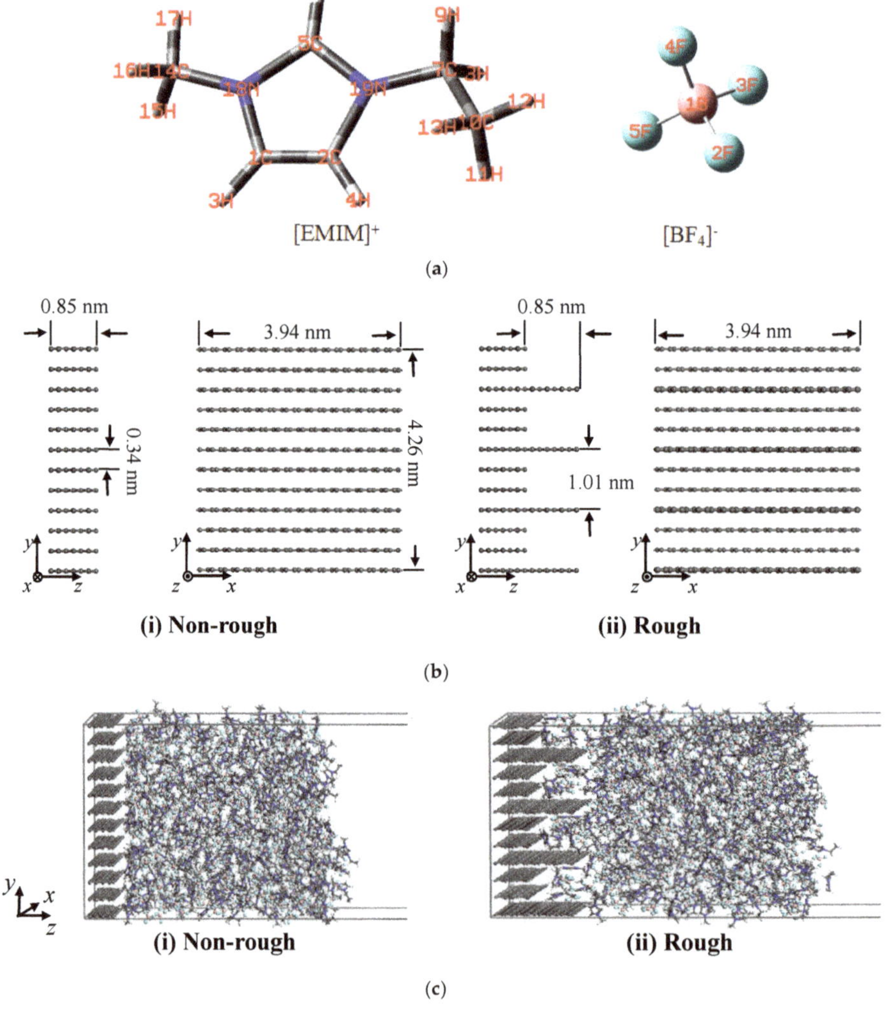

Figure 1. The detailed illustration of the simulation system: (**a**) ion structure of [EMIM]$^+$ and [BF$_4$]$^-$, (**b**) stack structure of graphene electrode, (**c**) interface models of graphene/ion liquid with non-rough and rough graphene surface.

The simulations are conducted in the canonical ensemble (NVT) using a universal force field (UFF) [32]. For each [EMIM]$^+$ cation and [BF$_4$]$^-$ anion, the partial charge q on each atom are modified and assigned according to the previous studies [33] (as shown in Table 1) and remain unchanged during the simulations. Periodic boundary conditions are imposed in all three directions. Under these boundary conditions, particles leave the domain at one side and reenter the domain at the opposite side with the same velocity. During the simulation, the electrode atoms are fixed in their initial positions in all cases. A Nose thermostat [34] is used to maintain the system temperature at 300 K. Starting from a random distribution of [EMIM]$^+$/[BF$_4$]$^-$ ion pairs near the electrode surfaces, the systems above are firstly relaxed to eliminate the undesirable interactions between ions and electrode. In addition, the SMART method [35] is used in this step to minimize the system energies. Then, the equilibrium and production processes are continued for 2×10^6 steps with a time step of 1 fs, and the atom positions and micro-structure of the electrode/electrolyte interface are saved every 20 ps. The Ewald summation method [36] is used for Coulombic interactions with an accuracy of 10^{-3} kcal/mol, and the atom-based summation method [37] is used for van der Waals interactions with a cutoff of 1.2 nm. To simulate the charge/discharge process, different constant charge densities σ (the ratio between total charge and atom numbers at the electrode surface) are set on the electrode surface as $\sigma = 0$, ± 0.02, ± 0.05 and ± 0.08 e/atom. According to the results of the ion-concentration distributions normal to the graphene surface under different charge densities, the differential capacitances are finally calculated by solving the Poisson equation.

Table 1. Atom types and partial charges on the atoms of cation and anion.

Atoms	Atom Type	q(e)	Atoms	Atom Type	q(e)
1C	C_R	−0.167	14C	C_3	−0.166
2C	C_R	−0.192	15H	H_b	0.129
3H	C R	0.248	16H	H b	0.129
4H	H_b	0.259	17H	H_b	0.129
5C	C_R	0.058	18N	N_R	0.079
6H	H_b	0.205	19N	N_R	−0.010
7C	C_3	0.033	1B	B_3	0.828
8H	H_b	0.089	2F	F_	−0.457
9H	H_b	0.089	3F	F_	−0.457
10C	C_3	−0.079	4F	F_	−0.457
11H	H_b	0.056	5F	F_	−0.457
12H	H_b	0.056	-	-	-
13H	H_b	0.056	-	-	-

3. Results and Discussion

3.1. Ion Distributions in EDL

The distribution and migration of ions in the EDL, especially those ions next to the electrode surface, could directly affect the capacitance of the ionic liquid EDL. Therefore, firstly, Figure 2 intuitively illustrates the three-dimensional arrangement of ions in the innermost cation/anion layer on the electrode surface with various electrode charge densities (σ). For the non-rough electrode, when $\sigma = 0$ e/atom (i.e., neutral electrode), the [EMIM]$^+$ ions become uniformly distributed next to the electrode surface, and the alkyl tail of the [EMIM]$^+$ cation spreads toward the neutral electrode due to its small steric effect (see the inset in Figure 2a). [BF$_4$]$^-$ ions are capable of getting into the space among the larger [EMIM]$^+$ ions, as shown in the front view of region A-A of Figure 2a, which is because of the small size of the [BF$_4$]$^-$ ion and the electrostatic attractions between [BF$_4$]$^-$ ions and [EMIM]$^+$ ions. As shown in Figure 2b, when exposed to the electrode charged by the positive potential, the [BF$_4$]$^-$ ions tend to accumulate in the innermost layer near the electrode surface due to electrostatic attractions. Meanwhile, there are still some [EMIM]$^+$ ions located in the innermost layer. This can be explained by the charge delocalization phenomenon of cations, which weakens the electrostatic repulsion between the [EMIM]$^+$ ions and the positively charged electrode surface. In addition, the [EMIM]$^+$ cations adjust their arrangement by retrieving their alkyl tails owing to the extrusion of the [BF$_4$]$^-$ anions. As shown in Figure 2c, when exposed to the electrode charged by the negative potential, the [BF$_4$]$^-$ ions are apt to move away from the electrode by the electrostatic repulsion, while the [EMIM]$^+$ ions are moved toward the electrode by electrostatic attraction.

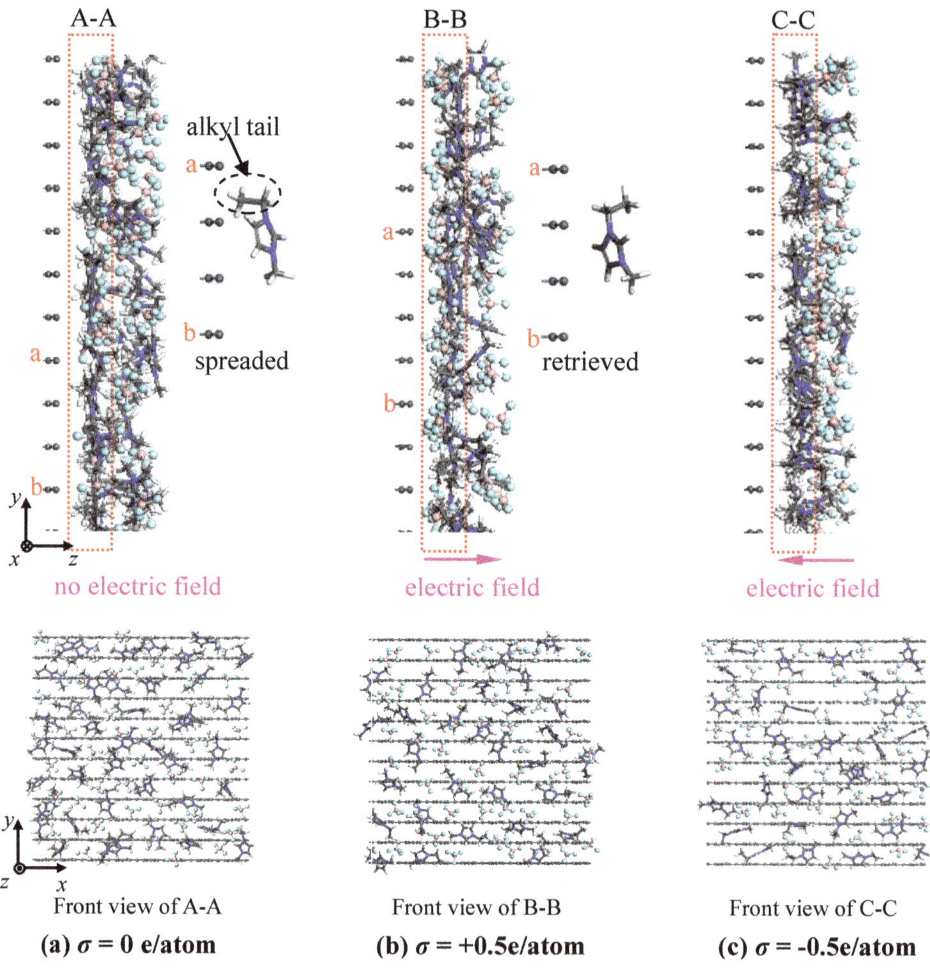

Figure 2. Snapshots of ions in the layer close to the non-rough electrode surface with different charges: (a) $\sigma = 0$ e/atom, (b) $\sigma = +0.5$ e/atom, (c) $\sigma = -0.5$ e/atom.

Compared with the ion distribution on the flat surface of the non-rough electrode, the cations and anions are wrapped in the grooves of the rough electrode, even on the neutral electrode surface, as marked by the blue dashed box in Figure 3. In particular, [EMIM]$^+$ ions are distorted in the grooves with their chain configuration aligned along the surface of the groove (see the inset in Figure 3a). In addition, the comparison between the ion distribution outside the grooves near the positively and negatively charged electrodes (see dashed box B-B in Figure 3b and the dashed box C-C in Figure 3c) indicates that the electrostatic force also has an important impact on the anions and cations on the rough electrode, and thus more $[BF_4]^-$ ions with small size and high mobility are repulsed into the layer region C-C outside the grooves with negative potential.

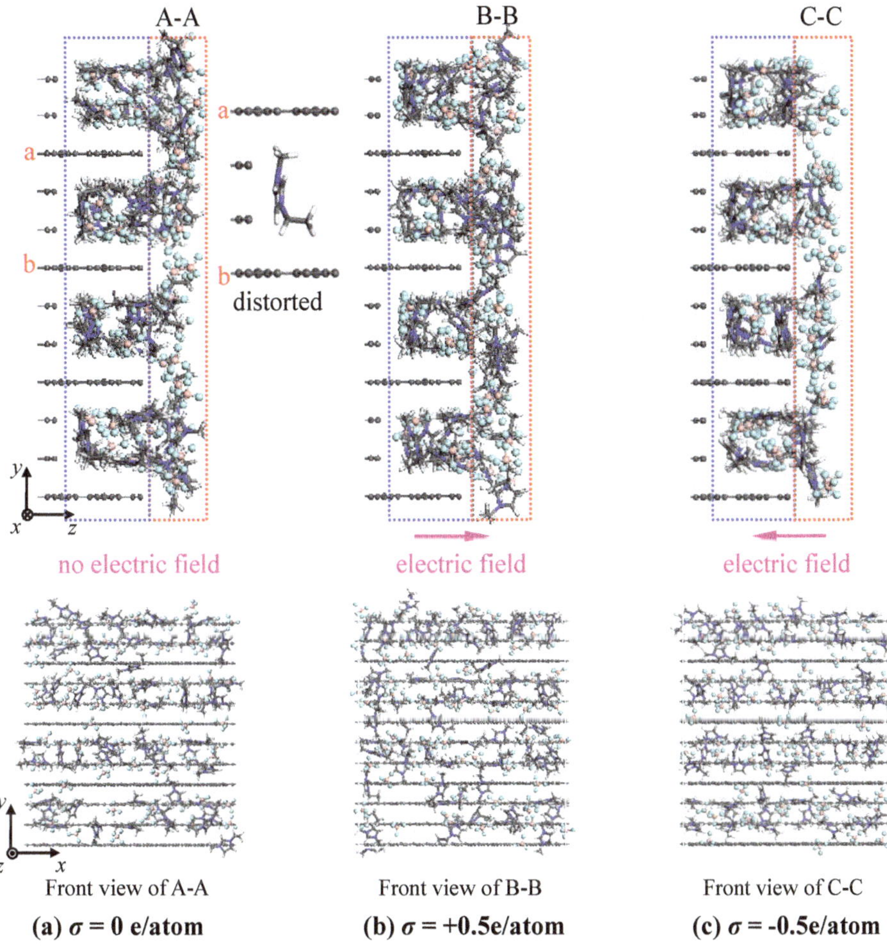

Figure 3. Snapshots of ions in the layer close to the rough electrode surface with different charges: (**a**) $\sigma = 0$ e/atom, (**b**) $\sigma = +0.5$ e/atom, (**c**) $\sigma = -0.5$ e/atom.

In order to further observe the distribution of ions, Figure 4 quantifies it by showing the concentration profiles of cations and anions along the z-direction perpendicular to the charged/uncharged electrode surface, where ρ_{lcl} is the local ion concentration along the z-direction and ρ_{avg} is the average concentration of ions in the simulation box. It can be seen that the concentration profiles of ions exhibit periodical oscillations with decaying amplitude along the z-direction, implying that anions and cations are distributed alternately in several consecutive layers outside the electrode. In addition, at the liquid–vacuum interface, the number of cations and anions is quickly reduced to zero and almost no ions are moved into the vacuum space because of the strong long-range electrostatic interactions between cations and anions. As shown in Figure 4a, the first peak of the $[EMIM]^+$ concentration profile corresponding to the first layer has only a little migration outside the charged electrode, relative to that in uncharged cases. This is due to the significant steric effect of the big $[EMIM]^+$ cation, which dramatically reduces its migration ability and thus weakens the electrostatic repulsion and attraction. However, when the electrode is positively and negatively charged, the alternate layers of $[BF_4]^-$ anions (i.e., the crests of the concentration profile) show obvious migration towards and away from the electrode relative to the case of the $\sigma = 0$ e/atom, respectively (see Figure 4b). This different phenomenon from that of the $[EMIM]^+$ cation can be mainly explained by the good mobility of the $[BF_4]^-$ anion under the same electrostatic repulsion or attraction, owing to its small size and steric effect. In addition, the small size and steric effect also lead to easy accumulation of the $[BF_4]^-$ anions, which produces higher peaks of the $[BF_4]^-$ anion concentration profile than those of the $[EMIM]^+$ cations.

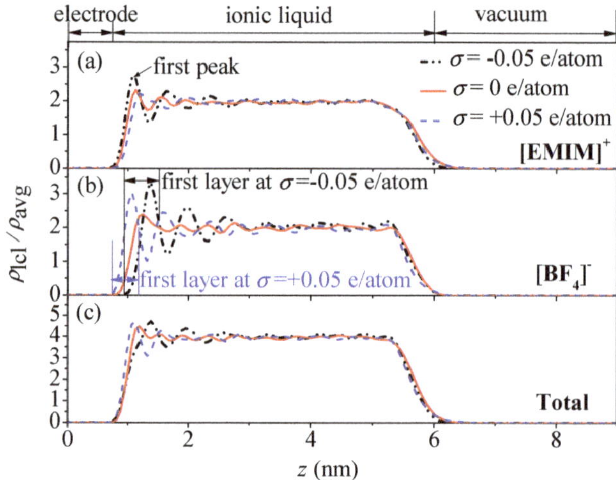

Figure 4. Concentration profiles of $[EMIM]^+$ cations and $[BF_4]^-$ anions near charged/uncharged non-rough electrodes: (**a**) $\sigma = 0$ e/atom, (**b**) $\sigma = +0.05$ e/atom, (**c**) $\sigma = -0.05$ e/atom.

As shown in Figure 5, the oscillations in ion concentration profiles are still observed on the rough electrode surface, which also reflects the presence of alternate stacked layers of cations and anions near the rough electrode surface. However, at the charged/uncharged rough electrode surface, only a limited number of $[EMIM]^+$ and $[BF_4]^-$ ions can move into the narrow grooves, which results in smaller values of the first peaks in the ion concentration profiles than on the non-rough electrode surface. In addition, the rough electrode induces a wider scope of oscillations in the ion concentration profiles, implying that the ion layers can be extended deeper into the bulk electrolyte solution by the larger interaction from the rough electrode.

What is more, Figure 6 indicates the ion–ion pair correlation functions [38] or radial distribution functions (RDF) for the $[EMIM]^+/[BF_4]^-$ ions near the non-rough neutral electrode surface, so as to further understand the alternate stacked layers of cations and anions in the EDL, where $g(r)$ is the probability of finding one kind of ion at a distance r from the reference ion with respect to that in the bulk phase, and where r is the distance between a pair of ions. The comparison among the RDFs of different ion pairs indicates that the highest first peak appears in the RDF of the $[EMIM]^+/[BF_4]^-$ pair at the smallest r, implying that the stacked layer of $[EMIM]^+$ cations is nearest to the corresponding layer of $[BF_4]^-$ anions due to the strong electric attraction between them. Meanwhile, the lowest first peak with the largest broadness emerges on the $[EMIM]^+/[EMIM]^+$ RDF at a larger r. This suggests a loose accumulation of $[EMIM]^+$ cations owing to their great steric effect and a greater distance between the two stacked layers of $[EMIM]^+$ cations in a pair. On the other hand, the RDF of $[BF_4]^-/[BF_4]^-$ exhibits a medium first peak at a value of r similar to that of $[EMIM]^+/[EMIM]^+$, which is attributed to the easier packing of $[BF_4]^-$ anions due to the smaller steric effect compared to the $[EMIM]^+$ cations. In addition, this also shows that the distance between the stacked layers of $[BF_4]^-$ anions in the pair approximates that between the stacked layers of $[EMIM]^+$ cations. Especially, it can be seen that the crests on the RDF curves of $[EMIM]^+/[EMIM]^+$ and $[BF_4]^-/[BF_4]^-$ appear to correspond to the valley on the RDF curves of $[EMIM]^+/[BF_4]^-$ and vice versa, which further supports the alternate distribution of stacked layers of cations and anions near the electrode surface.

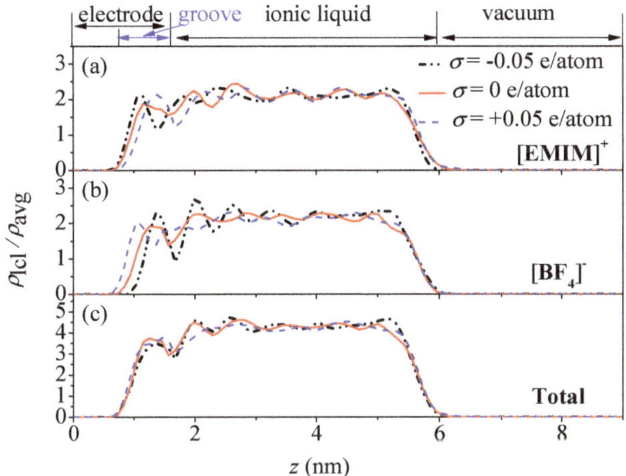

Figure 5. Concentration profiles of $[EMIM]^+$ cations and $[BF_4]^-$ anions near charged/uncharged rough electrodes. (**a**) $\sigma = 0$ e/atom, (**b**) $\sigma = +0.05$ e/atom, (**c**) $\sigma = -0.05$ e/atom.

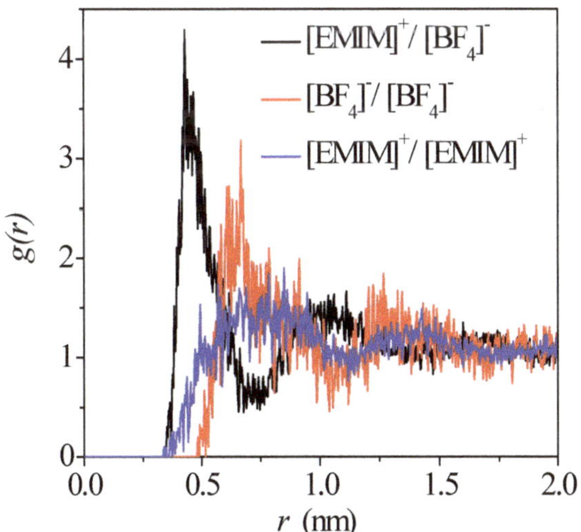

Figure 6. Ion–ion radial distribution functions (RDFs) for the $[EMIM]^+/[BF_4]^-$ electrolyte on the non-rough neutral electrode surface.

Additionally, it is worth noting from Figures 4 and 5 that even on the neutral electrode surface ($\sigma = 0$ e/atom) there also exists an obvious first layer of ions in the EDL, which is able to produce the potential on the uncharged electrode surface (as discussed in the following Section 3.2). In order to clarify this phenomenon, Figure 7 depicts the variations in potential energies of the [EMIM]$^+$ cation and [BF$_4$]$^-$ anion near the neutral electrode, where E is the potential energy of the ion and D represents the distance between the ions and the electrode. As shown, potential energy valleys are observed for both the [EMIM]$^+$ cation and the [BF$_4$]$^-$ anion, which establishes the potential energy difference to compensate the energy cost of the accumulation of ions in the location of the valleys. Therefore, the [EMIM]$^+$ cations and [BF$_4$]$^-$ anions are accumulated in the location of the valleys, forming the first layers on the neutral electrode. In addition, it can be seen that the potential energy valley of the [EMIM]$^+$ cation appears a little closer to the electrode than the [BF$_4$]$^-$ anion, which explains the occurrence of the first peak at the smaller z for the [EMIM]$^+$ cation on the neutral electrode surface ($\sigma = 0$ e/atom) (see insets (a) and (b) in Figures 4 and 5).

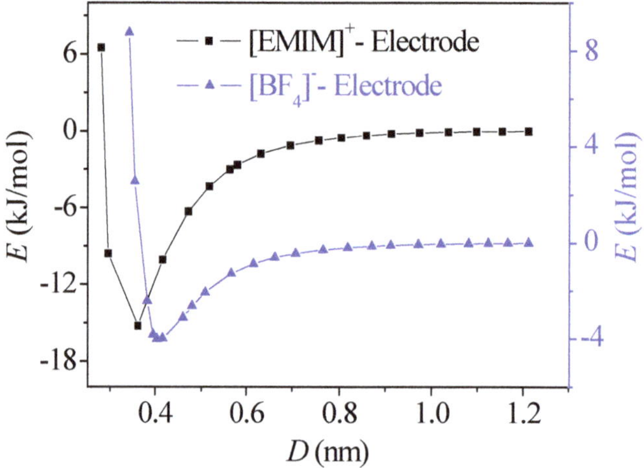

Figure 7. Total interaction energies of the [EMIM]$^+$ ion and [BF$_4$]$^-$ ion at different positions on the neutral electrode surface.

3.2. Differential Capacitance Calculation and Analysis

Apparently, the abovementioned different characteristics of ion distributions near the non-rough and rough electrodes could affect the capacitance properties of the EDL, which can be represented by the C–V curve (i.e., differential capacitance versus the potential drop across the EDL) of the EDL. To obtain the C–V curve, it is necessary to firstly find the total potential ($\Phi = \Phi_\sigma + \Phi_{IL}$) distribution of the whole simulation system along the z-direction. Herein, $\Phi_\sigma = \sigma z/\varepsilon$ is produced by the charge on the electrode, where ε is the effective dielectric constant of the electrolyte. Φ_{IL} is produced by the ionic liquid, which can be gained by integrating the Poisson equation [39]:

$$\frac{d^2\Phi_{IL}(z)}{dz^2} = -\frac{\rho_e(u)}{\varepsilon} \tag{1}$$

where $\rho_e(u) = e(\rho_{cat} - \rho_{ani})$ is the space charge density along the surface normal direction (ρ_{cat} and ρ_{ani} are the ion concentrations of [EMIM]$^+$ cautions and [BF$_4$]$^-$ anions, respectively, and e is the elementary charge), and $\varepsilon = \varepsilon_0\varepsilon_r$ is the effective dielectric constant of the electrolyte (ε_0 and ε_r are the permittivity of vacuum and relative dielectric constant, respectively). Considering that the dielectric constant in the compact layer can be nearly ten times smaller than that in the bulk phase [25,40], we chose $\varepsilon_r = 7.8$

for the $[EMIM]^+/[BF_4]^-$ electrolyte. With the boundary condition $\Phi(0) = 0$ and $d\Phi/dz = 0$ in the bulk phase of electrolyte, the total potential distribution along the z-direction can be expressed as:

$$\Phi(z) = -\frac{1}{\varepsilon}\int_0^z (z-u)\rho_e(u)du + \frac{\sigma}{\varepsilon}z \tag{2}$$

When the potential distribution is obtained according to Equation (2), the potential drop across the EDL (U_{EDL}) can be calculated as $U_{EDL} = \Phi_e - \Phi_b$, where Φ_e is the potential on the electrode and Φ_b is the potential in the middle of the electrolyte phase. In this way, U_{EDL} values on the non-rough electrode and the rough one are calculated as shown in Figure 8. Based on the data points of σ versus U_{EDL}, the σ-U_{EDL} curve is obtained by fitting these points with a fifth-order polynomial (see Figure 8). As shown, the potential drop across the EDL (U_{EDL}) is not zero but a positive value at $\sigma = 0$ e/atom. As discussed above, this phenomenon must be generated by the first layer of ions in the EDL on the neutral electrode surface.

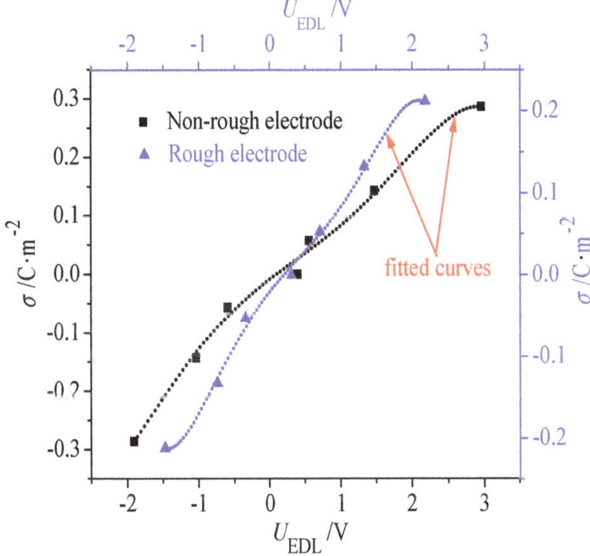

Figure 8. Correlation between the electrode charge density and the potential drop across the electrical double layers (EDLs) on the non-rough and rough electrodes.

Accordingly, the C–V curves can be obtained by analytically differentiating the polynomial according to the equation $DC = d\sigma/dU_{EDL}$, and are depicted in Figure 9 for the non-rough and rough electrodes. As shown, the C–V curves on the non-rough and rough electrodes both exhibit the double-hump shape. That is, a minimum point appears on the C–V curves near the potential of zero charge (PZC), while two maximum ones appear at lower and higher values of U_{EDL}. As discussed above, it is the accumulation of ions near the negatively and positively charged electrode surface that results in the maximum capacitance at negative U_{EDL} and the one at positive U_{EDL}. However, with the increasing concentration of ions near the charged electrode surface, the accumulation of ions at the charged electrode surface tends to be weakened when the electrode is further charged. As a result, the capacitance climbs over two maximum points and decreases in turn, forming the double-hump-shaped C–V curves. It is also observed that, despite whether the electrode is charged positively or negatively, there is a more significant increase in capacitance in the rough cases than in the non-rough cases. This phenomenon can be attributed to the fact that the rough electrode increases the effective contact area between the ions and electrode, which attracts more counterions to the electrode surface compared to the non-rough electrode. In addition, the double-hump-shaped C–V curve for the rough electrode is asymmetric with respect to the symmetric one for the non-rough electrode, where the capacitance is increased more significantly when the electrode is positively charged. This must be caused by the different impact of the grooved electrode surface on the accumulation of $[EMIM]^+$ cations and $[BF_4]^-$ anions. Clearly, when the electrode is charged, it is harder for the larger-sized $[EMIM]^+$ cations to accumulate in the narrow grooves on the rough electrode when compared with the smaller $[BF_4]^-$. Consequently, a smaller capacitance appears on the negatively charged rough electrode. The comparisons between the C–V curves of the EDLs on the non-rough and rough electrodes suggest that roughness on the electrode can increase the capacitance in the EDL, and thus be beneficial for improving the energy density of the EDL supercapacitor in real applications. We believe this work can provide useful information to improve the energy density of the EDL supercapacitor via optimizing the micro-structure of the electrode. In addition, the charging/discharging rate, which is a crucial parameter for the super-capacitor system, will be considered in our future work.

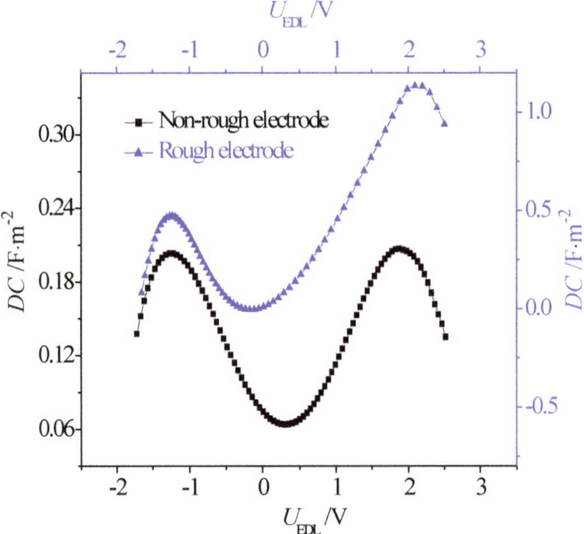

Figure 9. Correlation between the capacitance and the potential drop across the EDLs on the rough and non-rough electrodes.

4. Conclusions

In this work, based on the molecular dynamics method, an electrode–electrolyte model is established to numerically investigate the molecular-level structure in the electrical double layers of the $[EMIM]^+/[BF_4]^-$ ionic liquid on the vertically oriented graphene electrode. The detailed distribution and migration of the ions on the non-rough and rough electrode surfaces with different charge densities are compared and analyzed. In addition, the C–V curve of the EDL is obtained to clarify the effect of the electrode surface morphology on the capacitance of the EDL. The main conclusions can be summarized as follows:

(1)	The concentration profiles of ions exhibit periodical oscillations with decaying amplitude along the direction perpendicular to the charged/uncharged electrode surface, which suggests an alternate distribution of anions and cations in several consecutive layers in the EDL on the electrode surface. When the electrode is charged, the alternate layers of $[BF_4]^-$ anions experience more-significant migration than the layers of [EMIM]$^+$ cations, owing to the good mobility of the [BF$_4$]$^-$ anion due to its small size and steric effect. Additionally, these ion layers can be extended deeper into the bulk electrolyte solution by the stronger interaction of the rough electrode compared to the situation for the non-rough electrode.

(2) The potential energy valley of ions near the neutral electrode surface establishes a potential energy difference to compensate the energy cost of the ion accumulation. As a result, ions are able to accumulate in the location of the valley to form the first layers near the electrode, allowing the potential drop across the EDL on the uncharged electrode surface to be produced.

(3) Due to the greater effective contact area between the ions and electrode, the rough electrode possesses a larger capacitance than the non-rough one when it is negatively and positively charged. In addition, when the electrode is charged, it is harder for the larger-sized [EMIM]$^+$ cations to accumulate in the narrow grooves on the rough electrode when compared with the smaller [BF$_4$]$^-$. Consequently, when compared with the symmetric double-hump-shaped C–V curve for the non-rough electrode surface, the double-hump-shaped C–V curve for the rough electrode is asymmetric, where the capacitance is increased more significantly when the electrode is positively charged.

Acknowledgments: The authors gratefully acknowledge the support provided by the National Natural Science Foundation of China (51406175) and the Natural Science Foundation of Jiangsu Province (BK20140488).

Author Contributions: Xiangdong Liu provided the guidance and supervision. Pengfei Lu implemented the main research, discussed the results, and wrote the paper. Qiaobo Dai and Liangyu Wu checked and revised the paper. All authors read and approved the final manuscript.

Conflicts of Interest: The authors declare no conflict of interest.

References

1. Winter, M.; Brodd, R.J. What are batteries, fuel cells, and supercapacitors? *Chem. Rev.* **2004**, *104*, 4245–4269. [CrossRef] [PubMed]

2. Fronk, B.M.; Neal, R.; Garimella, S. Evolution of the transition to a world driven by renewable energy. *J. Energy Resour. Technol.* **2010**, *132*, 021009. [CrossRef]

3. Yang, W.; Li, J.; Ye, D.D.; Zhang, L.; Zhu, X.; Liao, Q. A hybrid microbial fuel cell stack based on single and double chamber microbial fuel cells for self-sustaining pH control. *J. Power Sources* **2016**, *306*, 685–691. [CrossRef]

4. Zhu, X.B.; Zhao, T.S.; Wei, Z.H.; Tian, P.; An, L. A high-rate and long cycle life solid-state lithium-air battery. *Energy Environ. Sci.* **2015**, *8*, 3745–3754. [CrossRef]

5. Demirocak, D.E.; Srimivasan, S.S.; Stefanakos, E.K. A Review on Nanocomposite Materials for Rechargeable Li-ion Batteries. *Appl. Sci.* **2017**, *7*, 731. [CrossRef]

6. Simon, P.; Gogotsi, Y.; Dunn, B. Where do batteries end and supercapacitors begin? *Science* **2014**, *343*, 1210–1211. [CrossRef] [PubMed]

7. Rao, C.N.R.; Sood, A.K.; Subrahmanyam, K.S.; Govindaraj, A. Graphene: The new two-dimensional nanomaterial. *Angew. Chem. Int. Ed.* **2009**, *48*, 7752–7777. [CrossRef] [PubMed]

8. Iro, Z.S.; Subramani, C.; Dash, S.S. A brief review on electrode materials for supercapacitor. *Int. J. Electrochem. Sci.* **2016**, *11*, 10628–10643. [CrossRef]

9. Rajput, N.N.; Monk, J.; Hung, F.R. Ionic liquids confined in a realistic activated carbon model: A molecular simulation study. *J. Phys. Chem. C* **2014**, *118*, 1540–1553. [CrossRef]

10. Yang, L.; Fishbine, B.H.; Migliori, A.; Pratt, L.R. Molecular simulation of electric double-layer capacitors based on carbon nanotube forests. *J. Am. Chem. Soc.* **2008**, *131*, 12373–12376. [CrossRef] [PubMed]

11. Feng, G.; Jiang, D.E.; Cummings, P.T. Curvature effect on the capacitance of electric double layers at ionic liquid/onion-like carbon interfaces. *J. Chem. Theory Comput.* **2012**, *8*, 1058–1063. [CrossRef] [PubMed]

12. Wen, Z.Q.; Li, M.; Zhu, S.J.; Wang, T. Three-dimensional (3D) nanocomposites of MnO$_2$-modified mesoporous carbon filled with carbon spheres/carbon blacks for supercapacitors. *Int. J. Electrochem. Sci.* **2016**, *11*, 23–33.

13. Shim, Y.; Kim, H.J. Nanoporous carbon supercapacitors in an ionic liquid: A computer simulation study. *ACS Nano* **2010**, *4*, 2345–2355. [CrossRef] [PubMed]

14. Hayes, R.; Warr, G.G.; Atkin, R. Structure and nanostructure in ionic liquids. *Chem. Rev.* **2015**, *115*, 6357–6426. [CrossRef] [PubMed]

15. Baldelli, S.; Bao, J.; Wu, W.; Pei, S.S. Sum frequency generation study on the orientation of room-temperaturre ionic liquid at the graphehe-ionic liquid interface. *Chem. Phys. Lett.* **2011**, *516*, 171–173. [CrossRef]

16. Vijayakumar, M.; Schwenzer, B.; Shutthanandan, V.; Hu, J.Z.; Liu, J.; Aksay, I.A. Elucidating graphehe-ionic liquid interfacial region: A combined expermental and computational study. *Nano Energy* **2014**, *3*, 152–158. [CrossRef]

17. Mo, Y.F.; Wan, Y.F.; Chou, A.; Huang, F.C. Graphene/ionic liquid composite films and ion exchange. *Sci. Rep.* **2014**, *4*, 5466. [CrossRef] [PubMed]

18. Vatamanu, J.; Bedrov, D. Capacitive energy storage: Current and future challenges. *J. Phys. Chem. Lett.* **2015**, *6*, 3594–3609. [CrossRef] [PubMed]

19. Lin, T.Y.; Kandlikar, S.G. A theoretical model for axial heat conduction effects during single-phase flow in microchannels. *J. Heat Transf.* **2012**, *134*, 020902. [CrossRef]

20. Li, W.M.; Qu, X.P.; Alam, T.; Yang, F.H.; Chang, W.; Khan, J.; Li, C. Enhanced flow boiling in microchannels through integrating multiple micro-nozzles and reentry microcavities. *Appl. Phys. Lett.* **2017**, *110*, 014104. [CrossRef]

21. Helmholtz, H. Studies of electrical interfaces. *Ann. Phys.* **1879**, *243*, 337–382. [CrossRef]

22. Bolt, G.H. Analysis of the validity of the Gouy-Chapman theory of the electric double layer. *J. Colloid Sci.* **1955**, *10*, 206–218. [CrossRef]

23. Oldham, K.B. A Gouy-Chapman-Stern model of the double layer at a (metal)/(ionic liquid) interface. *J. Electroanal. Chem.* **2008**, *613*, 131–138. [CrossRef]

24. Kornyshev, A.A. Double-layer in ionic liquids: paradigm change? *J. Phys. Chem. B* **2007**, *111*, 5545–5557. [CrossRef] [PubMed]

25. Feng, G.; Qiao, R.; Huang, J.S.; Dai, S.; Sumpter, B.J.; Meunier, V. The importance of ion size and electrode curvature on electrical double layers in ionic liquids. *Phys. Chem. Chem. Phys.* **2011**, *13*, 1152–1161. [CrossRef] [PubMed]

26. Li, S.; Feng, G.; Bauelos, J.L.; Rother, G.; Fulvio, P.F.; Dai, S.; Cummings, P.T. Distinctive nanoscale organization of dicationic versus monocationic ionic liquids. *J. Phys. Chem. C* **2013**, *117*, 18251–18257. [CrossRef]

27. Vatamanu, J.; Hu, Z.Z.; Bedrov, D. Increasing energy storage in electrochemical capacitors with ionic liquid electrolytes and nanostructured carbon electrodes. *J. Phys. Chem. Lett.* **2013**, *4*, 2829–2837. [CrossRef]

28. Fedorov, M.V.; Kornyshev, A.A. Ionic liquid near a charged wall: Structure and capacitance of electrical double layer. *J. Phys. Chem. B* **2008**, *112*, 11868–11872. [CrossRef] [PubMed]

29. Fedorov, M.V.; Georgi, N.; Kornyshev, A.A. Double layer in ionic liquids: The nature of the camel shape of capacitance. *Electrochem. Commun.* **2010**, *12*, 296–299. [CrossRef]

30. Vatamanu, J.; Cao, L.L.; Borodin, O.; Bedrov, D.; Smith, G.D. On the influence of surface topographyc double layer structure and differential capacitance of graphite/ionic liquid interfaces. *J. Phys. Chem. Lett.* **2011**, *2*, 2267–2272. [CrossRef]

31. Bo, Z.; Zhu, W.G.; Ma, W.; Wen, Z.H.; Shuai, X.R.; Chen, J.H. Vertically oriented graphene bridging active-layer/current-collector interface for ultrahigh rate supercapacitors. *Adv. Mater.* **2013**, *25*, 5799–5806. [CrossRef] [PubMed]

32. Rappé, A.K.; Casewit, C.J.; Colwell, K.S.; Goddard, W.A., III; Skiff, W.M. UFF, a full periodic table force field for molecular mechanics and molecular dynamics simulations. *J. Am. Chem. Soc.* **1992**, *114*, 10024–10035.

33. De Andrade, J.; Böes, E.S.; Stassen, H. Computational study of room temperature molten salts composed by 1-alkyl-3-methylimidazolium cations—Force-field proposal and validation. *J. Phys. Chem. B* **2002**, *106*, 13344–13351. [CrossRef]

34. Nosé, S. A unified formulation of the constant temperature molecular dynamics methods. *J. Chem. Phys.* **1984**, *81*, 511–519. [CrossRef]

35. Zhang, C.B.; Lu, P.F.; Chen, Y.P. Molecular dynamics simulation of electroosmotic flow in rough nanochannels. *Int. J. Heat Mass Transf.* **2014**, *59*, 101–105. [CrossRef]

36. Nymand, T.M.; Linse, P. Ewald summation and reaction field methods for potentials with atomic charges, dipoles, and polarizabilities. *J. Chem. Phys.* **2000**, *112*, 6152–6160. [CrossRef]
37. Nyden, M.R.; Stoliarov, S.I.; Westmoreland, P.R.; Guo, Z.X.; Jee, C. Applications of reactive molecular dynamics ot the study of the thermal decomposition of polymers and nanoscale structures. *Mater. Sci. Eng. A* **2004**, *365*, 114–121. [CrossRef]
38. Dell, A.E.; Tsang, B.; Jiang, L.X.; Granick, S.; Schweizer, K.S. Correlated two-particle diffusion in dense colloidal suspensions at early times: Theory and comparison to experiment. *Phys. Rev. E* **2015**, *92*, 052304. [CrossRef] [PubMed]
39. Outhwaite, C.W. Modified Poisson-Boltzmann equation in electric double layer theory based on the Bogoliubov-Born-Green-Yvon integral equations. *J. Chem. Soc.* **1978**, *74*, 1214–1221. [CrossRef]
40. Singh, T.; Kumar, A. Static dielectric constant of room temperature ionic liquids: internal pressure and cohesive energy density approach. *J. Phys. Chem. B* **2008**, *112*, 12968–12972. [CrossRef] [PubMed]

 *applied sciences*

Article

Dual Functionalized Freestanding TiO$_2$ Nanotube Arrays Coated with Ag Nanoparticles and Carbon Materials for Dye-Sensitized Solar Cells

Ho-Sub Kim [1,†], Myeung-Hwan Chun [1,†], Jung Sang Suh [1], Bong-Hyun Jun [2,*] and Won-Yeop Rho [1,2,*]

[1]   Department of Chemistry, Seoul National University, Seoul 151-747, Korea; hosub@snu.ac.kr (H.-S.K.); hwanmc@hanmail.net (M.-H.C.); jssuh@snu.ac.kr (J.S.S.)
[2]   Department of Bioscience and Biotechnology, Konkuk University, Seoul 143-701, Korea
*   Correspondence: bjun@konkuk.ac.kr (B.-H.J.); rho7272@gmail.com (W.-Y.R.); Tel.: +82-2-450-0521
†   These authors contributed equally to this work.

Academic Editors: Elias K. Stefanakos and Sesha S. Srinivasan
Received: 21 March 2017; Accepted: 30 May 2017; Published: 2 June 2017

Abstract: Highly ordered, freestanding TiO$_2$ nanotube arrays (TiO$_2$ NTAs) were prepared using an electrochemical method. The barrier layer was etched to open the bottom of each array, aptly named "open-ended TiO$_2$ NTAs". These arrays were coated with silver nanoparticles (Ag NPs) and/or carbon materials to enhance electron generation and transport. The energy conversion efficiency of the resulting dye-sensitized solar cells (DSSCs) with open-ended freestanding TiO$_2$ NTAs, when coated with Ag NPs, increased from 5.32% to 6.14% (by 15%) due to plasmonic interactions. Meanwhile, coating the open-ended freestanding TiO$_2$ NTAs with carbon materials increased the energy conversion efficiency from 5.32% to 6.07% (by 14%), due to π-π conjugation. When the Ag NPs and carbon materials were simultaneously applied to the open-ended freestanding TiO$_2$ NTAs, the energy conversion efficiency increased from 5.32% to 6.91%—an enhancement of 30%, due to the additive effects of plasmonics and π-π conjugation.

Keywords: dye-sensitized solar cells; carbon materials; Ag nanoparticles; freestanding TiO$_2$ nanotube arrays

---

1. Introduction

Since the initial development of dye-sensitized solar cells (DSSCs) in 1991 by the Grätzel group [1], global research has continued due to their low cost, ease of fabrication, and high power conversion efficiency [2,3]. Titanium dioxide nanoparticles (TiO$_2$ NPs) are typically used as the photoanode in DSSCs because they have a desirable direct band gap (3.2 eV) and a large surface area for adsorbing dyes; both help to generate electrons [4–7]. However, TiO$_2$ NPs are randomly networked, and the countless grain boundaries within them lead to material defects and charge recombinations that inhibit smooth electron transport [8–10].

In recent years, TiO$_2$ nanotube arrays (NTAs) have been explored as an alternative to TiO$_2$ NPs [11–13]. The TiO$_2$ NTAs can be fabricated using an electrochemical method (i.e., anodization) [11,14], and their well-ordered, vertically aligned tubular structures serve as direct electron pathways; this enhances not only electron transport, but also charge separation [15,16]. However, despite their merits, the barrier layer on the bottom of the TiO$_2$ NTAs may impede charge transfer and electrolyte diffusion. To overcome this problem, we recently removed the bottom layer of TiO$_2$ NTAs using argon ion (Ar$^+$) milling, which resulted in improved electron transport and improved electrolyte diffusion [17].

There have been an increasing number of studies that add carbon to TiO_2 NTAs in order to improve the charge separation and transfer of electrons; this is due to carbon's superior electrical properties by π-π conjugation [18–20]. Many researchers have explored the application of carbon materials in solar cell technologies. Carbon 60 (C_{60} or "fullerene") and carbon nanotubes (CNTs) are well known for their roles as electron acceptors and charge separators in organic solar cells [21,22]. When incorporated in organic solar cells, CNTs act as exciton dissociation sites and hopping centers for hole transport [23], and in DSSCs, graphene mixed with TiO_2 NPs plays a role in promoting charge separation and movement [24]. As mentioned, TiO_2 NTAs were developed as alternatives to TiO_2 NPs. While it is not simple to blend carbon materials with TiO_2 NTAs, we recently reported a method for enriching freestanding TiO_2 NTAs with carbon for use in DSSCs. When a small amount of carbon was deposited on TiO_2 NTAs, compared to those without carbon enrichment, the energy conversion efficiency increased by approximately 22.4% [25]. We tentatively attributed this to an improved efficiency of electron transport by the π-π conjugation introduced through carbon enrichment.

A plasmonic effect triggered by metal NPs (such as silver and gold) can be used to enhance photoabsorption in solar cells [26–28]. When incident photons pass by Ag NPs, they cause electron vibration and photo scattering of the nanoparticles, which facilitates photon control more efficiently [29]. The metal NPs were incorporated through mixing with TiO_2 sol in the DSSCs, or with precursors of the active layer in organic solar cells. However, it is difficult to insert metal NPs into the channels of TiO_2 NTAs, as the fabrication of DSSCs based on TiO_2 NTAs requires. We recently devised a simple method for the complete formation of Ag NPs in the channels of TiO_2 NTAs using ultraviolet (UV) irradiation [13,30]. These NPs helped the dyes to generate electrons, as was demonstrated by a high current density in the DSSCs.

To date, we have confirmed the separate effects of both carbon enrichment and the incorporation of Ag NPs in previous studies. However, the effects of adding both carbon and Ag NPs remain unknown. Here, we report such effects on the performance of DSSCs—in terms of enhanced electron transport and plasmonic effects—when enriching freestanding TiO_2 NTAs with one or both materials. Carbon materials were synthesized by chemical vapor deposition (CVD) and deposited on the wall of TiO_2 NTAs. The Ag NPs were formed using UV irradiation within the channels of TiO_2 NTAs.

2. Materials and Methods

2.1. Preparation of Closed- and Open-Ended TiO$_2$ NTAs

To fabricate TiO_2 NTAs, titanium foils (Alfa Aesar, 99.7% purity, 2.5 cm $\times$ 4.0 cm $\times$ 320 µm) were prepared and anodized using an electrochemical method. The electrolyte was composed of 0.8 wt % NH_4F and 2 vol % H_2O in ethylene glycol. Carbon rods served as the cathode material. A 60 V DC potential was supplied to the titanium foils at 25 °C for 2 h. Later, the anodized titanium foils were annealed in a tube furnace (at 450 °C for 1 h), and a second anodization process was then conducted on the samples (at 30 V for 10 min). After the second anodization, the sample was immersed in 10% H_2O_2 for 24 h to detach the TiO_2 NTAs from the titanium foils. Ion milling with Ar^+ bombardment was used to make open-ended tips by removing the bottom of the TiO_2 NTAs [25].

2.2. Preparation of Photoanodes for DSSCs Based on the TiO$_2$ NTAs

Fluorine-doped tin oxide (FTO) glass was washed and sonicated in ethanol and acetone to remove impurities. Titanium diisopropoxide bis(acetylacetonate) (5 wt % in *n*-butanol) was spin-coated on the clean FTO glass to form a compact TiO_2 blocking layer after annealing at 450 °C for 1 h. A TiO_2 paste (Solaronix, T/SP) was applied to the FTO glass using a doctor blade method, in order to attach the closed- or open-ended TiO_2 NTAs. Finally, the samples were annealed in a furnace at 450 °C for 30 min.

2.3. Synthesis of Ag NPs on the TiO_2 NTAs by UV Irradiation

The samples were placed in a 0.3 mM $AgNO_3$ aqueous solution. Ag NPs were synthesized in the channels of closed- or open-ended TiO_2 NTAs using a 254 nm UV lamp for 3 min.

2.4. Synthesis of Carbon Materials on the TiO_2 NTAs by CVD

The samples were placed in a quartz tube furnace filled with nitrogen (200 standard cubic centimeter per minute (sccm)). Hydrogen gas (30 sccm) and ethylene gas (40 sccm) were flowed into the tube furnace at 450 °C for 30 s.

2.5. Fabrication of Dye-Sensitized Solar Cells

All DSSC samples were post-treated with 10 mM $TiCl_4$ solution at 50 °C for 30 min, then annealed at 450 °C for 1 h. These steps not only enhanced the photocurrent, but also prevented the dissolution of the Ag NPs upon contacting the iodine-iodide electrolyte. Each treated sample was stained using dye molecules in ethanol at 50 °C for 8 h; here the dye molecules were 0.5 mM solutions of N719 ($(Bu_4N)_2Ru(dobpyH)_2(NCS)_2$, Solaronix). Following this treatment, samples were washed with ethanol to eliminate physisorbed dye molecules. To fabricate counter electrodes, chloroplatinic acid (H_2PtCl_6) in ethanol was drop-casted onto clean FTO glass and annealed in a tube furnace at 400 °C for 1 h.

The electrolyte used to separate the electrodes contained 0.7 M 1-butyl-3-methyl-imidazolium iodide (BMII), 0.03 M I_2, 0.1 M guanidium thiocyanate (GSCN), and 0.5 M 4-*tert*-butyl pyridine (TBP) in a mixture of acetonitrile and valeronitrile (85:15 v/v). A 60-μm-thick hot-melt Surlyn spacer (Solaronix) was placed between the photoanode and counter electrode; the electrolyte was injected into the space formed.

2.6. Characterization of Dye-Sensitized Solar Cells

The structures of TiO_2 NTAs on FTO glass were confirmed using a field emission scanning electron microscope (FE-SEM, JSM-6330F, JEOL Inc., Tokyo, Japan) The existence of Ag NPs in the channels of TiO_2 NTAs was verified by the high-angle annular dark-field (HAADF) imaging technique using a scanning transmission electron microscope (TEM, JEM-2200FS, JEOL Inc., Tokyo, Japan). Raman spectra were measured using a Raman spectrometer (LabRAM HV Evolution spectrometer, HORIBA, Tokyo, Japan). The ultraviolet-visible (UV-Vis) spectra were recorded using a UV-Vis spectrophotometer (NEOSYS-2000, SCINCO, Seoul, Korea). Current density-voltage measurements were carried out using an electrometer (Keithley 2400) and a solar simulator (1 kW Xenon with AM 1.5 filter, PEC-L01, Peccel Technologies, Kanagawa, Japan). Electrochemical impedance spectroscopy (EIS) data were collected using a potentiostat (Solartron 1287) equipped with a frequency response analyzer (Solartron 1260) between 10^{-2} and 10^6 Hz under AM 1.5 light illumination, and analyzed using Z-View software (Solartron Analytical). The applied bias voltage and AC amplitude were set at the open circuit voltage (V_{oc}) of the DSSCs and at 10 mV.

3. Results and Discussion

The fabrication of DSSCs based on freestanding TiO_2 NTAs is shown in Figure 1. The bottom layer was present in the closed-ended freestanding TiO_2 NTAs, but was removed by ion milling in the open-ended TiO_2 NTAs. The DSSCs were fabricated from both the open- and closed-ended freestanding TiO_2 NTAs to compare energy conversion efficiencies. In both cases, the freestanding TiO_2 NTAs were attached to the FTO glass with TiO_2 paste, and Ag NPs were synthesized using UV irradiation (Figure 1a). Carbon materials were synthesized using CVD (Figure 1b). By using the UV irradiation and CVD, Ag NPs and carbon materials were deposited in the channel of highly ordered TiO_2 NTAs without any distortion. The dye (N719) was adsorbed onto both types of freestanding TiO_2 NTAs (Figure 1c). Finally, DSSCs were fabricated by assembling the working electrode (freestanding TiO_2 NTAs on FTO glass) and the counter electrode (Pt on FTO glass), as shown in Figure 1d.

FE-SEM images of TiO_2 NTAs are shown in Figure 2. The top view (Figure 2a) shows a pore size of approximately 100 nm after having applied the electrochemical method. The bottom of the TiO_2 NTAs before ion milling ("closed-ended TiO_2 NTAs") is shown in Figure 2b, with a total bottom pore size of approximately 100 nm (including that of the wall thickness). However, when the bottom was removed by ion milling to produce the "open-ended TiO_2 NTAs" (Figure 2c), the bottom pore was reduced to 30 nm in size, while the wall thickness was approximately 35 nm. An HAADF image of Ag NPs in the channels of TiO_2 NTAs is shown in Figure 2d, and the diameter of Ag NPs was approximately 30 nm. This allowed the Ag NPs to be successfully immobilized inside the channel of TiO_2 NTAs by UV irradiation, and the resulting plasmonic interactions may have affected all the surface areas. A side view of TiO_2 NTAs attached to the FTO glass by TiO_2 paste after being sintered at 450 °C is shown in Figure 2e. The main role of the TiO_2 paste is to connect the TiO_2 NTAs with the FTO glass surface. The thickness of the TiO_2 film layer was 3 µm, and the length of the TiO_2 NTAs was approximately 18 µm.

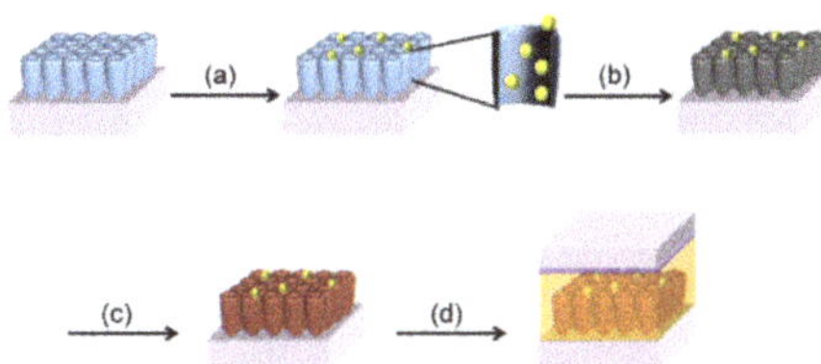

Figure 1. Overall scheme of the fabrication of dye-sensitized solar cells (DSSCs), based on freestanding TiO_2 nanotube arrays (TiO_2 NTAs) coated with silver nanoparticles (Ag NPs) and carbon materials. (**a**) Synthesis of Ag NPs in the channel of TiO_2 NTAs, (**b**) deposition of carbon materials, (**c**) dye adsorption, and (**d**) fabrication of the DSSC.

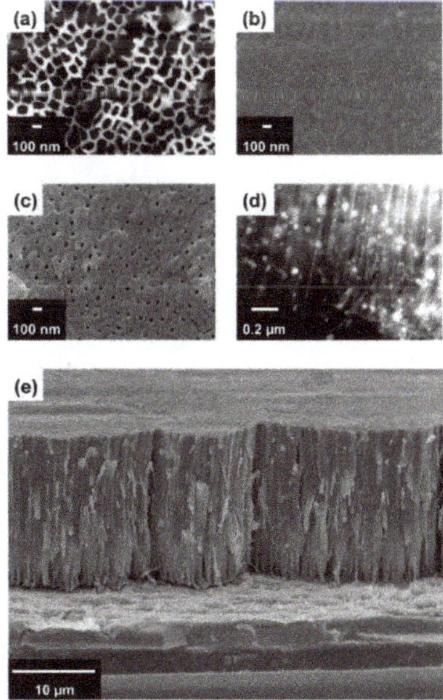

Figure 2. Field emission scanning electron microscope (FE-SEM) images of TiO_2 NTAs: (**a**) top view, (**b**) bottom view, (**c**) bottom view after ion milling, (**d**) high-angle annular dark-field (HAADF) image of Ag NPs in the channel of TiO_2 NTAs, and (**e**) side view of TiO_2 NTAs on fluorine-doped tin oxide (FTO) glass.

Carbon materials on the TiO_2 NTAs were synthesized by CVD, and Figure 3 shows their structure as confirmed by Raman spectroscopy (TEM images of TiO_2 NTAs were shown in Figure S1). In a previous publication, we reported the optimization of TiO_2 NTAs for DSSCs using carbon materials [25]. The B_{1g} (397 cm^{-1}), A_{1g} (518 cm^{-1}), and E_g (641 cm^{-1}) peaks indicated that the TiO_2 NTAs were in the form of anatase TiO_2, as shown in Figure 3a [31]. When carbon materials were synthesized on the TiO_2 NTAs using CVD, the G band at 1600 cm^{-1} represented graphite, while the D band at 1384 cm^{-1} was due to the disorderly network of sp^2 and sp^3 sites in the carbon materials (Figure 3b). The sp^2 sites of the carbon materials resulted in a π-π conjugation that improved the efficiency of electron transport across the TiO_2 NTAs.

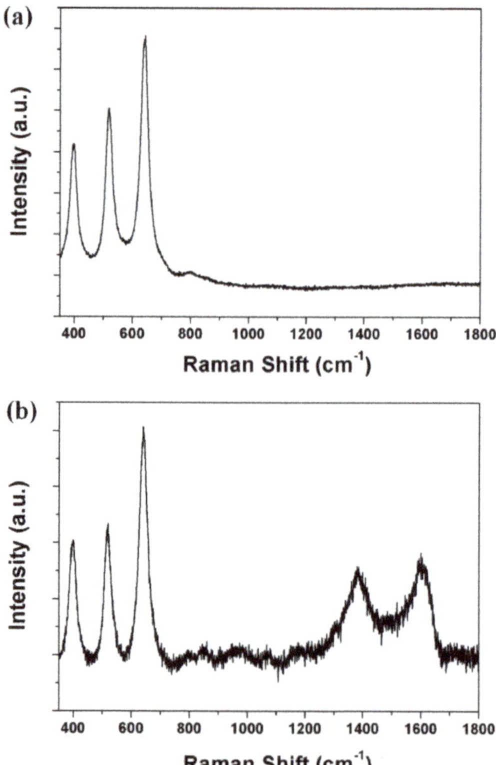

Figure 3. Raman spectra of TiO_2 NTAs: (**a**) without carbon materials, and (**b**) with carbon materials.

Ag NPs were synthesized on the TiO_2 NTAs using UV irradiation, and this was confirmed by the UV-Vis spectrum. Using the HAADF image shown in Figure 2d, the size of Ag NPs was confirmed to be approximately 30 nm. An absorption peak centered at 405 nm was also observed (Figure 4). Our previous paper reported on the optimization of TiO_2 NTAs using Ag NPs [25]. Other researchers have reported that Ag NPs with sizes of approximately 30 nm had UV-Vis absorption peaks at 420 nm. However, in this case, the Ag NPs were synthesized using UV irradiation (at 254 nm) without the addition of any stabilizing or reducing agents. As such, the Ag NPs were immobilized in the TiO_2 NTAs, which would affect absorption in the UV-Vis spectrum. The absorption band of Ag NPs is within the same range as that of the dye N719 (*cis*-diisothiocyanato-bis(2,2′-bityridyl-4,4′-dicarboxylato) ruthenium(II) bis(tetrabutylammonium), 390–530 nm), which led to enhanced electron generation from the dye by means of plasmonic interactions.

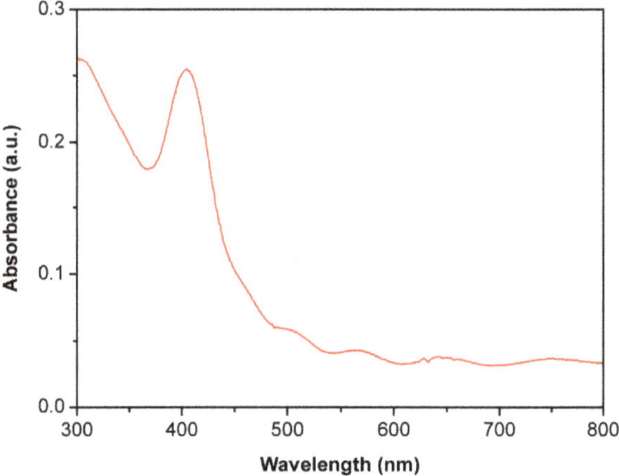

Figure 4. Ultraviolet-visible (UV-Vis) spectrum of Ag NPs on the TiO_2 NTAs.

The current density-voltage curves of DSSCs using closed-ended TiO_2 NTAs both with and without modification were measured under air-mass (AM) 1.5 sunlight, and the results are presented in Figure 5. The V_{oc}, short-circuit current density (J_{sc}), fill factor (ff), and energy conversion efficiency (η) of the DSSCs are summarized in Table 1. For the DSSCs without any treatment, the energy conversion efficiency was 4.10%, which increased to 5.73% when Ag NPs were embedded via UV irradiation (corresponding to an overall increase of 40%). When carbon materials were added to the closed-ended TiO_2 NTAs via CVD, the energy conversion efficiency improved to 5.69%, corresponding to a 39% increase. With both Ag NPs and carbon materials, the energy conversion efficiency further improved to 6.36%, corresponding to an overall increase of 55%. Note that when Ag NPs were treated with $TiCl_4$, the core-shell type $Ag@TiO_2$ nanoparticles were formed. Because the dye is adsorbed on $Ag@TiO_2$, the amount of dye loading may not be significantly reduced (Table 1). As previously reported, a large amount of carbon doping materials could lower the conversion efficiency by decreasing dye loading [25]. However, in this case, only a trace amount of carbon material was deposited, which did not significantly decrease the dye loading.

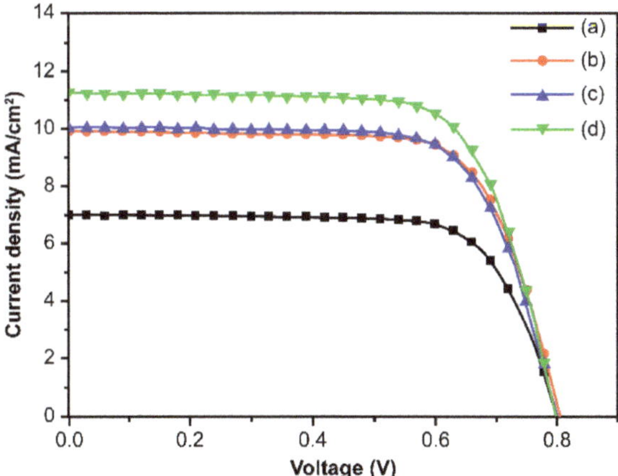

Figure 5. Current density-voltage curves of DSSCs based on: (a) unmodified closed-ended TiO_2 NTAs, (b) embedded Ag NPs, (c) applied carbon materials, and (d) both Ag NPs and carbon materials.

Table 1. Photovoltaic properties of dye-sensitized solar cells (DSSCs) based on closed-ended TiO$_2$ nanotube arrays (TiO$_2$ NTAs) with Ag nanoparticles (NPs) and/or carbon materials.

DSSCs Based on Closed-Ended TiO$_2$ NTAs Decorated	J_{sc} (mA/cm^2)	V_{oc} (V)	ff	η (%)	Dye Loading (nmol/cm^2)
without Ag NPs and carbon materials	7.02	0.81	0.72	4.10 ± 0.28	144
with Ag NPs	9.92	0.81	0.72	5.73 ± 0.31	142
with carbon materials	10.03	0.80	0.71	5.69 ± 0.26	139
with Ag NPs and carbon materials	11.25	0.80	0.71	6.36 ± 0.34	141

The current density-voltage curves for DSSCs based on open-ended TiO$_2$ NTAs with or without modification were also measured under AM 1.5 sunlight, and the results are presented in Figure 6. The V_{oc}, J_{sc}, ff and η values of these DSSCs are summarized in Table 2. When unmodified TiO$_2$ NTAs were used, DSSCs based on open-ended TiO$_2$ NTAs had higher energy conversion efficiency (5.32%) compared to those based on closed-ended TiO$_2$ NTAs (4.10%). The closed-end barrier of the TiO$_2$ NTA disturbs electron transport between the TiO$_2$ layer and the electrode [17,25].

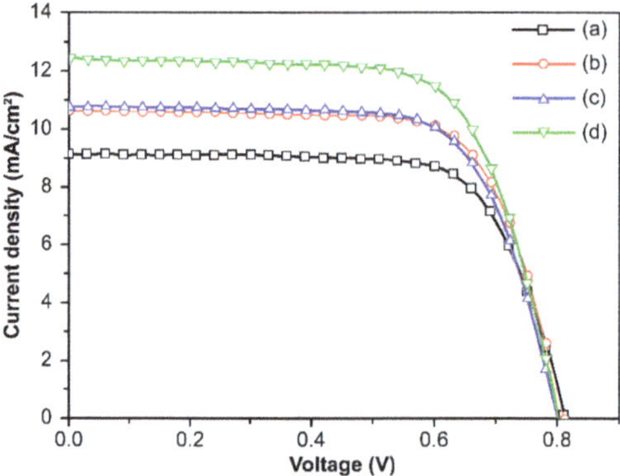

Figure 6. Current density-voltage curves of DSSCs based on: (a) unmodified open-ended TiO$_2$ NTAs, (b) embedded Ag NPs, (c) applied carbon materials, and (d) both Ag NPs and carbon materials.

Table 2. Photovoltaic properties of DSSCs based on open-ended TiO$_2$ NTAs with Ag NPs and/or carbon materials.

DSSCs Based on Open-Ended TiO$_2$ NTAs Decorated	J_{sc} (mA/cm^2)	V_{oc} (V)	ff	η (%)	Dye Loading (nmol/cm^2)
without Ag NPs and carbon materials	9.12	0.81	0.72	5.32 ± 0.36	153
with Ag NPs	10.61	0.81	0.71	6.14 ± 0.46	151
with carbon materials	10.73	0.80	0.71	6.07 ± 0.30	147
with Ag NPs and carbon materials	12.41	0.80	0.69	6.91 ± 0.41	149

When Ag NPs were embedded in the open-ended TiO$_2$ NTAs, the energy conversion efficiency improved from 5.32% to 6.14%, corresponding to a 15% enhancement. In this case, electron generation in the DSSCs was enhanced by the plasmonics from the NPs, despite the slightly diminished dye loading (from 153 to 151 nmol/cm^2). When carbon materials alone were applied to TiO$_2$ NTAs, the energy conversion efficiency improved to 6.07% (a 14% increase). In this case, electron transport was improved due to the π-π conjugation across the small quantity of carbon materials in spite of a diminished dye load (153 to 147 nmol/cm^2, which was even less than with Ag NPs). Here carbon materials were distributed to interact with the TiO$_2$ and the dye, making up for the loss of dye loading in terms of the energy conversion efficiency. When Ag NPs and carbon materials were both applied to the open-ended TiO$_2$ NTAs, the energy conversion efficiency improved to 6.91%, corresponding to a 30% enhancement when compared to the unmodified open-ended TiO$_2$ NTAs. In this case, the Ag NPs and carbon materials produced additive effects with their respective plasmonics and π-π conjugations; this was in spite of a slightly reduced dye loading of 149 nmol/cm^2. Comparing the performance parameters in Table 2, the V_{oc} and *ff* decreased with treatment; the conduction band of the TiO$_2$ NTAs shifted down as shown in Figure S2, which in turn affected the V_{oc} and the charge recombination through electron density suppressing the *ff*. However, the J_{sc} was increased by the plasmonic activity in conjunction with π-π, which improved the energy conversion efficiency of the DSSCs.

The DSSCs based on open-ended TiO$_2$ NTAs were characterized by EIS across the frequency range from 10^{-2} to 10^6 Hz, as shown in Figure 7. The applied bias voltage was set at the V_{oc} with 10 mV of AC amplitude. The data were analyzed using an equivalent circuit (Figure 7 inset), and the fit parameters are listed in Table 3. The ohmic series resistance (R$_s$) was due to the sheet resistance corresponding to the x-axis value where the first semicircle begins (on the left-hand side of Figure 7). The value of R$_s$ was similar with or without Ag NPs or carbon materials, indicating that the additional deposits did not affect the sheet's resistance to FTO or the current collector. The R$_1$ value is given by the sum of the small semicircle, which at high frequency was assigned to the parallel combination of resistances and the capacitances at the Pt-FTO/electrolyte and the FTO/TiO$_2$ interfaces. The R$_2$ value is given by the sum of the large semicircle at low frequency (associated with the resistance) and the capacitance at the dye-adsorbed TiO$_2$/electrolyte interface, as well as the transport resistance. The values of R$_1$ without and with Ag NPs were approximately 5.58 and 5.54 Ω, respectively. However, the value of R$_2$ with Ag NPs (36.90 Ω) was much lower than without Ag NPs (61.12 Ω). More electrons were generated by plasmonic activities than were produced at the dye-adsorbed TiO$_2$/electrolyte interface. As a result, the R$_2$ value was reduced in the presence of Ag NPs. The value of R$_1$ with carbon materials (5.07 Ω) was less than for without or with Ag NPs (5.58 and 5.54 Ω, respectively), whereas the value of R$_2$ (36.40 Ω) was less than without Ag NPs (61.12 Ω). Electrons were better transported by π-π conjugation affected by the FTO/TiO$_2$ and dye-adsorbed TiO$_2$/electrolyte interfaces. Hence, the values of both R$_1$ and R$_2$ decreased in the presence of carbon materials. In the presence of both Ag NPs and carbon materials, the values of R$_1$ (4.88 Ω) and R$_2$ (24.55 Ω) were the lowest. In this case, more electrons were generated and better transported by a combination of plasmonics and the π-π conjugation affecting the FTO/TiO$_2$ and dye-adsorbed TiO$_2$/electrolyte interfaces. Therefore, the values of R$_1$ and R$_2$ were reduced and the parameters were determined by EIS, as shown in Table S1.

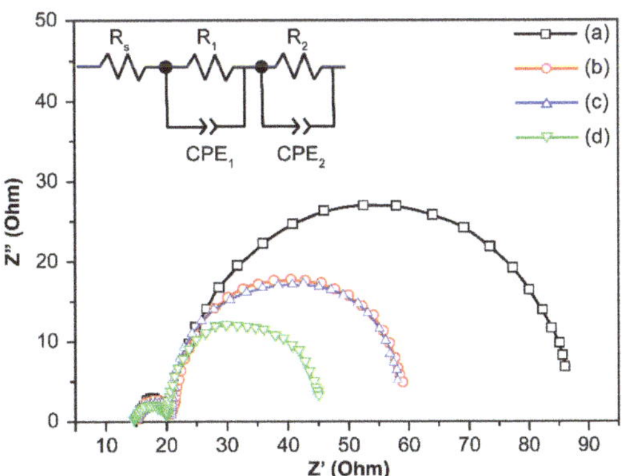

Figure 7. Electrochemical impedance spectroscopy (EIS) data of DSSCs based on: (a) unmodified open-ended TiO$_2$ NTAs, (b) embedded Ag NPs, (c) applied with carbon materials, and (d) with both Ag NPs and carbon materials.

Table 3. EIS fitting results for DSSCs with open-ended TiO_2 NTAs.

DSSCs Based on Open-Ended TiO_2 NTAs Decorated	R_s (Ω)	R_1 (Ω)	CPE_1 (F)	R_2 (Ω)	CPE_2 (F)
without Ag NPs and carbon materials	15.50	5.58	6.91×10^{-6}	61.12	1.99×10^{-3}
with Ag NPs	15.52	5.54	8.65×10^{-6}	36.90	2.10×10^{-3}
with carbon materials	15.56	5.07	1.62×10^{-5}	36.40	2.03×10^{-3}
with Ag NPs and carbon materials	14.99	4.88	1.16×10^{-6}	24.55	2.99×10^{-3}

The incident photon-to-current efficiency (IPCE) of DSSCs based on the open-ended TiO_2 NTAs is shown in Figure 8. Plasmon is a type of quasiparticle consisting of free electrons collectively vibrating within the metal. At the interface of a metal with a negative dielectric constant and a medium with a positive dielectric constant, surface plasmon resonance (SPR) combines a spreading electromagnetic wave (from visible to near-infrared frequency) with the plasmon. This combination generates plasmon-polariton, which leads to optical absorption; a strong electric field is also generated in some parts. During SPR, the light energy accumulates on the metal nanoparticle surface, and optical control is possible in the frequency range below the optical diffraction limit. Therefore, the intensity of DSSC based on open-ended TiO_2 NTAs embedded with Ag NPs was higher than without embedded Ag NPs. This may mean that more electrons were generated by the plasmonic activities, which increased the short circuit current. The current intensity in the DSSCs based on open-ended TiO_2 NTAs with carbon materials was also higher than in those without. This may mean that electrons were better transported by π-π conjugation, which also increased the short circuit current. Moreover, the current intensity was the strongest in the presence of both Ag NPs and carbon materials. In this case, electrons were generated in large quantities and were better transported by plasmonic activities and π-π conjugation.

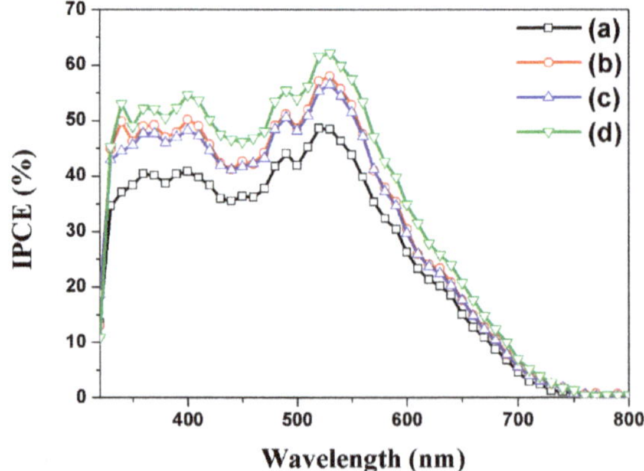

Figure 8. Incident photon-to-current efficiency (IPCE) of DSSCs based on: (a) unmodified open-ended TiO_2 NTAs, (b) embedded with Ag NPs, (c) applied with carbon materials, and (d) with both Ag NPs and carbon materials.

4. Conclusions

We deposited Ag NPs and carbon materials in the channels of closed- and open-ended TiO_2 NTAs using UV irradiation and CVD, respectively. These modifications improved the energy conversion efficiency of the corresponding DSSCs; the electron generation was enhanced by plasmonics from the Ag NPs, while the resistance of TiO_2 NTAs was suppressed via the π-π conjugation from the carbon materials. DSSCs made of freestanding TiO_2 NTAs coated with both Ag NPs and carbon materials had the best energy conversion efficiency, due to the combination of these two factors. Comparing the open-ended and closed-ended TiO$_2$ NTAs (both with Ag NPs and carbon materials), the energy conversion efficiency of the DSSCs was higher for the former.

Supplementary Materials: Supplementary Materials are available online at http://www.mdpi.com/2076-3417/7/6/576/s1.

Acknowledgments: This work was supported by the Bio & Medical Technology Development Program of the National Research Foundation (NRF), and funded by the Korean government (MSIP & MOHW) (2016-A423-0045).

Author Contributions: H.-S. Kim, M.-H. Chun, J.S. Suh, B.-H. Jun, and W.-Y. Rho conceived and designed the experiments. H.-S. Kim and W.-Y. Rho performed the experiments. H.-S. Kim, and M.-H. Chun analyzed the data. H.-S. Kim, B.-H. Jun, and W.-Y. Rho wrote the manuscript. All authors read and approved the final manuscript.

Conflicts of Interest: The authors declare no conflict of interest.

References

1. Oregan, B.; Grätzel, M. A low-cost, high-efficiency solar cell based on dye-sensitized colloidal TiO$_2$ films. *Nature* **1991**, *353*, 737–740. [CrossRef]
2. Grätzel, M.J. Dye-sensitized solar cells. *Photochem. Photobiol. C Photochem. Rev.* **2003**, *4*, 145–153. [CrossRef]
3. Hardin, B.E.; Snaith, H.J.; McGehee, M.D. The renaissance of dye-sensitized solar cells. *Nat. Photonics* **2012**, *6*, 162–169. [CrossRef]
4. Sang, L.; Zhao, Y.; Burda, C. TiO$_2$ Nanoparticles as functional building blocks. *Chem. Rev.* **2014**, *114*, 9283–9318. [CrossRef] [PubMed]
5. Hara, K.; Sato, T.; Katoh, R.; Furube, A.; Ohga, Y.; Shinpo, A.; Suga, S.; Sayama, K.; Sugihara, H.; Arakawa, H. Molecular design of coumarin dyes for efficient dye-sensitized solar cells. *J. Phys. Chem. B* **2003**, *107*, 597–606. [CrossRef]
6. Galoppini, E. Linkers for anchoring sensitizers to semiconductor nanoparticles. *Coord. Chem. Rev.* **2004**, *248*, 1283–1297. [CrossRef]
7. Nazeeruddin, M.K.; Pechy, P.; Renouard, T.; Zakeeruddin, S.M.; Humphry-Baker, R.; Comte, P.; Liska, P.; Cevey, L.; Costa, E.; Shklover, V.; et al. Engineering of efficient panchromatic sensitizers for nanocrystalline TiO$_2$-based solar cells. *J. Am. Chem. Soc.* **2001**, *123*, 1613–1624. [CrossRef] [PubMed]
8. Grätzel, M. Photoelectrochemical cells. *Nature* **2001**, *414*, 338–344. [CrossRef] [PubMed]
9. Katoh, R.; Furube, A.; Yoshihara, T.; Hara, K.; Fujihashi, G.; Takano, S.; Murata, S.; Arakawa, H.; Tachiya, M. Efficiencies of electron injection from excited N3 into nanocrystalline semiconductor (ZrO$_2$, TiO$_2$, ZnO, Nb$_2$O$_5$, SnO$_2$, In$_2$O$_3$) films. *J. Phys. Chem. B* **2004**, *108*, 4818–4822. [CrossRef]
10. Du, L.; Furube, A.; Yamamoto, K.; Hara, K.; Katoh, R.; Tachiya, M. Plasmon-induced charge separation and recombination dynamics in gold–TiO$_2$ nanoparticle systems: Dependence on TiO$_2$ particle size. *J. Phys. Chem. C* **2009**, *113*, 6454–6462. [CrossRef]
11. Mor, G.K.; Varghese, O.K.; Paulose, M.; Shankar, K.; Grimes, C.A. A review on highly ordered, vertically oriented TiO$_2$ nanotube arrays: Fabrication, material properties, and solar energy applications. *Sol. Energy Mater. Sol. Cells* **2006**, *90*, 2011–2075. [CrossRef]
12. Shin, Y.; Lee, S. Self-organized regular arrays of anodic TiO$_2$ nanotubes. *Nano Lett.* **2008**, *8*, 3171–3173. [CrossRef] [PubMed]
13. Rho, W.-Y.; Jeon, H.; Kim, H.-S.; Chung, W.-J.; Suh, J.S.; Jun, B.-H. Ag Nanoparticle–functionalized open-ended freestanding TiO$_2$ nanotube arrays with a scattering layer for improved energy conversion efficiency in dye-sensitized solar cells. *J. Nanomater.* **2016**, *6*, 117. [CrossRef] [PubMed]
14. Ruan, C.M.; Paulose, M.; Varghese, O.K.; Mor, G.K.; Grimes, C.A. Fabrication of highly ordered TiO$_2$ nanotube arrays using an organic electrolyte. *J. Phys. Chem. B* **2005**, *109*, 15754–15759. [CrossRef] [PubMed]
15. Martinson, A.B.; Hamann, T.W.; Pellin, M.J.; Hupp, J.T. New architectures for dye-sensitized solar cells. *Chemistry* **2008**, *14*, 4458–4467. [CrossRef] [PubMed]
16. Chen, Q.W.; Xu, D.S. Large-scale, noncurling, and free-standing crystallized TiO$_2$ nanotube arrays for dye-sensitized solar cells. *J. Phys. Chem. C* **2009**, *113*, 6310–6314. [CrossRef]
17. Rho, C.; Min, J.H.; Suh, J.S. barrier layer effect on the electron transport of the dye-sensitized solar cells based on TiO$_2$ nanotube arrays. *J. Phys. Chem. C* **2012**, *116*, 7213–7218. [CrossRef]

18. Zhang, H.; Lv, X.; Li, Y.; Wang, Y.; Li, J. P25-graphene composite as a high performance photocatalyst. *ACS Nano* **2010**, *4*, 380–386. [CrossRef] [PubMed]

19. Wang, W.; Bando, Y.; Zhi, C.; Fu, W.; Wang, E.; Golberg, D. Aqueous noncovalent functionalization and controlled near-surface carbon doping of multiwalled boron nitride nanotubes. *J. Am. Chem. Soc.* **2008**, *130*, 8144–8145. [CrossRef] [PubMed]

20. Guo, X.; Baumgarten, M.; Müllen, K. Designing π-conjugated polymers for organic electronics. *Prog. Polym. Sci.* **2013**, *38*, 1832–1908. [CrossRef]

21. Schulze, K.; Uhrich, C.; Schüppel, R.; Leo, K.; Pfeiffer, M.; Brier, E.; Reinold, E.; Baeuerle, P. Efficient vacuum-deposited organic solar cells based on a new low-bandgap oligothiophene and fullerene C60. *Adv. Mater.* **2006**, *18*, 2872–2875. [CrossRef]

22. Cheng, Y.-J.; Yang, S.-H.; Hsu, C.-S. Synthesis of conjugated polymers for organic solar cell applications. *Chem. Rev.* **2009**, *109*, 5868–5923. [CrossRef] [PubMed]

23. Pradhan, B.; Batabyal, S.K.; Pal, A.J. Functionalized carbon nanotubes in donor/acceptor-type photovoltaic devices. *Appl. Phys. Lett.* **2006**, *88*, 3106. [CrossRef]

24. Roy-Mayhew, J.D.; Aksay, I.A. Graphene materials and their use in dye-sensitized solar cells. *Chem. Rev.* **2014**, *114*, 6323–6348. [CrossRef] [PubMed]

25. Rho, W.-Y.; Kim, S.-H.; Kim, H.-M.; Suh, J.S.; Jun, B.-H. Carbon-doped freestanding TiO_2 nanotube arrays in dye-sensitized solar cells. *New J. Chem.* **2017**, *41*, 285–289. [CrossRef]

26. Lu, L.; Luo, Z.; Xu, T.; Yu, L. Cooperative plasmonic effect of Ag and Au nanoparticles on enhancing performance of polymer solar cells. *Nano Lett.* **2012**, *13*, 59–64. [CrossRef] [PubMed]

27. Pillai, S.; Catchpole, K.; Trupke, T.; Green, M. Surface plasmon enhanced silicon solar cells. *J. Appl. Phys.* **2007**, *101*, 093105. [CrossRef]

28. Nakayama, K.; Tanabe, K.; Atwater, H.A. Plasmonic nanoparticle enhanced light absorption in GaAs solar cells. *Appl. Phys. Lett.* **2008**, *93*, 121904. [CrossRef]

29. Bhattacharyya, D.; Sarswat, P.K.; Islam, M.; Kumar, G.; Misra, M.; Free, M.L. Geometrical modifications and tuning of optical and surface plasmon resonance behaviour of au and ag coated tio 2 nanotubular arrays. *RSC Adv.* **2015**, *5*, 70361–70370. [CrossRef]

30. Rho, W.-Y.; Kim, H.-S.; Lee, S.H.; Jung, S.; Suh, J.S.; Hahn, Y.-B.; Jun, B.-H. Front-illuminated dye-sensitized solar cells with Ag nanoparticle-functionalized freestanding TiO_2 nanotube arrays. *Chem. Phys. Lett.* **2014**, *614*, 78–81. [CrossRef]

31. Yan, J.; Wu, G.; Guan, N.; Li, L.; Li, Z.; Cao, X. Understanding the effect of surface/bulk defects on the photocatalytic activity of TiO_2: Anatase versus rutile. *Phys. Chem. Chem. Phys.* **2013**, *15*, 10978–10988. [CrossRef] [PubMed]

 *applied sciences*

Article

Hydrogeochemical Modeling to Identify Potential Risks of Underground Hydrogen Storage in Depleted Gas Fields

Christina Hemme * and Wolfgang van Berk

TU Clausthal, Department of Hydrogeology, Institute of Disposal Research, Leibnizstrasse 10, 38678 Clausthal-Zellerfeld, Germany; wolfgang.van.berk@tu-clausthal.de
* Correspondence: christina.hemme@tu-clausthal.de; Tel.: +49-5323-72-2142

Received: 23 October 2018; Accepted: 15 November 2018; Published: 19 November 2018

Abstract: Underground hydrogen storage is a potential way to balance seasonal fluctuations in energy production from renewable energies. The risks of hydrogen storage in depleted gas fields include the conversion of hydrogen to $CH_{4(g)}$ and $H_2S_{(g)}$ due to microbial activity, gas–water–rock interactions in the reservoir and cap rock, which are connected with porosity changes, and the loss of aqueous hydrogen by diffusion through the cap rock brine. These risks lead to loss of hydrogen and thus to a loss of energy. A hydrogeochemical modeling approach is developed to analyze these risks and to understand the basic hydrogeochemical mechanisms of hydrogen storage over storage times at the reservoir scale. The one-dimensional diffusive mass transport model is based on equilibrium reactions for gas–water–rock interactions and kinetic reactions for sulfate reduction and methanogenesis. The modeling code is PHREEQC (pH-REdox-EQuilibrium written in the C programming language). The parameters that influence the hydrogen loss are identified. Crucial parameters are the amount of available electron acceptors, the storage time, and the kinetic rate constants. Hydrogen storage causes a slight decrease in porosity of the reservoir rock. Loss of aqueous hydrogen by diffusion is minimal. A wide range of conditions for optimized hydrogen storage in depleted gas fields is identified.

Keywords: hydrogen storage; porous media; bacterial sulfate reduction; methanogenesis; gas loss; diffusion; reactive transport modeling; PHREEQC

---

1. Introduction and Aims

Underground hydrogen storage (UHS) is used to store large amounts of hydrogen generated from renewable energy sources (such as wind and solar) to compensate for seasonal fluctuations in the supply and demand of energy [1,2]. Large amounts of hydrogen can be stored in depleted oil and gas fields, in salt caverns, and in aquifers [1,3]. Nevertheless, practical experience of underground hydrogen storage is still rare. In the US and UK, hydrogen is currently stored in salt caverns [3–6], but hydrogen storage in depleted oil and gas fields is still under research and discussion. Depleted oil and gas fields have a huge storage capacity, are well known from former exploration and production, and qualify therefore for hydrogen storage. However, the existing underground gas storages (UGS) are designed for the storage of natural gas, which does not contain hydrogen (or only very low amounts). Because the chemical and physical properties of hydrogen are different to those of methane (CH_4)—the main component of natural gas—the effects of hydrogen on the reservoir rock, cap rock, and storage facilities must be analyzed before injecting hydrogen into these storages [2].

Compared to methane, hydrogen has a higher diffusivity in pure water. The diffusion coefficient for hydrogen is 5.13×10^{-9} m^2 s^{-1} and for CH_4 it is 1.85×10^{-9} m^2 s^{-1}, both in pure water at 25 °C. The data are given in the database phreeqc.dat, that is provided by the US Geological Survey.

Generally, the diffusion process of an aqueous species is described by its diffusion coefficient in pure water (m^2 s^{-1}). However, in any porous system, where solids are available, the influence of porosity on the diffusion must be considered by scaling the diffusion coefficient with tortuosity [7], resulting in an effective diffusion coefficient (Equation (1)), where d'_m is the effective diffusion coefficient, d_m is the diffusion coefficient in pure water, and τ is the tortuosity. The tortuosity is the ratio of the effective diffusion mass fluxes in a porous medium to ideal diffusion mass fluxes in solutions without a rock matrix.

$$d'_m = \frac{d_m}{\tau^2} \tag{1}$$

Therefore, the effective diffusion coefficient for methane is $2.35{-}2.49 \times 10^{-10}$ m^2 s^{-1} in clayey host rocks at 21 °C [8]. The effective diffusion coefficient for hydrogen is 3.0×10^{-11} m^2 s^{-1} in clayey host rocks saturated with water at 25 °C [9]. However, there is a lack of data regarding effective diffusion coefficients of hydrogen at underground storage conditions with higher temperatures and pressures [10].

Compared to methane, hydrogen has a lower viscosity. The viscosity of hydrogen at the two-phase boundary at 17 MPa and 350 K is 1.01×10^{-5} kg m^{-1} s^{-1} [11] and methane fluid at 20 MPa and 350 K is 1.81×10^{-5} kg m^{-1} s^{-1} [12]. Furthermore, hydrogen has a lower density than methane. At normal conditions, the density of hydrogen is eight times smaller than that of methane and the difference increases by 20–30% under storage conditions [2]. The solubility of $H_{2(g)}$ in pure water is smaller than the solubility of $CH_{4(g)}$. At 25 °C and 1.0 atm, 7.9×10^{-4} mol kgw^{-1} $H_{2(g)}$ and 1.4×10^{-3} mol kgw^{-1} $CH_{4(g)}$ dissolve in pure water (data based on phreeqc.dat).

The properties of hydrogen—being different from those of methane—could lead to risks when storing hydrogen in depleted gas fields, which were constructed for storing methane (underground gas storage). Potential risks include (i) the conversion of hydrogen to CH_4 and H_2S due to microbial activity, (ii) chemical reaction of hydrogen with the minerals of the reservoir rock/cap rock and thus potential resulting porosity changes, and (iii) the loss of aqueous $H_{2(aq)}$ by diffusion through the cap rock.

The first risk (i) arises from microbial activities in the underground. The two processes relevant here are bacterial sulfate reduction (2) and methanogenesis (3), where hydrogen is microbially catalyzed by bacteria and is converted to $H_2S_{(aq)}$ or $CH_{4(aq)}$. The activity of the different bacteria depends on the availability of the different electron acceptors such as sulfate or carbon dioxide [1]. Other possible microbial processes are acetogenesis and iron(III)-reduction, summarized by Hagemann et al. [13].

$$\text{Bacterial sulfate reduction (BSR): } SO_4{}^{2-}{}_{(aq)} + 5H_{2(aq)} \rightarrow H_2S_{(aq)} + 4H_2O \tag{2}$$

$$\text{Methanogenesis: } CO_{2(aq)} + 4H_{2(aq)} \rightarrow CH_{4(aq)} + 2H_2O \tag{3}$$

In aqueous anoxic environments, the sulfate-reducing bacteria (SRB) use sulfate as an electron acceptor to oxidize hydrogen and generate sulfide-S which could be available as aqueous $S^{2-}{}_{(aq)}$, $HS^-{}_{(aq)}$, and $H_2S_{(aq)}$ and as gaseous $H_2S_{(g)}$ as well. The sulfate is derived from the aqueous dissolution of mineral phases like anhydrite ($CaSO_{4(s)}$). The energy gained from sulfate reduction is used by the SRB for cell growth [14]. SRB prefer temperatures around 38 °C [15] and near-neutral pH conditions [14], but are active even at extreme habitat conditions such as at great depths and temperatures, ranging from 0 °C to 60–80 °C [16–18] and in some cases up to 110 °C [19]. SRB activity was observed, for example, in hydrocarbon reservoirs [18] and in the underground storage of town gas [20]. The risk arising from bacterial sulfate reduction lies in the produced H_2S, which is corrosive towards the storage facilities, toxic if inhaled, and can pose a risk to the environment.

Typical environments for methanogenic bacteria are, for example, anoxic sediments and flooded soils [21]. However, the existence of methanogenic bacteria was also observed in town gas storages [22]. Methanogenic bacteria prefer temperatures of 30–40 °C for growth [10], but they also have been found at higher temperatures of 80 °C [23,24] and up to 97 °C [10,25]. With these bacteria, $H_{2(aq)}$ is the electron donor and $CO_{2(aq)}$ is the electron acceptor, which is reduced to form $CH_{4(aq)}$. The methanogenic bacteria obtain their energy for cell growth from this conversion [21,22]. The problem associated with methanogenesis lies in the loss of hydrogen and the related energy loss [2]. This phenomenon has already been observed in the underground storage of town gas. A well-known example is the Czech town gas storage at Lobodice, where the 54 vol. % injected $H_{2(g)}$ diminished to 37 vol. % $H_{2(g)}$. Concurrently, $CH_{4(g)}$ increased from 21.9 vol. % to 40.0 vol. % within a time span of 7 months [22].

Another risk of underground hydrogen storage (ii) is the reaction of $H_{2(g/aq)}$ with the minerals of the reservoir rock and the cap rock, which can lead to mineral dissolution and precipitation and resulting porosity changes as known from carbon capture and storage e.g., [26–29]. Changes in the porosity of the cap rock can either improve or deteriorate its sealing capacity. Furthermore, precipitation of minerals at the well equipment may cause scaling [2].

In depleted gas fields, the high diffusivity of hydrogen could be the reason for hydrogen loss (iii). The hydrogen is dissolved in the formation water of the cap rock and diffuses through the cap rock [2]. Reitenbach et al. [2] stated that a significant loss of hydrogen can be expected. The high diffusivity, low viscosity, and low density of hydrogen leads to a high mobility and therefore the potential loss due to leakage should be considered [1].

A natural analog for an unintended and unpredictable leakage of hydrogen can be found in hydrogen anomalies associated with faults. These so called natural hydrogen seeps have been described by Larin et al. [30], Sato et al. [31], Wakita et al. [32], Ware et al. [33], and Zgonnik et al. [34]. In these cases, faults act as "fluid conduits" [30]. An increased CH_4 concentration is also observed within these anomalies because of the increased microbial activity stimulated by the increased amount of hydrogen [30]. If hydrogen reaches the surface, it could inhibit the growth of trees, underbrush, and grass [30].

There are experimental studies concerning kinetics of pyrite and pyrrhotite reduction by hydrogen at high temperatures [35], the reactivity of hydrogen in sandstones [36], and the petrographic and petrophysical variation in reservoir sandstones due to hydrogen storage [37]. However, modeling studies of hydrogen storage to predict "long-term" behaviors are still rare. Therefore, this study is performed to model the loss of hydrogen (and related energy loss) as a combination of (i) the conversion by bacteria, (ii) gas-water-rock interactions in the reservoir and cap rock, which are connected with porosity changes, and (iii) loss by aqueous diffusion along a gradient of decreasing pressure and temperature conditions.

These processes are retraced by a one-dimensional reactive mass transport model to simulate gas–water–rock interactions resulting from hydrogen storage in depleted gas fields. The aim of this study is to investigate the basic mechanisms of BSR and methanogenesis in an integrated way over storage times at the reservoir scale and to describe qualitatively and quantitatively which reservoir rock and cap rock minerals dissolve or precipitate because of hydrogen storage as well as the related porosity changes. Furthermore, the parameters that influence the loss of hydrogen are determined.

Therefore, a model is presented, in which the hydrogeochemical mechanisms of underground hydrogen storage are simulated in a reference scenario (Sections 3.1–3.3). In further scenarios, single parameters will be changed in the model to show their effects on the modeling results and to identify the parameters that are most sensitive for underground hydrogen storage (Section 3.4).

2. Methodology

2.1. Modeling Tools

The modeling program for the one-dimensional reactive mass transport (1DRMT) model is PHREEQC version 3 (PHREEQC = pH-REdox-EQuilibrium written in the C programming language), provided by the US Geological Survey [38]. PHREEQC has capabilities to simulate (i) speciation and saturation-index calculations, (ii) batch-reaction and 1D transport calculations, and (iii) inverse modeling [38]. The used database is phreeqc.dat extended by dawsonite, nahcolite,

$CH_{4(g)}$, $H_2S_{(g)}$, $N_{2(g)}$, which are taken from llnl.dat. The 1DRMT model considers equilibrium reactions for gas–water–rock interactions, and kinetic reactions for sulfate reduction and methanogenesis. The equilibrium calculations are based on the mass action law including all species used in this study (Al, Ba, C, Ca, Cl, Fe, K, Mg, N, Na, S, Si) and their corresponding equilibrium constants. The equilibrium phases, mass action equations, and equilibrium constants used in the model are summarized in Table 1. For detailed information about PHREEQC, see Parkhurst and Appelo [38].

Table 1. Equilibrium phases, mass action equations, and equilibrium constants used in the model. Data from phreeqc.dat, except for dawsonite, nahcolite, $CH_{4(g)}$, $H_2S_{(g)}$, $N_{2(g)}$, which are from llnl.dat [38].

Equilibrium Phase	Equilibrium Reaction	log K at 25 °C, 1 bar
K-feldspar	$KAlSi_3O_8 + 8H_2O = K^+ + Al(OH)_4^- + 3H_4SiO_4$	-20.573
Albite	$NaAlSi_3O_8 + 8H_2O = Na^+ + Al(OH)_4^- + 3H_4SiO_4$	-18.002
Kaolinite	$Al_2Si_2O_5(OH)_4 + 6H^+ = H_2O + 2H_4SiO_4 + 2Al^{3+}$	7.435
Quartz	$SiO_2 + 2H_2O = H_4SiO_4$	3.98
Calcite	$CaCO_3 = CO_3^{2-} + Ca^{2+}$	8.48
Pyrite	$FeS_2 + 2H^+ + 2e^- = Fe^{2+} + 2HS^-$	-18.479
Illite	$K_{0.6}Mg_{0.25}Al_{2.3}Si_{3.5}O_{10}(OH)_2 + 11.2H_2O = 0.6K^+ + 0.25Mg^{2+} + 2.3Al(OH)^{4-} + 3.5H_4SiO_4 + 1.2H^+$	-40.267
Dawsonite	$NaAlCO_3(OH)_2 + 3H^+ = Al^{3+} + HCO_3^- + Na^+ + 2H_2O$	4.35
Mackinawite	$FeS + H^+ = Fe^{2+} + HS^-$	-4.648
Dolomite	$CaMg(CO_3)_2 = Ca^{2+} + Mg^{2+} + 2CO_3^{2-}$	-17.09
Nahcolite	$NaHCO_3 = HCO_3^- + Na^+$	-0.11
Anhydrite	$CaSO_4 = Ca^{2+} + SO_4^{2-}$	-4.39
Halite	$NaCl = Cl^- + Na^+$	1.570
Gypsum	$CaSO_4 \cdot 2H_2O = Ca^{2+} + SO_4^{2-} + 2H_2O$	-4.58
Sulfur [a]	$S + 2H^+ + 2e^- = H_2S$	4.882
Barite	$BaSO_4 = Ba^{2+} + SO_4^{2-}$	-9.97
Goethite	$FeOOH + 3H^+ = Fe^{3+} + 2H_2O$	-1.0
$H_{2(g)}$	$H_2 = H_2$	-3.1050
$CO_{2(g)}$	$CO_2 = CO_2$	-1.468
$CH_{4(g)}$	$CH_4 = CH_4$	-2.8502
$H_2S_{(g)}$	$H_2S = H^+ + HS^-$	-7.9759
$N_{2(g)}$	$N_2 = N_2$	-3.1864

[a] Sulfur = elemental sulfur.

2.2. Model Setup

Gaseous hydrogen ($H_{2(g)}$) is injected into the reservoir rock where residual gas from the previous natural gas reservoir is still available and the reservoir brine is in equilibrium with these gases and the reservoir rock. When $H_{2(g)}$ is injected, the available brine is saturated with H_2. With ongoing injection of $H_{2(g)}$, the initial reservoir brine is displaced and $H_{2(g)}$ builds up a plume. The only available water in the reservoir rock is then the irreducible water. On contact with the cap rock, $H_{2(g)}$ dissolves into the cap rock brine and diffuses through the cap rock brine. These processes are retraced with a hydrogeochemical, 1D reactive mass transport modeling approach, which is of semi-generic nature. The conditions assumed in the model are based on conditions of a depleted gas field with a temperature of 40 °C and a reservoir pressure of 40 atm. The model is divided into a column of 1488 cells, with a cell height of 1.0 m each (in the z-direction). The cap rock is assumed to have a height of 182 m (cells 1–182), the reservoir rock consists of 667 m (cells 183–850), and the underlying rock of 637 m (cells 851–1488; Figure 1). For the sake of simplicity, a constant partial pressure in the $H_{2(g)}$ plume is assumed and the partial pressure is set to be equal to the hydrostatic pressure. Even during production, an amount of hydrogen remains in the reservoir.

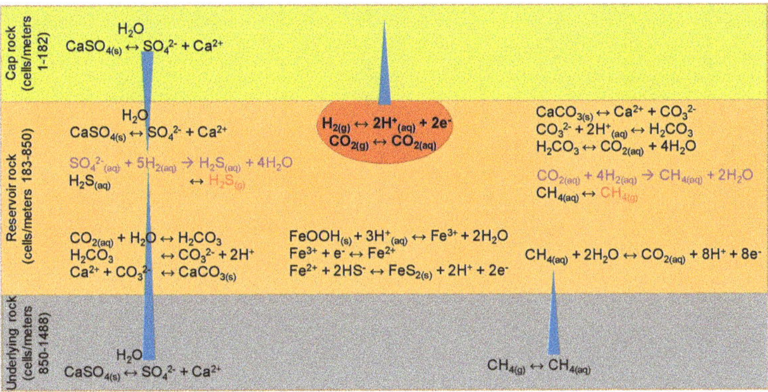

Figure 1. Selected reactions and processes associated with bacterial sulfate reduction and methanogenesis (purple). Single arrow = kinetic-controlled reactions; double arrow = equilibrium reactions; blue triangles = time-dependent diffusive transport of aqueous components; bold = injected gas for storage.

The mineralogical compositions of the cap rock, the reservoir rock, and the underlying rock are based on data of the Röt Formation [39,40], Buntsandstein Formation [41], and Zechstein Formation [42,43] and are summarized in Table 2. The reactive amount of each mineral phase is calculated in moles per kg of pore water (mol kgw^{-1}) with the consideration of the specific density of each mineral phase (g cm^{-3}). Dawsonite, mackinawite, elemental sulfur, albite (and pyrite) are considered as potential secondary phases, which may form at saturation. The clay minerals are considered as follows: illite is defined as a primary mineral phase, and the cation exchange capacity of chlorite, illite, montmorillonite, and kaolinite are estimated based on the data given by Appelo and Postma [44]. An initial small amount of CO_2 (pCO_2 = partial pressure of CO_2 = 1.0 atm) is assumed to be available in the cap rock, reservoir rock, and underlying rock from burial. The possible outgassing of $CH_{4(g)}$, $N_{2(g)}$, $CO_{2(g)}$, $H_2S_{(g)}$, and $H_{2(g)}$ is assumed in all cells.

The reservoir rock (cells 183–850) has an initial porosity of 10% ($n = 0.1$), 2.5% ($n = 0.025$) of the pore space is filled with irreducible water, which is equal to 1 L H_2O, and 7.5% ($n = 0.075$) of the pore space is filled with gas. The total volume is 40 L (1 L/0.025) and therefore, 3 L are filled with gas (40 L × 0.075). The porosity is filled with 3 L gas and 1 L H_2O, so 36 L are occupied by solids in each cell of the reservoir rock. The gas consists of 10% residual gas (that was not extracted during natural gas production) and 90% stored hydrogen gas. The composition of the residual gas in the reservoir rock is based on data from Bischoff and Gocht [45] with $pCH_{4(g)}$ = 3.56 atm, $pN_{2(g)}$ = 0.4 atm, and $pCO_{2(g)}$ = 0.04 atm. The stored hydrogen gas is composed of 96% $H_{2(g)}$ ($pH_{2(g)}$ = 34.56 atm) and 4% $CO_{2(g)}$ ($pCO_{2(g)}$ = 1.44 atm), these values are modified from data presented by Panfilov [5]. For modeling purposes, the stored hydrogen gas is additionally induced as a tracer gas with the same chemical properties as $H_{2(g)}$. It is used instead of $H_{2(g)}$ because $H_{2(g/aq)}$ is induced and used as the reactant in the calculation kinetics. The amount of tracer gas corresponds to 4.3 mol L^{-1} of gas in each cell of the reservoir. The summed volume of all gases in the reservoir rock is 3 L per cell and represents the stored gas volume at 40 atm under the assumed reservoir conditions. The sum of the partial pressure of all available gases (including tracer gas) is equivalent to the total pressure (40 atm) in the reservoir so that the generated gases ($H_2S_{(g)}$, $CH_{4(g)}$) are released as gas bubbles.

Table 2. Mineralogical composition of the reservoir rock, cap rock, and underlying rock. Dawsonite, nahcolite, mackinawite, sulfur, albite (and pyrite) are potential secondary phases.

Primary Minerals	Weight Percent (wt. %)	Amount (mol kgw^{-1})
Cap rock		
Halite	5.0	76.74
Quartz	50.0	746.42
Illite	20.0	46.73
Dolomite	5.0	24.32
Anhydrite	15.0	131.77
Reservoir rock		
K-feldspar	30.0	103.90
Kaolinite	1.0	3.73
Quartz	55.0	882.43
Calcite	0.5	4.82
Dolomite	0.5	0.03
Anhydrite	0.5	0.132 [a]
Illite	11.5	28.88
Barite	0.5	0.0009
Goethite	0.5	0.002
Underlying rock		
Halite	50.0	758.46
Quartz	8.0	118.04
Calcite	6.0	53.14
Dolomite	10.0	0.03
Pyrite	1.0	7.39
Anhydrite	25.0	66.11

[a] Reactive amount of anhydrite.

The cap rock (cells 1–182) is defined by an initial porosity of 5.0% ($n = 0.05$), with 1 L irreducible water and 19 L solid. The underlying rock (cells 851–1488) is defined by an initial porosity of 10% ($n = 0.1$), divided into 2.5% ($n = 0.025$) irreducible water, which is equal to 1 L H_2O, and 7.5% ($n = 0.075$) gas. The porosity is filled with 3 L gas and 1 L H_2O, so 36 L are occupied by solids in each cell of the underlying rock. The assumed natural gas composition in the underlying rock is $pCH_{4(g)} = 81.1$ atm, $pN_{2(g)} = 4.3$ atm, $pH_2S_{(g)} = 10.7$ atm, and $pCO_{2(g)} = 10.7$ atm [45].

The initial brine compositions of the cap rock, reservoir rock, and underlying rock are pre-calculated in a separate transport model to simulate the million-years-long interaction of the different brines with each other. The cap rock brine is composed of a 0.5 M Na^+/Cl^--dominated solution equilibrated with the mineral phases of the cap rock under cap rock pressure and temperature conditions. The temperature and pressure conditions change along the diffusive pathway of the aqueous species through the cap rock, starting with 40 °C and 40 atm at the reservoir depth (data from Pudlo et al. [46]). A gradient of decreasing temperature and pressure conditions according to the geothermal gradient (33.3 °C km^{-1} depth) and a pressure gradient of 100 atm km^{-1} depth under hydrostatic conditions is assumed. Therefore, each cell is defined by a specific temperature and pressure condition (e.g., cell 182: 39.967 °C and 39.9 atm; cell 181: 39.934 °C and 39.8 atm; and so on). The initial reservoir rock brine composition is a 0.5 M Na^+/Cl^--dominated solution, which is in equilibrium with the reservoir rock and the initial gas composition at 40 atm and 40 °C. The underlying rock is dominated by a 6.7 M Na^+/Cl^--dominated solution equilibrated with the underlying rock and gas composition under underlying rock pressure and temperature conditions (106.7 atm and 60 °C). To simulate the million-years-long interaction, molecular diffusion of all aqueous species through the cap rock brine, reservoir rock brine, and underlying rock brine is simulated over one million years. The brine compositions are summarized in Table 3.

The high Na^+ and Cl^- concentrations and high ionic strength in the brines require the use of the Pitzer database. However, the Pitzer database does not include Al^{3+}, which is important when modeling hydrogen storage in depleted gas fields. Therefore, the database phreeqc.dat is used. The validation that PHREEQC using phreeqc.dat simulates correct results with high Na^+ and Cl^- concentrations and high ionic strength is shown by Hemme and van Berk [47]. Furthermore, Parkhurst and Appelo [48] stated that models are reliable at higher ionic strength if the system is sodium chloride dominated.

Table 3. Composition of the initial irreducible water in the reservoir rock, the cap rock brine, and the underlying rock brine.

Parameter	Cap Rock Brine	Irreducible Water in the Reservoir Rock	Underlying Rock Brine
pH	6.4	6.4	5.9
Temperature (°C)	37.0	40.0	60.0
Elements	Concentration (mol kgw^{-1})	Concentration (mol kgw^{-1})	Concentration (mol kgw^{-1})
Al	1.31×10^{-7}	2.209×10^{-8}	1.776×10^{-8}
Ba	5.85×10^{-7}	3.922×10^{-7}	2.206×10^{-5}
C_{tot} [a]	3.11×10^{-2}	1.762×10^{-2}	7.405×10^{-3}
Ca	3.63×10^{-2}	1.186×10^{-2}	1.562×10^{-2}
Cl	1.27	1.123	5.396
Fe	5.69×10^{-2}	6.572×10^{-2}	4.263×10^{-11}
K	5.28×10^{-1}	6.151×10^{-1}	4.604×10^{-1}
Mg	8.73×10^{-4}	2.579×10^{-3}	1.142×10^{-2}
N	5.61×10^{-2}	6.485×10^{-2}	5.081×10^{-2}
Na	1.27	1.123	5.396
S_{tot} [b]	1.01	1.143	7.540×10^{-1}
Si	8.73×10^{-5}	9.878×10^{-5}	1.509×10^{-4}

[a] Summed concentration of aqueous CH_4 and C(+IV) species; [b] Summed concentration of aqueous S(+VI) and S(−II) species.

Molecular diffusion of all aqueous species is the only mass transport accounted for by the model. Multicomponent diffusion is calculated where each "solute can be given its own diffusion coefficient, allowing it to diffuse at its own rate, but with the constraint that overall charge balance is maintained" [38,49,50]. Fluid flow is not considered in the model due to the lack of total hydraulic head differences in depleted gas reservoirs. A homogeneous distribution of the relevant parameters in the reservoir and cap rock such as mineralogical composition is assumed. Furthermore, no fractures or natural faults are considered. Because of the lack of data, the storage time of hydrogen in depleted oil and gas fields is based on the equipment lifetime of 30 years [51], but is varied in an additional scenario (Section 3.4.1).

The oxidation of hydrogen by SO_4^{2-} and/or CO_2 in the model is kinetically controlled. The reaction kinetics describe the time-dependent changes of reaction products and educts of a chemical reaction. To calculate the irreversible reactions, catalyzed by microorganisms like sulfate-reducing bacteria and methanogenic bacteria, the Monod equation (Equation (4)) is used. Bacterial concentrations are kept constant in the model with the aim to model the "worst-case" scenario.

$$\psi^{growth} = \psi_{max}^{growth} \left(\frac{c^s}{\alpha + c^s} \right) \tag{4}$$

where ψ_{max}^{growth} is the maximum specific growth rate (mol L^{-1} s^{-1}), c^s is the concentration of the limiting substrates (mol L^{-1}), and α is the half-velocity constant. The maximum specific growth rate that is assumed in the model is 2.30×10^{-9} mol kgw^{-1} s^{-1} for methanogenesis (at 37 °C) [22] and 9.26×10^{-8} mol kgw^{-1} s^{-1} for BSR (at 30 °C) [52]. For bacterial sulfate reduction, the limiting substrate is sulfate and for methanogenesis it is carbon dioxide.

Panfilov [5] defined a new model to describe the substrate-limited growth models (Equation (5)), where t_e is the "individual time of eating" (s) and n_{max} is the maximum population size (m^{-3}). Due to the lack of data regarding the characteristic time of eating and the maximum population size, the Panfilov model is not considered.

$$\psi^{\text{growth}} = \frac{1}{t_e} \frac{n}{1 + \frac{n^2}{n_{\max}^2}} \left(\frac{c^s}{\alpha + c^s} \right) \tag{5}$$

The lower boundary of the column (in the underlying rock) and the upper boundary of the column (in the cap rock) are defined as diffusive flux. To check the numerical accuracy of the results, discretization studies for one scenario are performed, in which the number of shifts and the number of cells are refined. The model calculation for one scenario is rerun and results are compared [38].

3. Results and Discussion

3.1. Loss of $H_{2(g/aq)}$ by Bacterial Conversion to $CH_{4(g)}$ and $H_2S_{(g)}$

The modeling results of the reference scenario (defined in Section 2.2; input file S1) show that the maximum amount of bacterial-converted $H_{2(g/aq)}$ to $CH_{4(g)}$ and $H_2S_{(g)}$ depends on the amount of available (and reactive) electron acceptors, carbon dioxide and sulfate, and on the amount of available $H_{2(g/aq)}$. Figure 1 shows selected reactions and processes that are coupled to bacterial sulfate reduction and methanogenesis.

When CO_2 and SO_4^{2-} are fully consumed, methanogenesis and BSR will not proceed. The stored gas is composed of 96% hydrogen and 4% carbon dioxide (residual gas is still available). Therefore, CO_2 is available as electron acceptor for the methanogenesis and $CH_{4(g)}$ is generated. After less than 3 years of storage, the injected $CO_{2(g)}$ is completely consumed, but an ongoing supply of CO_2 is provided by the dissolution of carbonate-bearing minerals (such as calcite, for example). Consequently, the amount of generated $CH_{4(g)}$ is limited by the assumed storage time of 30 years and the time-dependent kinetics of methanogenesis.

The available sulfate for BSR in the reservoir rock results from anhydrite dissolution. If anhydrite is still present in a depleted natural gas reservoir, it is not reactive (e.g., it is coated by other minerals). Otherwise, the sulfate would have been catalyzed by bacterial sulfate reduction, with methane—the main component of natural gas—as electron donor. The same applies to hydrogen storage in depleted natural gas reservoirs. However, the creation of new fractures and natural faults or the dissolution of carbonate cements (pore-filling cements) leads to contact of reactive anhydrite with the stored hydrogen. To simulate this "worst-case" scenario, a small amount of reactive anhydrite is assumed in this model. The dissolution of anhydrite provides $SO_4^{2-}{}_{(aq)}$ for $H_2S_{(g)}$ generation and $Ca^{2+}{}_{(aq)}$ for calcite precipitation. Additionally, the dissolution of anhydrite creates porosity, which in turn can be superseded by the precipitation of other minerals (see Section 3.2). The reactive anhydrite is in solution equilibrium with the brine. The consumption of sulfate$_{(aq)}$ by BSR is accompanied by ongoing anhydrite dissolution and more sulfate becomes available for BSR until the reactive anhydrite is completely dissolved.

As a product of BSR and methanogenesis, small amounts of water (0.007 kg) are formed (Equations (2) and (3)). This increases the water content in the reservoir rock (1.0 kg initial water + 0.007 kg generated water = 1.007 kg total water mass per cell after 30 years). The water will probably be suppressed by the gas along the reservoir/cap rock boundary. However, the small increase in the mass of water could increase the anhydrite dissolution, resulting in a higher amount of sulfate ions that are available for BSR. Additional sulfate is delivered by diffusion from the cap rock and the underlying rock, where the dissolution of anhydrite acts as a sulfate source. The amount of generated $H_2S_{(g)}$ depends on the storage time, the amount of available and reactive anhydrite, the diffusion, and the time-dependent kinetic rate constant. However, the last three mentioned factors depend on pressure and/or temperature.

Two hotspot regions with higher amounts of generated $H_2S_{(g)}$ (related to a higher loss of hydrogen) are identified: the contact area of the reservoir rock and the cap rock and the contact area of the reservoir rock and the underlying rock. These are the regions where additional sulfate is delivered into the reservoir rock by diffusion. However, the highest $CH_{4(g)}$ concentrations can be found at the contact area of the reservoir rock and the underlying rock, where additional CO_2 is delivered by diffusion of the gas in the underlying rock into the reservoir rock.

The 1DRMT modeling results show an average loss of 3.3 mol $H_{2(g/aq)}$ after 30 years in each cell of the reservoir rock (cells 183–850). Over the same period, on average, 0.2 mol L^{-1} of gas $CH_{4(g)}$ and 0.005 mol L^{-1} of gas $H_2S_{(g)}$ are generated and 0.06 mol L^{-1} of gas $CO_{2(g)}$ is consumed, which corresponds to the total amount of $CO_{2(g)}$ that was available as injected and residual gas. The changes in gas composition in the reservoir rock in cell 183, located at the contact with the cap rock, with ongoing time are shown in Figure 2a. The mass of lost $H_{2(g/aq)}$ that is not converted to $CH_{4(g)}$ or $H_2S_{(g)}$ is bound in aqueous species (OH^-, CH_4, HCO_3^-, $Al(OH)_4^-$, $CaOH^+$, $CaHCO_3^+$, $MgOH^+$, $MgHCO_3^+$, H_2, NH_4^+, NH_3, H_2S, HS^-, H_4SiO_4, and $H_3SiO_4^-$) or in mineral phases (see Section 3.2).

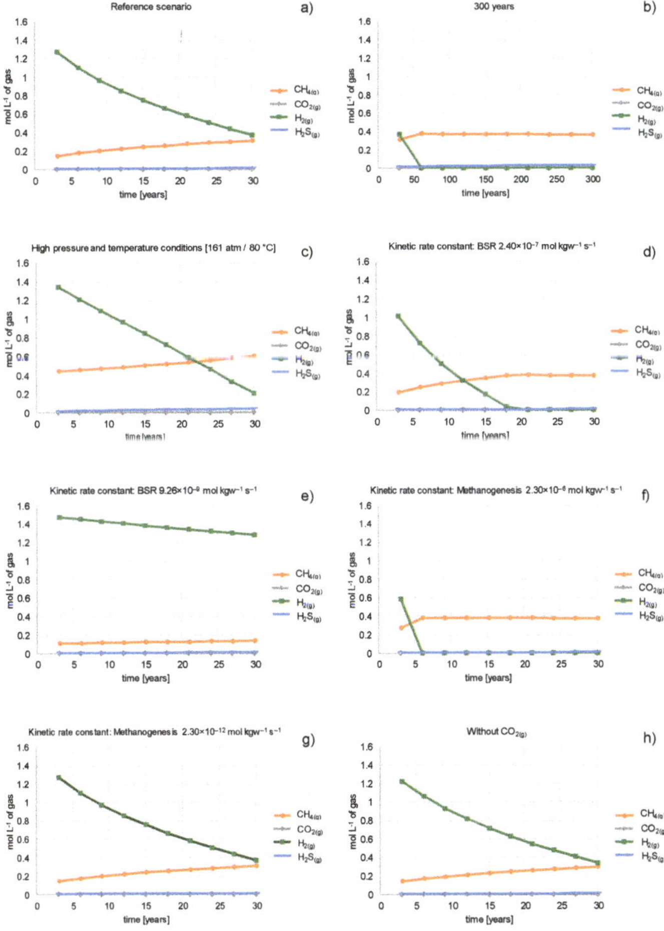

Figure 2. Changes in gas composition in the reservoir rock with ongoing time in (**a**) reference scenario and influenced by (**b**) increased storage time of 300 years, (**c**) higher pressure and temperature conditions, (**d**) higher kinetic rate constant for BSR of 2.40×10^{-7} mol kgw^{-1} s^{-1}, (**e**) lower kinetic rate constant for BSR of 9.26×10^{-9} mol kgw^{-1} s^{-1}, (**f**) higher kinetic rate constant for methanogenesis of 2.30×10^{-6} mol kgw^{-1} s^{-1}, (**g**) lower kinetic rate constant for methanogenesis of 2.30×10^{-12} mol kgw^{-1} s^{-1}, (**h**) varied stored gas composition (without $CO_{2(g)}$).

The competition between methanogenic microorganisms and sulfate-reducing bacteria is widely discussed in the literature [53–56] and is not the object of this study. However, in PHREEQC, BSR occurs prior to methanogenesis. This reaction order of $H_{2(g/aq)}$ with $SO_4^{2-}{}_{(aq)}$ (BSR) and $CO_{2(aq)}$ (methanogenesis) is tested in a separate batch model. It includes $H_2O_{(aq)}$, $CO_{2(aq)}$, and $SO_4^{2-}{}_{(aq)}$ (in equal amounts) and a sufficient amount of hydrogen. $H_{2(g/aq)}$ reacts primarily with $SO_4^{2-}{}_{(aq)}$ to form $S^{2-}{}_{(aq)}$ until the total amount of the initial $SO_4^{2-}{}_{(aq)}$ is consumed. This is followed by the reaction of $H_{2(g/aq)}$ with $CO_{2(aq)}$ to $CH_{4(aq)}$. Thereby, the equilibrium constant ($\log K$) is the crucial value. In a separate model, the $\log K$ value for SO_4^{2-} (33.65; (6)) is reduced to below the $\log K$ value of CO_2 (16.861; (7); data from phreeqc.dat). In this case, the methanogenesis takes place before BSR.

$$SO_4^{2-}{}_{(aq)} + 9H^+{}_{(aq)} + 8e^- = HS^-{}_{(aq)} + 4H_2O \tag{6}$$

$$CO_3^{2-}{}_{(aq)} + 2H^+{}_{(aq)} = CO_{2(aq)} + H_2O \tag{7}$$

3.2. Hydrogeochemical Effects of Hydrogen Storage on Reservoir Rock and Cap Rock

Storage of hydrogen in depleted gas fields can lead to dissolution and precipitation of minerals, which in turn could change the porosity of the reservoir rock and cap rock. An increase in the cap rock porosity would decrease the sealing capacity and increase the risk of pathways for the stored hydrogen to reach, for example, overlying aquifers. On the contrary, a decrease of the cap rock porosity would increase its sealing capacity.

Truche et al. [57] argued that the influence of hydrogen on sulfur-bearing minerals like framboidal pyrite is high but the effect on other minerals available in claystones like clay minerals, quartz and calcite is minor at "low temperatures". By contrast, Reitenbach et al. [2] stated that the addition of hydrogen causes abiotic reactions of hydrogen with minerals of the reservoir and cap rocks which leads to dissolution of sulfate and carbonate-bearing minerals, clay minerals of the chlorite group and feldspars and to precipitation of iron-sulfide bearing minerals, illite, and pyrrhotite. However, both studies agree that pyrite is a potential oxidant for hydrogen and is reduced to pyrrhotite (FeS1 + x) [2,57]. The 1DRMT modeling results of the reference scenario of this study show that at 40 °C and 40 atm the storage of hydrogen induces K-feldspar, kaolinite, and dolomite precipitation in the reservoir rock. Quartz, calcite, and illite are dissolved. The reactive amounts of barite, goethite, and anhydrite are completely consumed during the storage time of 30 years. After the complete consumption of reactive anhydrite, the only sulfate source comes from the cap rock and the underlying rock by diffusion and consequently this limits the loss of hydrogen by bacterial sulfate reduction. High concentrations of $S(-II)_{(aq)}$ species in the reservoir brine can be observed from the modeling results. The assumed reactive amount of goethite is completely consumed in less than 3 years. The goethite dissolution buffers the effect of $H_2S_{(g)}$ generation because the amount of Fe(+II), which has been reduced from Fe(+III), reacts with the aqueous sulfide to form pyrite (a potential secondary phase in the model) so that aqueous sulfide is no longer available for $H_2S_{(g)}$ generation. The modeling results show also that the higher pH conditions (pH increases from initial 6.4 to 8.2–8.7) and the high sodium concentrations in the brine induce weak albitization in the hotspot areas (the contact area of the reservoir rock and the cap rock and the contact area of the reservoir rock and the underlying rock). Except for pyrite and albite, the other potential secondary phases, dawsonite, mackinawite, nahcolite, and sulfur, are not formed under reservoir conditions. Elemental sulfur does not form, even if mackinawite and pyrite are not considered as potential secondary phases. Small amounts of calcite are precipitated because the dissolution of anhydrite contributes $Ca^{2+}{}_{(aq)}$ into the reservoir brine. However, at the same time, calcite is dissolved and delivers CO_2 for methanogenesis so that the total amount of calcite decreases. The volume changes of the mineral phases are shown in Figure 3. The changes in porosity are calculated from the PHREEQC modeling results. The porosity in the reservoir rock decreases from initially 10% to 9.79–9.95%. This means that the available pore space for hydrogen storage decreases over 30 years such that (i) the same amount of hydrogen is moved to greater distances from the bore hole or (ii) less hydrogen can be stored in future injection phases (max. 0.2%).

The mineralogical changes in the cap rock, induced by diffusion of aqueous $H_{2(aq)}$ from the reservoir rock, are minimal and cause no changes to the initial porosity. The reasons for that are the short storage time of 30 years and the slow diffusion process. Furthermore, the hydrogen reacts with the mineralogical assemblage of the reservoir rock and brine and it is converted by BSR and methanogenesis.

The used modeling program, PHREEQC, has no capabilities to simulate gas transport between the different cells (multiphase flow). To overcome this limitation, a separate transport model provides an initial amount of $H_{2(g)}$ in the cap rock for the kinetic calculation to simulate the influence of $H_{2(g)}$ on the cap rock properties in the case of a gas loss from the reservoir rock by, for example, new fractures or natural faults. The initial amount corresponds to the maximum available amount of $H_{2(g)}$ that could be available per cell in the cap rock and represents a worst-case scenario. Modeling results indicate an increase in pH from 6.4 to 7.8 in the cap rock brine after 30 years of storage. The intense contact of hydrogen with the cap rock results in a total loss in porosity, which increases the sealing capacity of the cap rock. This decrease in porosity is caused mainly by albitization, whereas the dissolution of halite, quartz, illite, and anhydrite counteracts a stronger porosity decrease. Precipitation of dolomite is negligible. The potential secondary phase pyrite is precipitated whereas dawsonite, mackinawite, and sulfur are not formed.

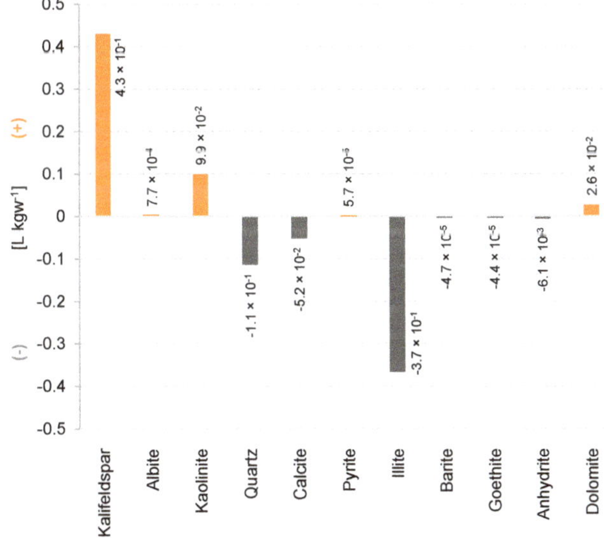

Figure 3. Volume changes of mineral phases (L kgw^{-1}) in the reservoir rock after 30 years of hydrogen storage (reference scenario).

3.3. Loss of Aqueous $H_{2(aq)}$ by Diffusion through the Cap Rock

By assuming specific diffusion coefficients for the different aqueous species, the diffusion of hydrogen and related products is modeled. A non-reactive tracer, with the same diffusion coefficient as hydrogen of 5.13×10^{-9} m^2 s^{-1}, shows that significant amounts of hydrogen (greater than 4.0×10^{-7} mol kgw^{-1}) were calculated only for distances less than 4 m through the cap rock (in the z-direction) over 30 years, assuming an initial cap rock porosity of 5%, no faults, and non-reactive hydrogen (Figure 4a). However, hydrogen reacts with the mineralogical assemblage of the reservoir rock and brine and is converted by BSR and methanogenesis. The loss of aqueous hydrogen by diffusion is minor. Dissolved hydrogen and corresponding aqueous species concentrations (OH$^-$,

CH_4, HCO_3^-, $Al(OH)_4^-$, $CaOH^+$, $CaHCO_3^+$, $MgOH^+$, $MgHCO_3^+$, H_2, NH_4^+, NH_3, H_2S, HS^-, H_4SiO_4, and $H_3SiO_4^-$) in the cap rock brine show no significant changes. The most significant difference has been calculated for the NH_4^+ concentration, which rises to around 4×10^{-4} mol kgw^{-1} in the first meter of the cap rock (cell 182). Therefore, the effect of hydrogen storage on the cap rock is only visible in the first meter of the cap rock, which is in contact with the reservoir rock (cell 182), after 30 years. Even though each aqueous species migrates with a different rate along its concentration gradient, as described by Buzek et al. [58], the difference in migration is too small to see deviations in the model over the modeled time of 30 years. The loss of the non-reactive hydrogen tracer by diffusion into the cap rock is 25% and represents the maximal loss of aqueous hydrogen by diffusion over 30 years.

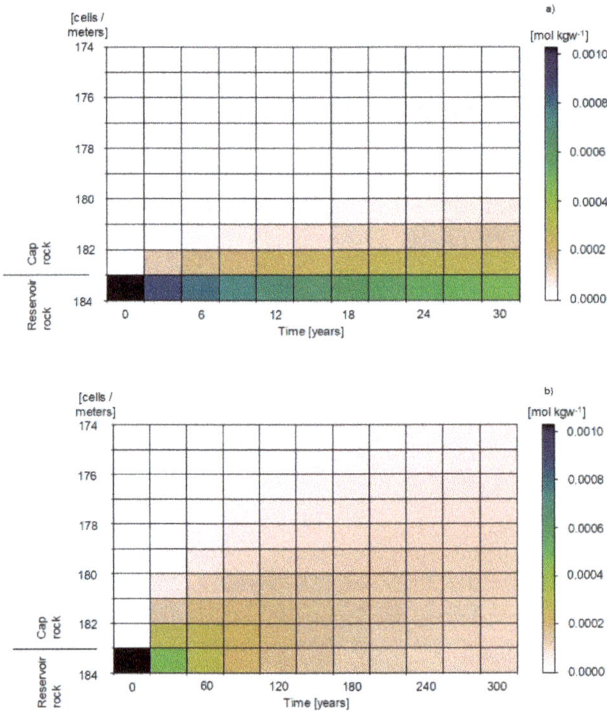

Figure 4. Diffusion of a non-reactive hydrogen tracer through the cap rock over (**a**) 30 years and (**b**) 300 years.

In summary (Sections 3.1–3.3), the loss of hydrogen gas by bacterial conversion, gas–water–rock interactions, and aqueous diffusion is 76% over 30 years.

3.4. Influencing Factors

To identify the controlling factors of hydrogen loss and the parameters that affect the modeling results, simulation of generic model scenarios was performed.

3.4.1. Storage Time

Because of the lack of data, the storage time has been chosen based on the equipment lifetime of 30 years [51]. To show the influence of longer storage times on the system environment, in this scenario a longer storage time of 300 years is modeled. With increasing storage time, the dissolved hydrogen has more time to diffuse into higher regions of the cap rock. After 300 years, the non-reactive tracer diffuses 10 m through the cap rock (Figure 4b), whereas the reactive hydrogen affects just the first 5 m of the cap rock.

When considering longer storage times, it becomes clear that the mass of generated water is increased to 1.03 kg (1 kg + 0.3 kg = 1.03 kg water in total) in the reservoir rock because the water-producing processes, BSR and methanogenesis (Equations (2) and (3)), are time-dependent. In the reservoir rock, the pH is increased to 8.7. The porosity loss from 10% to 9.19–9.99% is greater than that after 30 years of storage. The reason for the greater decrease is the intensified dissolution of quartz and illite. Calcite is completely consumed in the hotspot areas, the contact areas of the cap rock to the reservoir rock and of the underlying rock to the reservoir rock. In all other cells, calcite is still available, but it is only partly dissolved. On the other hand, the albitization process is intensified in the hotspot areas and more kaolinite is formed. The reactive amounts of anhydrite and goethite are completely consumed. The only sulfate source for BSR is delivered by diffusion from the underlying rock and the cap rock into the reservoir rock. The mineralogical changes in the cap rock are minimal and show that even after 300 years of hydrogen storage, the effects of diffusion of aqueous hydrogen are small.

When the storage time is increased to 300 years (under the assumption that no new gas mixture of $H_{2(g)}$ and $CO_{2(g)}$ is injected), the total amount of newly generated $CH_{4(g)}$ is increased to an average of 0.25 mol L^{-1} of gas and the $H_2S_{(g)}$ concentration is increased to an average of 0.006 mol L^{-1} of gas (Figure 2b). After 300 years, the total amount of stored $H_{2(g)}$ is consumed.

In the case of storage cycles, including gas injection, storage, and production phases, the newly injected hydrogen gas is again accompanied by a new amount of $CO_{2(g)}$, which is available as an electron acceptor for methanogenic bacteria. This aspect is modeled in a separate transport model. The total loss of hydrogen by conversion to $CH_{4(g)}$ is higher if the storage cycle has a higher frequency, because $CO_{2(g)}$ is co-injected and is available for methanogenesis. With a total storage time of 30 years and a dwelling time of 6 months, a maximum of 60 injections of "fresh gas" (composed of $H_{2(g)}$ and $CO_{2(g)}$) can be stored in the reservoir. This leads to a total generation of 6.72 mol L^{-1} of gas $CH_{4(g)}$ after 30 years. However, the loss of hydrogen after 6 months, the shorter storage period, is of course smaller (0.112 mol L^{-1} of gas in 6 months) than after 30 years of storage. Consequently, a shorter storage period leads to a lower amount of generated $CH_{4(g)}$ (and a lower loss of hydrogen) because methanogenesis is a time-dependent process. However, over the total time, the loss of hydrogen sums to a higher amount. This holds true only for methanogenesis. For bacterial sulfate reduction, the amount of generated $H_2S_{(g)}$ depends not only on the amount of available sulfate but also on the time-dependent kinetic rate constant.

To summarize, it is clear that longer storage times increase the risk of loss of hydrogen. On the other hand, the reactive anhydrite and calcite will be consumed after a few years of storage and safer storage conditions arise because BSR and methanogenesis will be limited by the amount of sulfate and carbon dioxide delivered by diffusion (and diffusion is a slow process). In this case, the hotspot areas gain in importance. However, assuming shorter storage periods of 6 months, the $CH_{4(g)}$ concentration in the stored gas will be lower than after a storage time of 30 years, but the total loss of hydrogen sums to higher amounts regarding shorter storage periods (60 injections in 30 years).

3.4.2. Pressure and Temperature Conditions in Gas Reservoirs

Each depleted gas field has specific pressure and temperature conditions. To analyze the influence of pressure and temperature on the modeling results of hydrogen storage, the pressure and temperature conditions are varied.

The higher pressure and temperature conditions chosen are 161 atm and 80 °C. At even higher temperature conditions, the sulfate-reducing bacteria and methanogenic bacteria are mostly no longer active. The maximum temperature for SRB and methanogenic bacteria is 80 °C [19,24]. At higher pressure and temperature conditions, thermochemical sulfate reduction, as known from deeply buried hydrocarbon reservoirs, must be considered.

At 161 atm and 80 °C (at invariable kinetic rate constants for BSR and methanogenesis), the amount of consumed $H_{2(g/aq)}$ is slightly lower (75%) than that in the reference scenario (76%; modeled at 40 atm and 40 °C). The increased pressure results in a decreased gas volume and in higher gas concentrations. The $H_2S_{(g)}$ concentration is increased to an average of 0.009 mol L^{-1} of gas and the $CH_{4(g)}$ concentration is increased to an average 0.54 mol L^{-1} of gas after 30 years (Figure 2c).

The porosity loss (from initial 10% to 9.52–9.86%) in the reservoir rock is, on average, higher than in the reference scenario. This can be explained by the stronger albitization at this higher pressure and temperature conditions and the pyrite formation increases as well. On the other hand, the dissolution of quartz, calcite, and illite is higher. However, the amount of precipitated minerals is higher than the amount of dissolved minerals and this causes the decrease in porosity. The pH in the reservoir rock brine is 7.5–8.4, which is lower than in the reference scenario. In addition, the amount of the aqueous sulfide sulfur (S(–II)) in the reservoir brine is decreased at these conditions. The mass of generated water is, despite BSR and methanogenesis, decreased from 1 kg to 0.96 kg. A reason for this could be the stronger albitization process, binding the water.

The changed pressure and temperature conditions affect the equilibrium constants of the mass action laws of gases, minerals, and aqueous species. Here, a modeling limitation is achieved because the phreeqc.dat database does not contain pressure dependencies for the equilibrium constants for all mineral phases that are used in the model. The pressure dependency of the following mineral phases in the model are included: calcite, dolomite, anhydrite, barite, quartz, and halite.

As in the reference scenario, the temperature and pressure conditions change along the diffusive pathway of the aqueous species through the cap rock according to the geothermal gradient of 33.3 °C km^{-1} depth and 100 atm km^{-1} depth under hydrostatic conditions, starting with 80 °C and 161 atm at the reservoir depth. Each cell is defined by a specific temperature and pressure condition. The influence of these changing pressure and temperature conditions on the cap rock is minor. The concentration of aqueous sulfate sulfur S(+VI) is higher and aqueous sulfide sulfur (S(–II)) is lower than in the reference scenario. Different from the reference scenario, albitization occurs at these higher pressure and temperature conditions even in the cap rock, and dawsonite precipitation and halite dissolution are slightly increased.

To summarize, at higher pressure and temperature conditions (while the kinetic rate constants are not changed), the concentration of $CH_{4(g)}$ and $H_2S_{(g)}$ is increased due to a lower gas volume. Furthermore, the loss in porosity in the reservoir rock is increased. However, the kinetic rate constant is pressure- and temperature-dependent; therefore, this influence is tested in a further scenario (Section 3.4.3).

3.4.3. Kinetic Rate Constant

Special consideration is given to the influence of the kinetic rate constants of bacterial sulfate reduction and methanogenesis on the loss of hydrogen during storage. Therefore, the rate constants are varied. As a reminder, the initial kinetic rate constant in the reference scenario for BSR is 9.26×10^{-8} mol kgw^{-1} s^{-1} and 2.30×10^{-9} mol kgw^{-1} s^{-1} for methanogenesis.

By increasing the kinetic rate constant for BSR slightly to 2.40×10^{-7} mol kgw^{-1} s^{-1}, based on data from Herrera et al. [52], the total amount of stored hydrogen is lost in less than 20 years. The amount of generated $H_2S_{(g)}$ does not change because the reactive amount of anhydrite in the reservoir rock is completely consumed and the only sulfate source comes from the cap rock and underlying rock by diffusion and diffusion is a slow process. By increasing the storage time to 300 years (with a kinetic rate constant of 2.40×10^{-7} mol kgw^{-1} s^{-1} for BSR), the effect of diffusion is recognizable and the $H_2S_{(g)}$ generation increases in the hotspot areas. The amount of generated $CH_{4(g)}$ increases from 0.2 mol L^{-1} of gas to an average of 0.25 mol L^{-1} of gas in each cell of the reservoir rock after 30 years (Figure 2d). The reason for this could be that methanogenesis starts as soon as all the sulfate for BSR is consumed. Because sulfate is consumed faster, due to the increased kinetic rate constant, the timeframe in which the methanogenesis can take place is expanded. On decreasing the kinetic rate constant for BSR to 9.26×10^{-9} mol kgw^{-1} s^{-1}, the amount of generated $H_2S_{(g)}$ does not change because the reactive amount of anhydrite in the reservoir rock is completely consumed. Less $H_{2(g/aq)}$ is consumed (14% loss) and a smaller amount of $CH_{4(g)}$ (on average 0.013 mol L^{-1} of gas) is generated (Figure 2e).

At a higher kinetic rate constant for methanogenesis of 2.30×10^{-6} mol kgw^{-1} s^{-1}, the total amount of stored hydrogen is lost in less than 6 years (Figure 2f). $CH_{4(g)}$ production increases to an average of 0.25 mol L^{-1} of gas and the mass of water rises to 1.03 kg per cell in the reservoir rock after 30 years of storage. The amount of $CO_{2(g)}$ from the injected gas mixture and the residual gas is completely consumed in less than 3 years of storage and is not the limiting factor for methanogenesis. If no $CO_{2(g)}$ is available anymore from the injected gas mixture and residual gas, the dissolution of carbonate-bearing minerals begins and delivers CO_2 for methanogenesis. With a smaller kinetic rate constant for methanogenesis of 2.30×10^{-12} mol kgw^{-1} s^{-1}, the loss of $H_{2(g/aq)}$ is slightly lower (75.7%). Less $CH_{4(g)}$ is generated (but the differences are in the third decimal place, only from 0.192 mol L^{-1} of gas in the reference scenario to 0.191 mol L^{-1} of gas) and the $H_2S_{(g)}$ generation is constant (Figure 2g). The produced mass of water and the pH conditions show no changes compared to the reference scenario.

It can be concluded that the kinetic rate constants are important factors controlling the loss of hydrogen storage. Knowledge of the kinetic rate constants for BSR and methanogenesis at elevated temperature and pressure are required. Nevertheless, the maximal amount of generated $H_2S_{(g)}$ is, if all reactive sulfate-bearing minerals (e.g., anhydrite) are consumed, limited by diffusion. On the other hand, the maximal amount of generated $CH_{4(g)}$ depends on the kinetic rate constant and the storage time.

3.4.4. Stored Gas Composition

In the "POWER-to-GAS-to-POWER" concept for the storage and usage of hydrogen, the hydrogen is stored purely in underground storage systems [13]. As a consequence of pure $H_{2(g)}$ injection and storage (no $CO_{2(g)}$ is co-injected), the amount of generated $CH_{4(g)}$ decreases to an average of 0.32 mol L^{-1} of gas after 30 years. $H_2S_{(g)}$ generation is constant with an average of 0.005 mol L^{-1} of gas after 30 years (Figure 2h). The loss of stored $H_{2(g/aq)}$ is slightly increased to 77% after 30 years of storage.

An alternative CO_2 source for methanogenesis is the residual gas that was not removed during natural gas production. Only if the residual gas is consumed or not reactive, does carbonate-bearing mineral dissolution increase and delivers CO_2 for methanogenesis, which is even more pronounced at a longer storage time of 300 years. Finally, the amount of available and reactive carbonate-bearing minerals (here calcite and dolomite) limits the process of methanogenesis if no $CO_{2(g)}$ is stored with $H_{2(g/aq)}$. On the other hand, a separate test model shows that a higher amount of co-injected $CO_{2(g)}$ (14.8 atm; 10 times higher than in the reference scenario) increases the amount of generated $CH_{4(g)}$ to an average of 1.0 mol L^{-1} of gas. If the amount of available $CO_{2(g)}$ would be infinitely large, the time-dependent kinetic rate constant and the storage time limit the loss of hydrogen by conversion to $CH_{4(g)}$.

Without $CO_{2(g)}$ co-injection, the porosity in the reservoir rock decreases from 10% to 9.82–9.98%. This loss in porosity is slightly smaller than in the reference scenario in which 4% $CO_{2(g)}$ is co-injected. The changes in mineral dissolution and precipitation are minor in comparison to the reference scenario. The amounts of precipitated kaolinite, K-feldspar, and dolomite are slightly smaller than in the reference scenario and less calcite and illite are dissolved. No changes in the pH are observable. The smaller amount of precipitated K-feldspar and kaolinite leads to a smaller amount of bound water in these minerals. Consequently, the mass of water increases to 1.025 kg in the reservoir rock and is higher than the increase in the reference scenario (1.007 kg). The changes in the mineralogical composition of the cap rock are minor.

4. Conclusions

The simulation results show that the underground storage of hydrogen in depleted gas fields entails the risk of hydrogen loss (and related energy loss) by bacterial conversion to $CH_{4(g)}$ and $H_2S_{(g)}$ and gas–water–rock interactions, which in turn lead to changes in the porosity of the reservoir rock. The modeling results of the one-dimensional reactive mass transport model identify the factors that control the loss of hydrogen:

- The loss of hydrogen by bacterial conversion to $CH_{4(g)}$ via methanogenesis is limited mainly by the amount of co-injected $CO_{2(g)}$, the reaction kinetics, and the connected maximal storage time of 30 years. Less co-injected $CO_{2(g)}$ will reduce $H_{2(g)}$ loss but cannot inhibit the conversion to $CH_{4(g)}$ if further CO_2 sources are available in the form of residual gas and carbonate-bearing minerals. The generation of $CH_{4(g)}$ by methanogenesis where CO_2 is only delivered by the dissolution of carbonate-bearing minerals is slower because the dissolution process limits the conversion to $CH_{4(g)}$.

- The loss of hydrogen by bacterial conversion to $H_2S_{(g)}$ via bacterial sulfate reduction is limited mainly by the amount of available sulfate in the reservoir. After complete consumption of reactive anhydrite, the only sulfate source comes from the cap rock and the underlying rock by diffusion and consequently limits the loss of hydrogen by bacterial sulfate reduction because the process of diffusion is slow.

- The mass of generated water, as a product of BSR and methanogenesis, increases the pressure in the system and diffusion can be intensified.

- The mineralogical changes in the reservoir rock result in a small decrease in porosity. As a consequence, the available pore space for hydrogen storage decreases over 30 years such that (i) the same amount of hydrogen is moved to greater distances from the bore hole or (ii) less hydrogen can be stored in future injection phases.

- The loss of aqueous hydrogen by diffusion and related effects on the cap rock mineralogy is negligibly small, with a storage time of 30 years, because hydrogen storage causes gas–water–rock interactions in the reservoir rock and brine and is converted by BSR and methanogenesis to a greater extent. Furthermore, the process of diffusion is slow.

- A longer storage period increases the loss of the stored hydrogen. Shorter storage periods lead to less hydrogen loss for each period, but over the total time the summed loss of hydrogen is higher because new $CO_{2(g)}$ is available for methanogenesis after each gas injection.

- At higher pressure and temperature conditions, the concentrations of $CH_{4(g)}$ and $H_2S_{(g)}$ increase due to a lower gas volume. Furthermore, the loss in porosity in the reservoir rock increases as well.

- Knowledge of kinetic rate constants for bacterial sulfate reduction and methanogenesis at elevated levels of pressure and temperature are required as accurately as possible.

It is recommended to choose depleted gas fields for hydrogen storage where the residual gas has low $CO_{2(g)}$ concentrations. The mineralogical composition of the reservoir rocks should contain low amounts of sulfate- and carbonate-bearing minerals but high amounts of reactive iron-bearing minerals. Reservoirs with low pressure and temperature conditions are recommended as well as storage gas compositions with low amounts of $CO_{2(g)}$. From the modeling results, it seems reasonable to implement a multicomponent transport model and to conduct a joint variation of several parameters as a next step.

Supplementary Materials: The following are available online at http://www.mdpi.com/2076-3417/8/11/2282/s1, Input file S1 for the transport model to calculate the reference scenario (for PHREEQC Version 3.1.4-8929; ready to copy, paste, and run).

Author Contributions: Conceptual model, C.H.; modeling, C.H. and W.v.B.; validation of results: C.H. and W.v.B.; analysis of results, C.H. and W.v.B.; writing—original draft preparation: C.H.; writing—review and editing: C.H. and W.v.B.; visualization: C.H.; supervision: W.v.B.

Appl. Sci. **2018**, *8*, 2282

Funding: This research received no external funding.

Acknowledgments: We thank Leo Fuhrmann for technical assistance. We would like to thank three anonymous reviewers for their constructive reviews.

Conflicts of Interest: The authors declare no conflict of interest.

References

1. Ebigbo, A.; Golfier, F.; Quintard, M. A coupled, pore-scale model for methanogenic microbial activity in underground hydrogen storage. *Adv. Water Resour.* **2013**, *61*, 74–85. [CrossRef]
2. Reitenbach, V.; Ganzer, L.; Albrecht, D.; Hagemann, B. Influence of added hydrogen on underground gas storage: A review of key issues. *Environ. Earth Sci.* **2015**, *73*, 6927–6937. [CrossRef]
3. Stone, H.B.J.; Veldhuis, I.; Richardson, R.N. Underground hydrogen storage in the UK. *Geol. Soc. Lond. Spec. Publ.* **2009**, *313*, 217–226. [CrossRef]
4. Henkel, S.; Pudlo, D.; Gaupp, R. Research sites of the H2STORE project and the relevance of lithological variations for hydrogen storage at depths. *Energy Procedia* **2013**, *40*, 25–33. [CrossRef]
5. Panfilov, M. Underground storage of hydrogen: In situ self-organisation and methane generation. *Transp. Porous Med.* **2010**, *85*, 841–865. [CrossRef]
6. Crotogino, F.; Donadei, S.; Bünger, U.; Landinger, H. Large-Scale Hydrogen Underground Storage for Securing Future Energy Supplies. In *18th World Hydrogen Energy Conference 2010—WHEC 2010 Proceedings*; Stolten, D., Grube, T., Eds.; Forschungszentrum IEF-3: Jülich, Germany, 2010.
7. Shen, L.; Chen, Z. Critical review of the impact of tortuosity on diffusion. *Chem. Eng. Sci.* **2007**, *62*, 3748–3755. [CrossRef]
8. Jacops, E.; Volckaert, G.; Maes, N.; Weetjens, E.; Govaerts, J. Determination of gas diffusion coefficients in saturated porous media: He and CH_4 diffusion in Boom Clay. *Appl. Clay Sci.* **2013**, *83*, 217–223. [CrossRef]
9. Krooss, B. *Evaluation of Database on Gas Migration through Clayey Host Rocks*; Belgian National Agency for Radioactive Waste and Enriched Fissile Material (ONDRAF-NIRAS); RWTH Aachen: Aachen, Germany, 2008.
10. Panfilov, M. Underground and pipeline hydrogen storage. In *Compendium of Hydrogen Energy: Volume 2: Hydrogen Storage, Distribution and Infrastructure*; Gupta, R.B., Basile, A., Veziroğlu, T.N., Eds.; Elsevier: Cambridge, UK; Waltham, MA, USA; Kidlington, UK, 2016; pp. 91–115.
11. McCarty, R.D.; Hord, J.; Roder, H.M. *Selected Properties of Hydrogen (Engineering Design Data)*; National Bureau of Standards Monograph 168—US Department of Commerce: Boulder, CO, USA, 1981.
12. Friend, D.G.; Ely, J.F.; Ingham, H. Thermophysical properties of methane. *J. Phys. Chem. Ref. Data* **1989**, *18*, 583–638. [CrossRef]
13. Hagemann, B.; Rasoulzadeh, M.; Panfilov, M.; Ganzer, L.; Reitenbach, V. Hydrogenization of underground storage of natural gas. *Comput. Geosci.* **2016**, *20*, 595–606. [CrossRef]
14. Cord-Ruwisch, R.; Kleinitz, W.; Widdel, F. Sulfate-reducing bacteria and their activities in oil production. *J. Pet. Technol.* **1987**, *39*, 97–106. [CrossRef]
15. Bernardez, L.A.; de Andrade Lima, L.R.; de Jesus, E.B.; Ramos, C.L.S.; Almeida, P.F. A kinetic study on bacterial sulfate reduction. *Bioprocess. Biosyst. Eng.* **2013**, *36*, 1861–1869. [CrossRef] [PubMed]
16. Postgate, J.R. *The Sulphate-Reducing Bacteria*, 2nd ed.; Cambridge University Press: Cambridge, UK, 1984.
17. Ehrlich, H.L. *Geomicrobiology*, 2nd ed.; Dekker: New York, NY, USA, 1990.
18. Machel, H.G. Bacterial and thermochemical sulfate reduction in diagenetic settings—Old and new insights. *Sediment. Geol.* **2001**, *140*, 143–175. [CrossRef]
19. Jorgensen, B.B.; Isaksen, M.F.; Jannasch, H.W. Bacterial sulfate reduction above 100 °C in deep-sea hydrothermal vent sediments. *Science* **1992**, *258*, 1756–1757. [CrossRef] [PubMed]
20. Postgate, J.R. *The Sulphate-Reducing Bacteria*; Cambridge University Press: Cambridge, UK, 1979.
21. Whitman, W.B.; Bowen, T.L.; Boone, D.R. The methanogenic bacteria. In *The Prokaryotes*; Dworkin, M., Falkow, S., Rosenberg, E., Schleifer, K.-H., Stackebrandt, E., Eds.; Springer: New York, NY, USA, 2006; pp. 165–207.
22. Šmigáň, P.; Greksák, M.; Kozánková, J.; Buzek, F.; Onderka, V.; Wolf, I. Methanogenic bacteria as a key factor involved in changes of town gas stored in an underground reservoir. *FEMS Microbiol. Lett.* **1990**, *73*, 221–224. [CrossRef]

23. Davydova-Charakhch'yan, I.A.; Kuznetsova, V.G.; Mityushina, L.L.; Belyaev, S.S. Methane-forming bacilli from oil fields of Tataria and western Siberia. *Microbiology* **1992**, *61*, 202.

24. Magot, M.; Ollivier, B.; Patel, B.K.C. Microbiology of petroleum reservoirs. *Antonie Leeuwenhoek* **2000**, *77*, 103–116. [CrossRef] [PubMed]

25. Gusev, M.V.; Mineeva, L.A. *Microbiology*; Moscow Lomonosov University: Moscow, Russia, 1992.

26. Bildstein, O.; Kervévan, C.; Lagneau, V.; Delaplace, P.; Crédoz, A.; Audigane, P.; Perfetti, E.; Jacquemet, N.; Jullien, M. Integrative modeling of caprock integrity in the context of CO_2 storage: Evolution of transport and geochemical properties and impact on performance and safety assessment. *Oil Gas Sci. Technol. Rev. IFP* **2010**, *65*, 485–502. [CrossRef]

27. Gaus, I.; Azaroual, M.; Czernichowski-Lauriol, I. Reactive transport modelling of the impact of CO_2 injection on the clayey cap rock at Sleipner (North Sea). *Chem. Geol.* **2005**, *217*, 319–337. [CrossRef]

28. Hemme, C.; van Berk, W. Change in cap rock porosity triggered by pressure and temperature dependent CO_2–water–rock interactions in CO_2 storage systems. *Petroleum* **2017**, *3*, 96–108. [CrossRef]

29. Mohd Amin, S.; Weiss, D.J.; Blunt, M.J. Reactive transport modelling of geologic CO_2 sequestration in saline aquifers: The influence of pure CO_2 and of mixtures of CO_2 with CH_4 on the sealing capacity of cap rock at 37 °C and 100bar. *Chem. Geol.* **2014**, *367*, 39–50. [CrossRef]

30. Larin, N.; Zgonnik, V.; Rodina, S.; Deville, E.; Prinzhofer, A.; Larin, V.N. Natural molecular hydrogen seepage associated with surficial, rounded depressions on the European Craton in Russia. *Nat. Resour. Res.* **2015**, *24*, 369–383. [CrossRef]

31. Sato, M.; Sutton, A.J.; McGee, K.A.; Russell-Robinson, S. Monitoring of hydrogen along the San Andreas and Calaveras faults in central California in 1980–1984. *J. Geophys. Res. Solid Earth* **1986**, *91*, 12315–12326. [CrossRef]

32. Wakita, H.; Nakamura, Y.; Kita, I.; Fujii, N.; Notsu, K. Hydrogen release: New indicator of fault activity. *Science* **1980**, *210*, 188–190. [CrossRef] [PubMed]

33. Ware, R.H.; Roecken, C.; Wyss, M. The detection and interpretation of hydrogen in fault gases. *Pure Appl. Geophys.* **1985**, *122*, 392–402. [CrossRef]

34. Zgonnik, V.; Beaumont, V.; Deville, E.; Larin, N.; Pillot, D.; Farrell, K.M. Evidence for natural molecular hydrogen seepage associated with Carolina bays (surficial, ovoid depressions on the Atlantic Coastal Plain, Province of the USA). *Prog. Earth Planet. Sci.* **2015**, *2*, 31. [CrossRef]

35. Truche, L.; Berger, G.; Destrigneville, C.; Guillaume, D.; Giffaut, E. Kinetics of pyrite to pyrrhotite reduction by hydrogen in calcite buffered solutions between 90 and 180 °C: Implications for nuclear waste disposal. *Geochim. Cosmochim. Acta* **2010**, *74*, 2894–2914. [CrossRef]

36. Yekta, A.E.; Pichavant, M.; Audigane, P. Evaluation of geochemical reactivity of hydrogen in sandstone: Application to geological storage. *Appl. Geochem.* **2018**, *95*, 182–194. [CrossRef]

37. Flesch, S.; Pudlo, D.; Albrecht, D.; Jacob, A.; Enzmann, F. Hydrogen underground storage—Petrographic and petrophysical variations in reservoir sandstones from laboratory experiments under simulated reservoir conditions. *Int. J. Hydrog. Energy* **2018**, *43*, 20822–20835. [CrossRef]

38. Parkhurst, D.L.; Appelo, C.A.J. *Description of Input for Phreeqc Version 3—A Computer Program for Speciation, Batch-Reaction, One-Dimensional Transport, and Inverse Geochemical Calculations*; U.S. Geological Survey: Denver, CO, USA, 2013.

39. Dersch-Hansmann, M.; Hug-Diegel, N.; Wonik, T. Ein vollständiges Röt-Profil (Oberer Buntsandstein) in Nordhessen—Lithostratigraphie, Sedimentfazies, Geochemie und Geophysik der Kernbohrung Fürstenwald. *Geol. Jahrb. Hessen* **2010**, *136*, 65–107.

40. Feist-Burkhardt, S.; Götz, A.E.; Szulc, J.; Borkhataria, R.; Geluk, M.; Haas, J.; Hormung, J.; Jordan, P.; Kempf, O.; Michalik, J. Triassic. In *The Geology of Central Europe*; McCann, T., Ed.; Geological Society: London, UK, 2008.

41. Soyk, D. Diagenesis and Reservoir Quality of the Lower and Middle Buntsandstein (Lower Triassic), SW Germany. Ph.D. Thesis, Ruprecht-Karls-Universität Heidelberg, University of Heidelberg, Heidelberg, Germany, 2015.

42. Biehl, B.C.; Reuning, L.; Schoenherr, J.; Lewin, A.; Leupold, M.; Kukla, P.A. Do CO_2-charged fluids contribute to secondary porosity creation in deeply buried carbonates? *Mar. Pet. Geol.* **2016**, *76*, 176–186. [CrossRef]

43. Schreiber, B.C.; Bąbel, M. *Evaporites through Space and Time*; Geological Society: London, UK, 2007; Volume 285.

44. Appelo, C.A.J.; Postma, D. *Geochemistry, Groundwater and Pollution*, 4th ed.; Balkema: Rotterdam, The Netherlands, 1999.

45. Bischoff, G.; Gocht, W. *Energietaschenbuch*, 2nd ed.; Vieweg+Teubner Verlag: Wiesbaden, Germany, 1984.

46. Pudlo, D.; Ganzer, L.; Henkel, S.; Kühn, M.; Liebscher, A.; De Lucia, M.; Panfilov, M.; Pilz, P.; Reitenbach, V.; Albrecht, D.; et al. The H2STORE Project: Hydrogen Underground Storage—A Feasible Way in Storing Electrical Power in Geological Media? In *Underground Storage of CO_2 and Energy*; Hou, M.Z., Xie, H., Yoon, J.S., Eds.; Springer: Berlin/Heidelberg, Germany, 2013; pp. 395–412.

47. Hemme, C.; van Berk, W. Potential risk of H_2S generation and release in salt cavern gas storage. *J. Nat. Gas Sci. Eng.* **2017**, *47*, 114–123. [CrossRef]

48. Parkhurst, D.L.; Appelo, C.A.J. *Users Guide to Phreeqc (Version 2)—A Computer Program for Speciation, Batch-Reaction, One-Dimensional Transport and Inverse Geochemical Calculations*; U.S. Geological Survey: Denver, CO, USA, 1999.

49. Appelo, C.A.J.; Wersin, P. Multicomponent Diffusion Modeling in Clay Systems with Application to the Diffusion of Tritium, Iodide, and Sodium in Opalinus Clay. *Environ. Sci. Technol.* **2007**, *41*, 5002–5007. [CrossRef] [PubMed]

50. Vinograd, J.R.; McBain, J.W. Diffusion of electrolytes and of the ions in their mixtures. *J. Am. Chem. Soc.* **1941**, *63*, 2008–2015. [CrossRef]

51. Lord, A.S.; Kobos, P.H.; Borns, D.J. Geologic storage of hydrogen: Scaling up to meet city transportation demands. *Int. J. Hydrog. Energy* **2014**, *39*, 15570–15582. [CrossRef]

52. Herrera, L.; Hernández, J.; Bravo, L.; Romo, L.; Vera, L. Biological process for sulfate and metals abatement from mine effluents. *Environ. Toxicol. Water Qual.* **1997**, *12*, 101–107. [CrossRef]

53. Adler, M.; Eckert, W.; Sivan, O. Quantifying rates of methanogenesis and methanotrophy in Lake Kinneret sediments (Israel) using pore-water profiles. *Limnol. Oceanogr.* **2011**, *56*, 1525–1535. [CrossRef]

54. Robinson, J.A.; Tiedje, J.M. Competition between sulfate-reducing and methanogenic bacteria for H_2 under resting and growing conditions. *Arch. Microbiol.* **1984**, *137*, 26–32. [CrossRef]

55. Kalyuzhnyi, S.V.; Fedorovich, V.V. Mathematical modelling of competition between sulphate reduction and methanogenesis in anaerobic reactors. *Bioresour. Technol.* **1998**, *65*, 227–242. [CrossRef]

56. Timmers, P.H.A.; Gieteling, J.; Widjaja-Greefkes, H.C.A.; Plugge, C.M.; Stams, A.J.M.; Lens, P.N.L.; Meulepas, R.J.W. Growth of anaerobic methane-oxidizing archaea and sulfate-reducing bacteria in a high-pressure membrane capsule bioreactor. *Appl. Environ. Microbiol.* **2015**, *81*, 1286–1296. [CrossRef] [PubMed]

57. Truche, L.; Jodin-Caumon, M.-C.; Lerouge, C.; Berger, G.; Mosser-Ruck, R.; Giffaut, E.; Michau, N. Sulphide mineral reactions in clay-rich rock induced by high hydrogen pressure. Application to disturbed or natural settings up to 250 degrees C and 30 bar. *Chem. Geol.* **2013**, *351*, 217–228. [CrossRef]

58. Buzek, F.; Onderka, V.; Vančura, P.; Wolf, I. Carbon isotope study of methane production in a town gas storage reservoir. *Fuel* **1994**, *73*, 747–752. [CrossRef]

 *applied sciences*

Article

The Effect of Magnetic Field on Thermal-Reaction Kinetics of a Paramagnetic Metal Hydride Storage Bed

Shahin Shafiee [1], Mary Helen McCay [1] and Sarada Kuravi [2,*]

[1] National Center for Hydrogen Research, Florida Institute of Technology, Melbourne, FL 32901, USA; sshafiee2011@my.fit.edu (S.S.); mmccay@my.fit.edu (M.H.M.)

[2] Department of Mechanical and Aerospace Engineering, New Mexico State University, Las Cruces, NM 88003, USA

* Correspondence: skuravi@nmsu.edu; Tel.: +1-575-646-1587

Received: 4 August 2017; Accepted: 7 September 2017; Published: 29 September 2017

Featured Application: The results of this research are applicable to metal hydride hydrogen storage systems and the onboard hydrogen storage system in the automobile industry.

Abstract: A safe and efficient method for storing hydrogen is solid state storage through a chemical reaction in metal hydrides. A good amount of research has been conducted on hydrogenation properties of metal hydrides and possible methods to improve them. Background research shows that heat transfer is one of the reaction rate controlling parameters in a metal hydride hydrogen storage system. Considering that some very well-known hydrides like lanthanum nickel ($LaNi_5$) and magnesium hydride (MgH_2) are paramagnetic materials, the effect of an external magnetic field on heat conduction and reaction kinetics in a metal hydride storage system with such materials needs to be studied. In the current paper, hydrogenation properties of lanthanum nickel under magnetism were studied. The properties which were under consideration include reaction kinetics, hydrogen absorption capacity, and hydrogenation time. Experimentation has proven the positive effect of applying magnetic fields on the heat conduction, reaction kinetics, and hydrogenation time of a lanthanum nickel bed. However, magnetism did not increase the hydrogenation capacity of lanthanum nickel, which is evidence to prove that elevated hydrogenation characteristics result from enhanced heat transfer in the bed.

Keywords: hydrogen storage systems; hydrogen absorption; thermochemical energy storage; metal hydride; magnetism; heat transfer enhancement

---

1. Introduction

Metal hydrides are materials which can chemically react with hydrogen to form a compound including hydrogen and at least one metal. A well-known group of metal hydrides are intermetallic compounds which are obtained by combining a stable-hydride-forming element and an unstable-hydride-forming element. They demonstrate good hydrogen storage properties in low pressures, while their volumetric densities are comparable with liquid hydrogen. Research on applications of metal hydrides has been in process for many years and includes investigating simple metal hydrides, such as magnesium hydride, and metal-hydride alloys, such as titanium-iron hydride.

A useful property of metal hydrides is their ability to absorb and desorb hydrogen at a constant pressure [1]. They also offer the advantage of hydrogen storage under moderate temperatures while safely performing as high volume efficient storage [2]. Another advantage of metal hydrides is that they can be used to store hydrogen and heat simultaneously. Their gravimetric and volumetric energy density is usually higher than other means of energy storage like Li-ion batteries while energy is stored at ambient temperature [3]. Metal hydride storage systems (like storage in LaNi5, for instance) usually have higher volumetric efficiency as compared to compressed gas technology, which is due to their near ambient operational pressures. A comparison of different hydrogen storage technologies is presented in Figure 1.

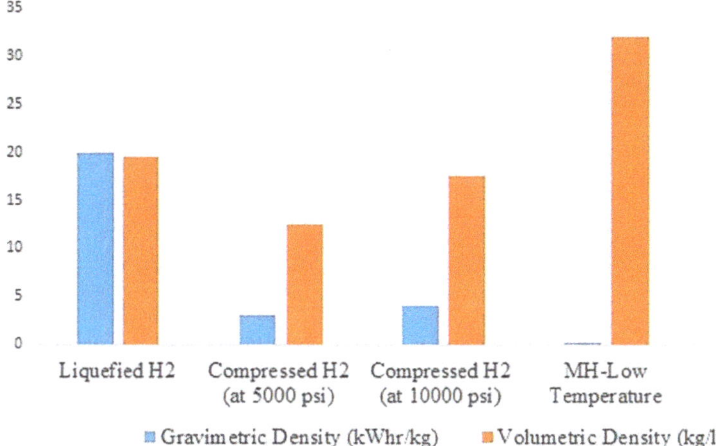

Figure 1. Comparison of kWhr/kg of hydrogen and total system volume for different hydrogen storage technologies.

The performance of a metal hydride system depends on the properties of the selected metal hydride. Some of the desired properties for a metal hydride are: highly specific gravimetric energy density, compact in design, long life and low performance degradation, and, of course, being economical. For a storage system, properties such as high hydrogen absorption capacity, high thermal conductivity, fast reaction kinetics, favorable equilibrium pressure, a simple activation process, and minimum degradation after cyclic operation are desired as well [4–8]. Some of the hydrogen release reaction characteristics are its thermochemistry, kinetics, phase changes, reagent requirements, catalyst morphology and amount, and species and levels of gaseous byproducts relative to hydrogen [9–12].

Metal hydrides can be categorized into different types including AB_5 (LaNi5), AB_2 (ZrV2), AB (FeTi), and A_2B (Mg2Ni) based on their chemical composition. Among all these types, AB_2 and AB_5 are the best known metal hydrides to absorb hydrogen. The AB_2 family of metal hydrides have an excellent hydrogen capacity (up to over 7 wt %) but have unacceptably slow kinetics of hydrogenation and dehydrogenation, even after extensive activation at 673 K (400 °C). The AB_5 group has an excellent hydrogenation performance at ambient temperatures but poor hydrogen capacity typically in the range of 1–1.5 wt % [13]. Metal hydrides useful for on-board hydrogen storage and can be broadly categorized as: (1) complex metal hydrides (e.g., $NaAlH_4$) that have low hydriding pressures (20–150 bar), relatively high hydriding temperatures (200–300 °C), and high volumetric capacities; and (2) high pressure metal hydrides (HPMH) that have high hydriding pressures and hydriding temperatures lower than 100 °C but have a low volumetric capacity (2–3 wt %) [14–16]. Both these types of metal hydrides require additional mechanisms such as external cooling, high pressure cylinders, etc., that increase the overall system weight and hence the reduced system capacity. In addition, carbon-based physisorption materials such as CNTs and MOFs show very high material storage capacities at low temperatures (77 K and 50–100 bar), and their hydrogen absorption capacity greatly decreases with an increase in temperatures [17,18]. Figure 2 shows the shows volumetric and gravimetric capacities of various storage materials and compares them to the best-known system capacities using these materials.

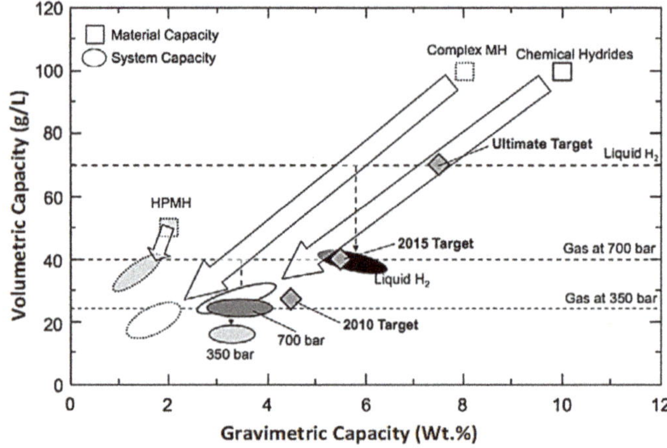

Figure 2. Capacity comparison of different storage techniques. Adapted with permission from [14], ©Elsevier, 2009.

In practical applications, the rate of the hydrogen absorption/desorption reaction is often limited by how quickly heat can be added to, or removed from, the hydride as opposed to the intrinsic kinetics of hydriding and dehydriding [1,19]. While metal hydrides are basically alloys with good thermal conductivity, they break down into micron size particles after a few hydrogenation/dehydrogenation cycles in a hydrogen system. As a result, thermal conductivity is greatly affected and decreases significantly; typically, the thermal conductivity is around 1 W/(m·K) for metal hydrides [20,21]. In addition, when hydrogen is introduced into a hydride bed, its atoms fill the pores in the bed and result in even lower conduction rates. The decreased rate of heat transfer in a bed will eventually result in extended hydriding times [14]. Figure 3 shows the effect of the thermal conductivity of the bed on its fill time. It can be observed that for the typical thermal conductivity of metal hydrides, i.e., 1 W/(m·K), the fill time is significantly higher and can be reduced if the thermal conductivity can be at least doubled.

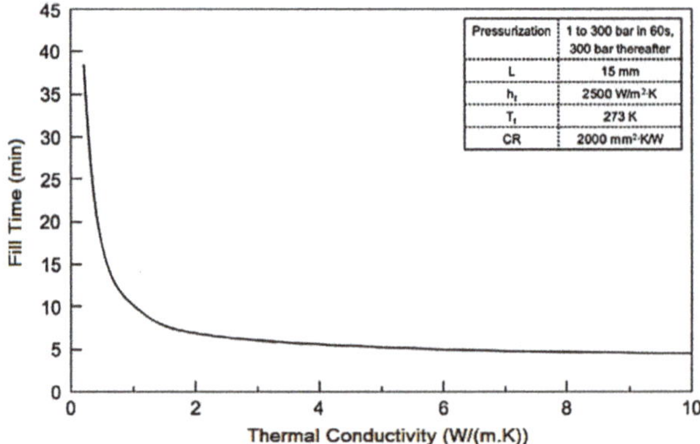

Figure 3. Effect of metal hydride thermal conductivity on hydrogen fill time [14]; fill time can be greatly reduced for thermal conductivity higher than 2 W/(m·K).

It is very important that the heat transfer is efficient in order for the hydriding reaction to proceed at a given pressure, and it can be achieved by using proper techniques for enhancing the thermal conductivity of the storage system. Enhancing heat conductivity is more a technical issue than a theoretical one, and measures have been presented experimentally to address this problem. Most of the methods are developed by adding solid matrices like fins or by compacting metal hydrides [22]. Some of these techniques are: using porous tubes/sheets for instantaneous hydrogen supply along the bed, porous bodies encapsulating the hydride powder, e.g., graphite structures, Porous Metal Hydride (PMH-compacts), metal-coated hydride particles, and the insertion of heat transfer enhancement matrices, e.g., metal screens and bands, metal foams, radial and axial metallic fins, and metallic wire nets [13,23–36]. The problem which arises by implementing any of these techniques is the increased thermal mass of the system, which affects the overall hydrogen storage capacity of the system.

Authors of this work also performed a study on the different reactor and heat exchanger configurations in metal hydride hydrogen storage systems [37] and recently proved experimentally that applying a magnetic field can enhance the effective thermal conductivity in porous beds [38,39]. Owing to the fact that metal hydride storage systems are similar to porous beds, we have tested the effect of the magnetic field on improving heat transfer and reaction kinetics in a metal hydride storage system. Considering the desired hydrogenation properties of AB_5 hydrides, and that $LaNi_5$ is a paradigm of this type of metal hydride, in addition to its magnetic properties, this material is selected for this study. The goal of this work is to test the effect of the magnetic field on improving the heat transfer, and hence the rate of hydrogenation, in a metal hydride hydrogen storage system. Though $LaNi_5$ does not meet the DOE hydrogen storage targets, this material was selected because of its commercial availability and magnetic properties.

Magnetic fields can exert a force on materials with magnetic properties and manipulate them without being in contact with those materials [40]. Manipulating magnetic particles by a magnetic field has a wide variety of applications including magnetohydrodynamic pumping in microchannels, fluid mixing, stabilizing or agitating magnetic particles containing fluid, and supporting bioreactions in microchannels [41].

When a magnetic field is applied on $LaNi_5$, nickel particles form in the compound and diffuse to the surface. Nickel is a strong paramagnetic material at room temperature and becomes ferromagnetic at 4.2 K [42–44]. Although experimental results state $LaNi_5$ as a paramagnetic material, calculations of Al Alam et al. [45] showed that, theoretically, it can exhibit properties of a very weak ferromagnetic material. Albertini et al. [46] and Marcos et al. [47] stated a significant magnetic entropy change in a nickel alloy ($Ni_{2+x}Mn_{1-x}Ga$) because of an increase in its nickel content.

Grechnev et al. [48] calculated and measured magnetic characteristics and the electron structure of three metal hydrides which are known as RNi_5 compounds. Their study included $LaNi_5$, YNi_5, and $CeNi_5$. According to their results, these hydrides do not demonstrate similar magnetic characteristics and each one has its individual magnetic properties. Thus, they cannot be categorized and characterized in the same series of compounds based on their magnetic characteristics.

In the present work, the heat transfer inside a metal hydride bed with $LaNi_5$ as the hydride material is studied. The difference in heat transfer with and without an external magnetic field is tested and the results are presented. The following sections describe the details.

2. Experimental Setup

2.1. Storage Material and Magnets

Lanthanum Nickel ($LaNi_5$) was selected for the experiments since it has fast and reversible sorption, a rather low operating pressure compared to pressurized gas at near ambient temperatures, and a good cycling life [49]. This metal hydride also has a low activation energy, good reaction kinetics, and the capability to store hydrogen at temperatures and pressures near normal conditions (1.7 to 20 bars for temperatures between 293 and 373 K, respectively) [1]. $LaNi_5$ also demonstrates properties of paramagnetic materials and shows segregation of lanthanum in a surface layer and the formation of ferromagnetic particles of nickel when exposed to a magnetic field [44]. Also, its hydrogen absorption capacity is 1.5 wt %. Particles of the powder employed in this work are flake formed with a density of 7.950 g/cm^3 [49].

Magnets utilized in this work are of N42 grade which is made of a composition of neodymium, iron, and boron, and is coated to avoid rusting. The magnets are axially magnetized disc magnets with a diameter of 5 cm. The thickness of magnets 1 and 2 is 1.27 and 2.54 cm, respectively. Nominal and measured magnetic intensities of the magnets are given in Table 1.

Table 1. Magnetic intensities of selected magnets.

Magnet Number	Thickness (cm)	Nominal Intensity (Gauss)
1	1.27	2952
2	2.54	4667

2.2. Metal Hydride Hydrogen Storage System Setup

The experimental device and its schematic is shown in Figure 4. The experimental setup consists of an argon cylinder and a hydrogen cylinder with a vacuum pump and hydrogen storing section. Argon is implemented to provide the required inert atmosphere for $LaNi_5$ when there are no experiments running. Hydrogen cylinder provides Ultra High Purity (UHP hydrogen of 99.999% purity) level for experiments. Both argon and hydrogen cylinders are connected to the rest of setup through a single pipe. Also, a vacuum pump provides a vacuum in connecting lines and a hydrogen storage section. The reactor is made of Stainless Steel containing 171 g of a $LaNi_5$ powder activated by successive absorption/desorption cycles. The particle size is about 60 μm and the porosity is about 0.5.

The whole hydrogen storage unit is submerged in water in a thermostatic bath. The bath heats and cools the hydrogen reservoir and Metal Hydride (MH) reactor during hydrogen desorption and absorption, respectively. During experiments, two different magnets (as shown in Table 1) are used to analyze the effect of change in magnetic field intensity on the heat transfer in the MH bed. In experiments with magnets, the magnet was placed at the bottom of the reactor and the temperature at the center of the reactor and on its wall, respectively, is measured. The reactor is connected to the reservoir using connecting tubes. The thermostatic bath is heated by a 1000 W heater with a build-in circulation pump. Deploying a thermostatic bath ensures consistency of boundary conditions throughout the experiments. It also ensures that adding the thermal mass of a magnet when the magnetic field is required does not impact the consistency of boundary conditions in the setup.

The pressure within the hydride bed is measured by installing a sensor between the hydrogen tank and the hydride bed. To measure the temperature at different locations, five thermocouples and thermal probes are utilized. Two thermal probes are inserted into the reservoir and reactor. Three thermocouples are measuring the temperature at the reactor wall, inside the thermostatic bath, and the room (ambient) temperature. Measured data from transducers and thermocouples is transferred to a computer by LabView through a data acquisition unit (DAQ). Thermocouple positions are presented in Table 2.

Table 2. Thermocouple positions in hydrogen storage setup.

Thermocouple	Position
T1	Metal hydride reactor
T2	Hydrogen reservoir
T3	Thermostatic bath
T4	Room temperature
T5	Reactor wall

For the absorption test, initially, the hydrogen tank is charged at the desired pressure and was kept in contact with the metal hydride reactor. The hydrogen pressure inside the tank decreases until it reaches a stable value. Because the absorption reaction is exothermic, the hydride bed is cooled by using water coming from a thermostatic bath.

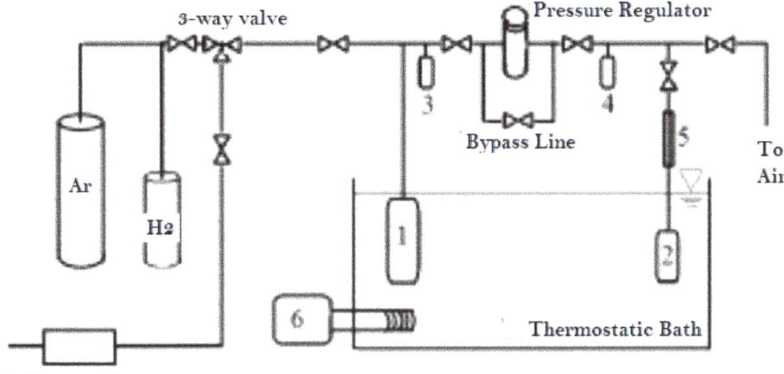

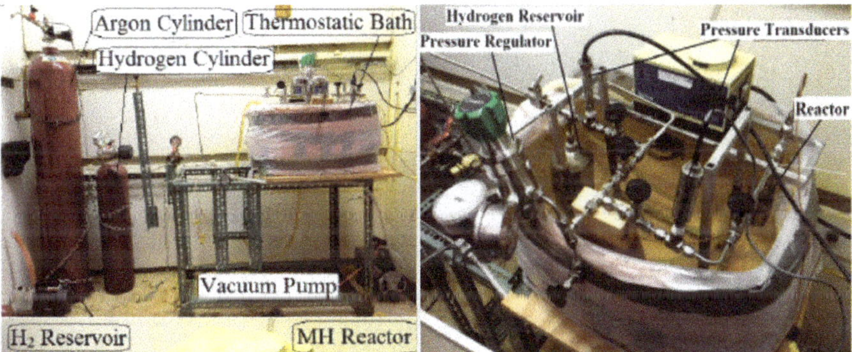

Figure 4. Hydrogen storage setup and its schematic diagram showing the components.

3. Results and Discussion

A series of experiments were performed to find the effect of magnetism on heat conduction in the lanthanum nickel bed. In these experiments, magnet 1 was placed at the bottom of the reactor and temperature was measured at thermocouples 1 and 5, which are at the center of the reactor and on its wall, respectively (Table 2). In these experiments, in order to estimate the improved thermal conductivity of the bed, no hydrogenation was performed. The improvement in thermal conductivity is calculated using the formula derived from Fourier's law of heat conduction under steady state conditions [42]:

$$\frac{k_{with\ magnet}}{k_{without\ magnet}} = \frac{(T_5 - T_1)_{without\ magnet}}{(T_5 - T_1)_{with\ magnet}}$$

where T_1 and T_5 are the temperatures of thermocouples at the center of the reactor and at the wall, respectively, and the temperature values were used once steady state is reached. When magnet 1 is used, the results showed around 8–9% improvement in the thermal conductivity of bed. The error margin of the thermocouples used in this work is 1–2.5% for temperatures below 973 K. Temperatures at the center of the reactor and its wall with and without magnet 1 are depicted in Figure 5. It can be observed that the steady state was reached quickly when the magnet was used, indicating a faster heat transfer rate. It can also be observed from the figure that the maximum temperature is lower inside the metal hydride tank (T_1) and on the reactor wall (T_5) when the magnetic field is applied. This shows that using the magnetic field can enhance the heat transfer in metal hydride beds.

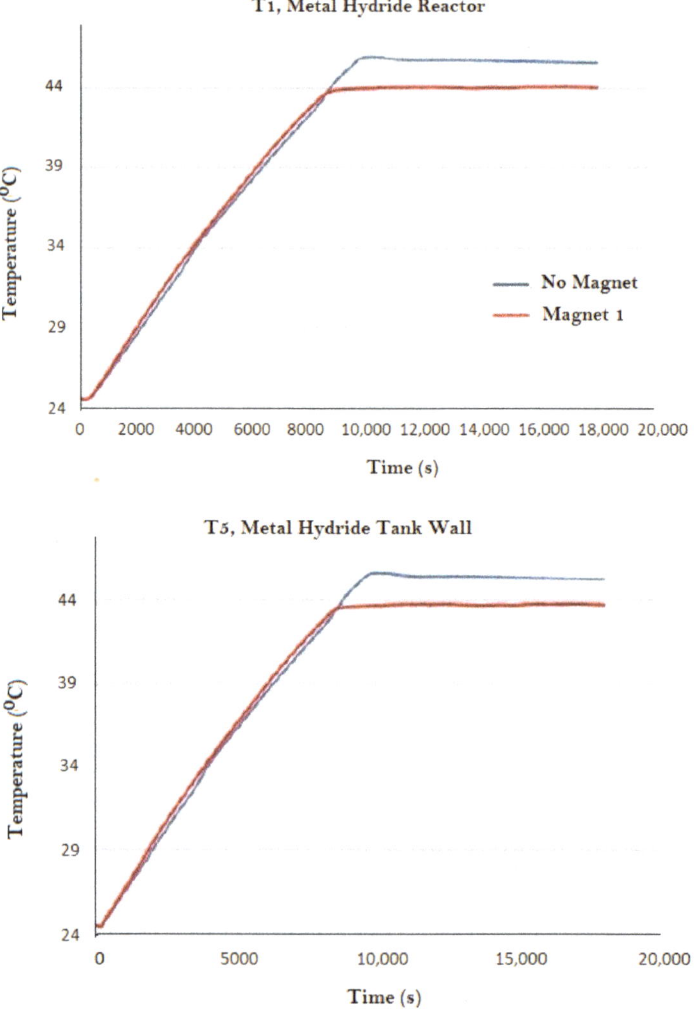

Figure 5. Temperature measurements of the reactor (Top) and wall (Bottom) with and without magnet in LaNi$_5$ bed during heat conduction experiments. Steady state is reached faster when magnet is present.

Considering that experimental results supported heat transfer enhancement in the LaNi$_5$ bed under magnetism, a set of experiments were run to check the hydrogen absorption properties of lanthanum nickel under magnetism. In these experiments, magnets 1 and 2 are used to provide magnetic fields inside the reactor. Initially, experiments were done to check the amount of hydrogen absorption in the bed without any magnetic fields, and then two series of experiments with magnets 1 and 2 were performed. Temperature distributions of two sample experiments during the hydriding process inside the reactor and on its wall are illustrated in Figure 6.

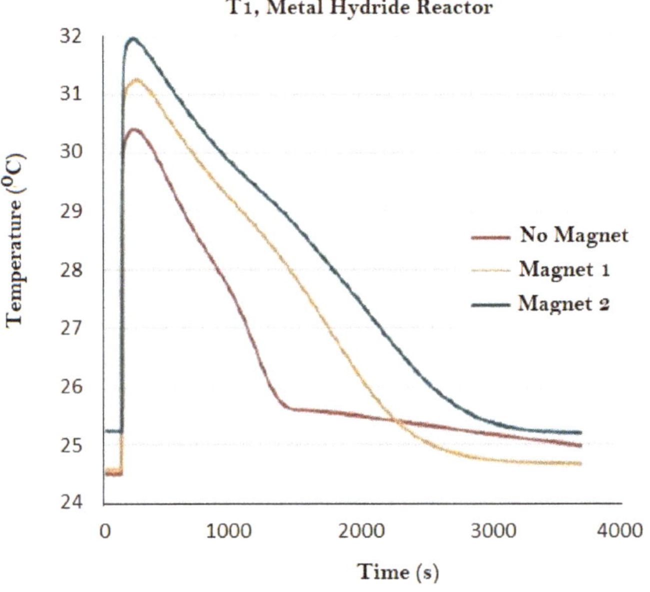

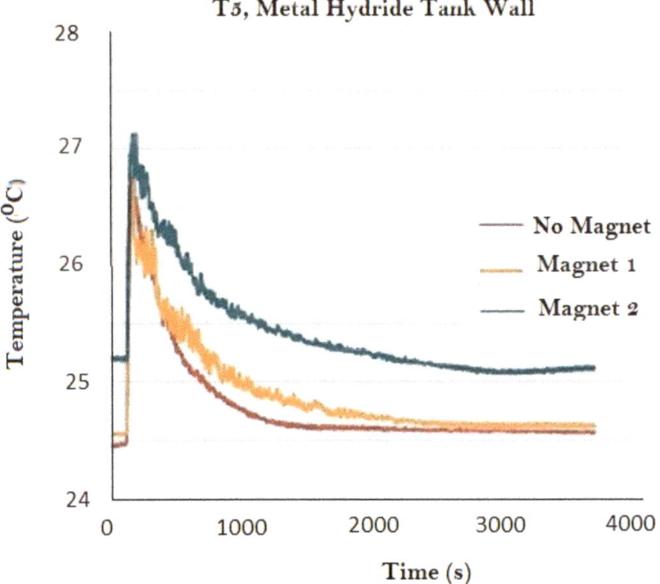

Figure 6. Temperature measurement inside reactor (**Top**) and on its wall (**Bottom**) during hydrogenation.

Few conclusions can be drawn from Figure 6. From the temperature measurement inside the metal hydride reactor, it is seen that the maximum temperature is recorded as soon as hydrogenation initiated. Also, the temperature peak increases with increased magnetic flux. Considering that the heat source in these experiments is the released heat through exothermic hydrogen absorption reaction, it is concluded that the higher magnetic intensity enhances the rate at which the reaction takes place.

In other words, the reaction kinetics are enhanced under the effect of the magnetic fields. Though the difference in temperatures under three cases is not significant before the reaction initiates, the difference in temperatures is more prominent during and after the reaction. This is due to the high energy released during the reaction and the dissipation of heat at different thermal conductivities in all the three cases.

Measured pressures in the reactor and reservoir are presented in Figure 7. From the final pressures in the reactor as shown in Figure 7, at the end of data recording time the equilibrium pressure of reaction experiences a slight increase. Also, the rate of decrease in pressure of the reservoir shows that the time required to absorb an identical amount of hydrogen in each case is decreased by implementing magnetic fields. Thus, applying magnetic fields enhances hydrogen absorption time in a LaNi$_5$ bed. It is postulated that the variation in equilibrium pressure and temperature is because of the decreased permeability and increased heat transfer area in the system, due to the compaction of particles when magnets are present.

The mass of absorbed hydrogen under each magnet is presented in Table 3. The mass is calculated by measuring the volume of absorbed hydrogen under normal conditions (T = 293 K & P = 101 kPa) and using the density of hydrogen under normal conditions (ρ = 0.089 kg/m^3). The driving potential for the hydriding reaction is the difference between the metal hydride tank pressure and equilibrium pressure; the larger this difference is, the faster is the reaction rate. The rate can be increased by either increasing the pressure or reducing the metal hydride temperature. If sufficient cooling is provided during a small pressure increment during the pressure ramp, the metal hydride temperature will be maintained below the equilibrium temperature, and the reaction continues as the pressure increases by a second increment. However, if the cooling is insufficient, the metal hydride temperature reaches the equilibrium temperature and the reaction stalls. From the calculated absorbed hydrogen for each experiment, it is evident that magnetism increases the amount of hydrogen which can be absorbed in lanthanum nickel. The hydrogen absorption amount (practical absorption capacity) should not be confused with theoretical absorption capacity, which is 1.5% for LaNi$_5$. The information presented in Table 3 is an indication of practical absorption capacity which can change by the change in the equilibrium condition of the chemical hydrogen absorption/desorption reactions. A change in the amount of absorbed hydrogen does not mean that there is a change to the theoretical hydrogen absorption capacity of the substance.

Table 3. Mass of absorbed hydrogen under different magnetic conditions.

Boundary Condition	Δm (gr)
No magnet	0.619
Magnet 1	0.818
Magnet 2	0.909

Thus far, the positive effect of magnetism on thermal conduction and hydrogen absorption capacity in LaNi$_5$ was tested. A series of experiments were performed to identify whether improved hydrogenation properties are because of the chemical reaction or because of the enhanced heat transfer rates. In these experiments, a hydriding process was initiated without magnetism on the setup. When hydriding reaches equilibrium, a magnet (magnet 2) is placed at the bottom of the reactor which causes the sudden temperature fluctuation shown in Figure 8. The temperature slope changes by adding the magnet (as presented in Figure 8), which supports the previous conclusion of changed equilibrium temperature and pressure in a magnetic field. However, as there is no significant change in the temperature pattern, no extra hydrogenation occurs by adding a magnet. Meanwhile, as illustrated in Figure 9, pressure does not have any changed slope when the magnetic field is applied. Consequently, no more hydrogen absorption occurs at this stage. From this experiment, it is concluded that the chemical reaction does not change by magnetism and the enhanced hydrogenation properties of the hydride are because of the enhanced heat transfer rate.

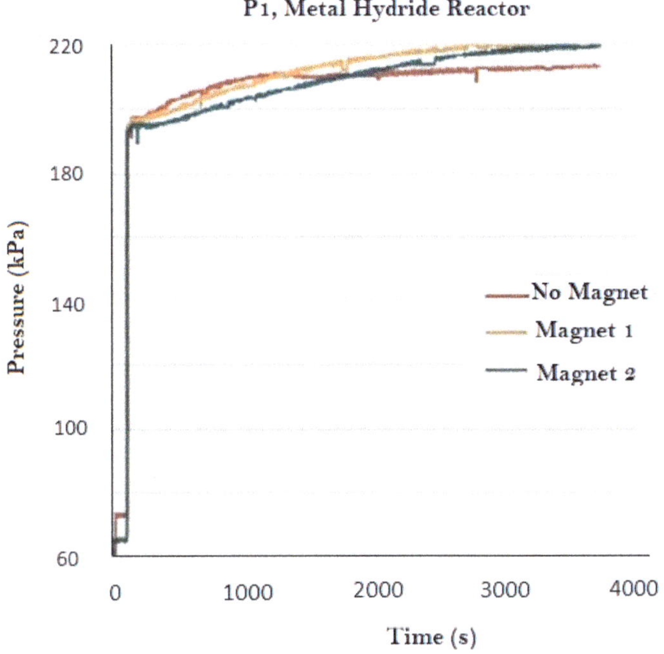

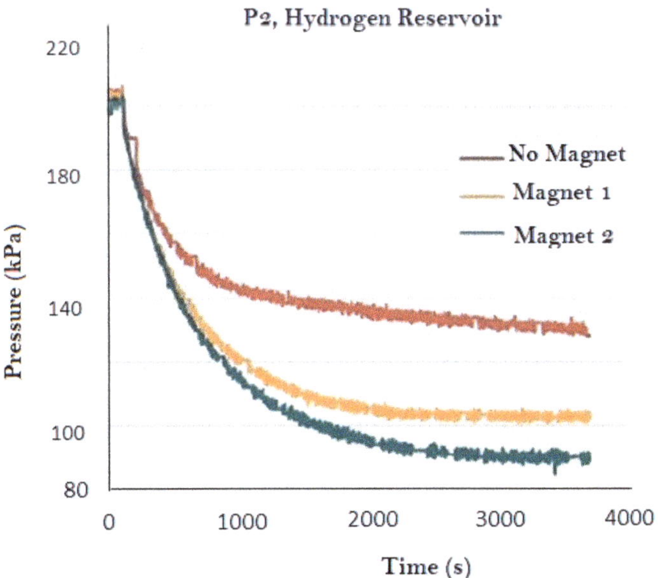

Figure 7. Pressure measurement inside reactor (**Top**) and reservoir (**Bottom**) during hydrogenation under different magnetic conditions.

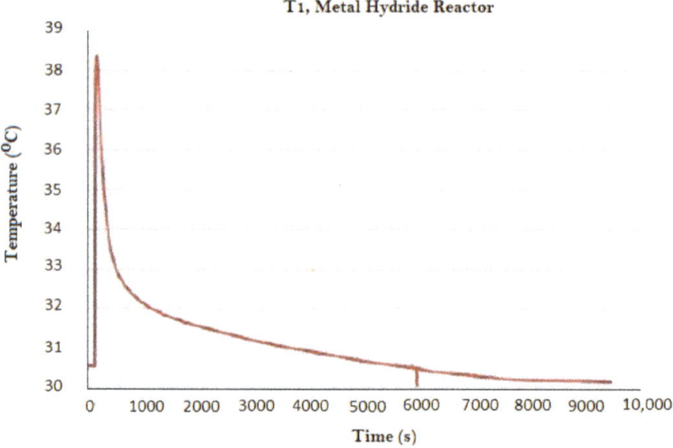

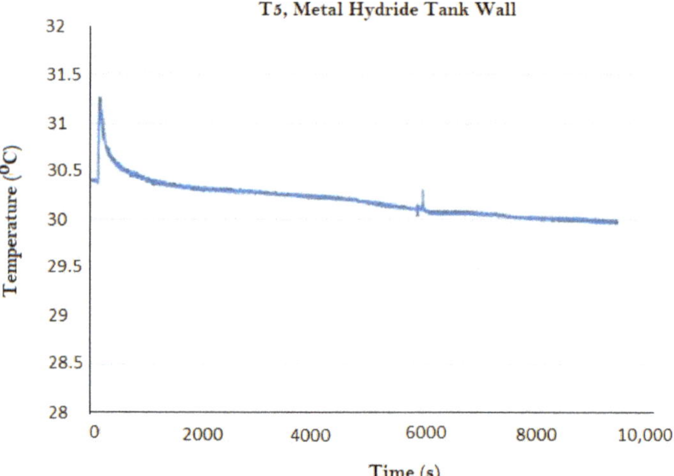

Figure 8. Temperature measurement inside reactor (**Top**) and the tank wall (**Bottom**) during hydrogenation.

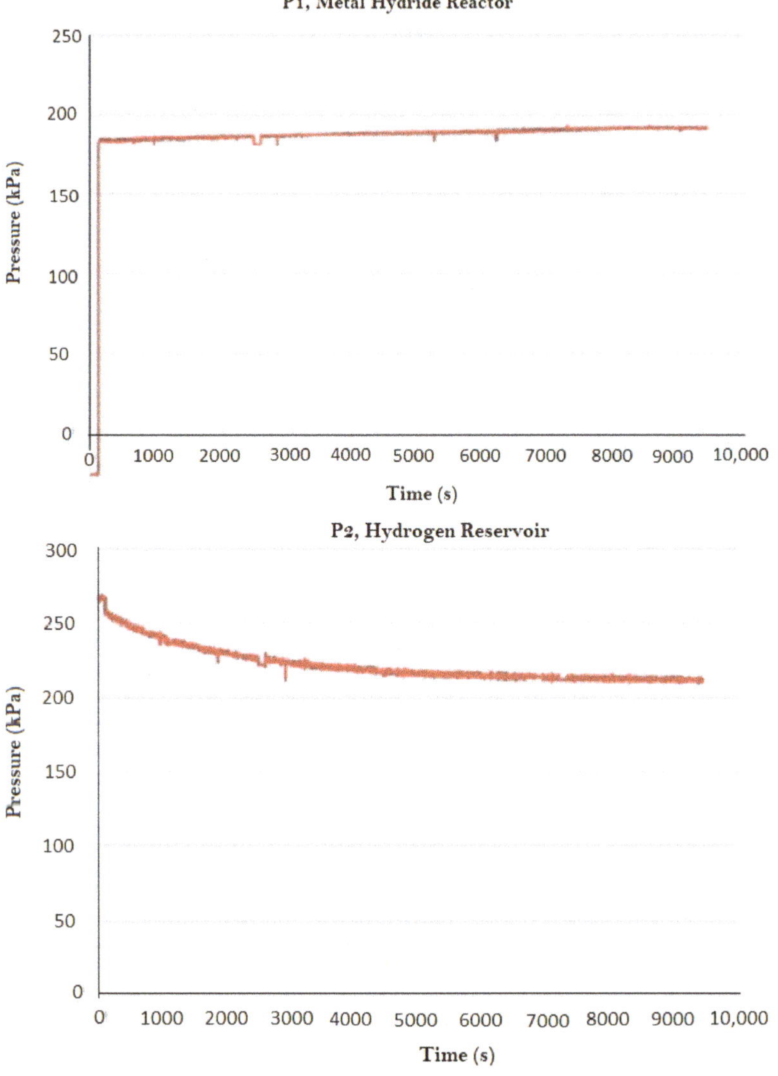

Figure 9. Pressure measurement inside reactor (**Top**) and reservoir (**Bottom**) during hydrogenation.

4. Conclusions

This paper studied a new heat transfer enhancement method of using an external magnetic field to improve heat conduction inside a metal hydride bed. Experiments have proven that implementing an external magnetic field improves effective thermal conductivity, reaction kinetics, and hydrogenation in a lanthanum nickel bed. Improved hydrogenation properties are requirements of hydrogen storage in metal hydrides. The selected metal hydride ($LaNi_5$) demonstrates magnetic properties which were utilized to apply the concept of increased heat transfer under magnetism, to improve hydrogenation properties of the metal hydride. However, magnetism did not increase the hydrogenation capacity of lanthanum nickel, which proves that elevated hydrogenation characteristics result from enhanced heat transfer in the bed. Considering that magnetic fields increase the heat transfer in both ferromagnetic and paramagnetic materials and many metal hydrides are paramagnetic, the same concept can be applied to any paramagnetic hydrogen absorbing medium.

Acknowledgments: Authors would like to thank the National Center for Hydrogen Research. The center supported this work by providing the funding and required facilities and equipment. The authors would also like to thank Johnathan Whitlow for his help in designing the experiments.

Author Contributions: Sarada Kuravi conceived the idea and designed the experiments. Shahin Shafiee performed the experiments and wrote the paper. Mary-Helen McCay contributed to materials and lab facilities. Mary-Helen McCay and Sarada Kuravi analyzed the data and proof-read the manuscript.

Conflicts of Interest: The authors declare no conflict of interest.

References

1. Harries, D.N.; Paskevivius, M.; Sheppard, D.A.; Price, T.E.C.; Buckley, C.E. Concentrating Solar Thermal Heat Storage Using Metal Hydrides. *Proc. IEEE* **2012**, *100*, 539–549. [CrossRef]
2. Sakintuna, B.; Lamari-Darkrim, F.; Hirscher, M. Metal hydride materials for solid hydrogen storage: A review. *Int. J. Hydrogen Energy* **2007**, *32*, 1121–1140. [CrossRef]
3. Durbin, D.J.; Malardier-Jugroot, C. Review of Hydrogen Storage Techniques for on Board Vehicle Applications. *Int. J. Hydrogen Energy* **2013**, *38*, 14595–14617. [CrossRef]
4. Gardiner, M.R.; Burke, A. Comparison of hydrogen storage technologies: A focus on energy required for hydrogen input. *ACS Div. Fuel Chem.* **2002**, *47*, 794–795. (Preprints).
5. Muthukumar, P.; Groll, M. Metal Hydride Based Heating and Cooling Systems: A Review. *Int. J. Hydrogen Energy* **2010**, *35*, 3817–3831. [CrossRef]
6. Werner, R.; Groll, M. Two-stage metal hydride heat transformer laboratory model: Results of reaction bed tests. *J. Less Common Met.* **1991**, *172*, 1122–1129. [CrossRef]
7. Groll, M. Reaction beds for dry sorption machines. *Heat Recovery Syst. CHP* **1993**, *13*, 341–346. [CrossRef]
8. Supper, W.; Groll, M.; Mayer, U. Reaction kinetics in metal hydride reaction beds with improved heat and mass transfer. *J. Less Common Met.* **1984**, *104*, 279–286. [CrossRef]
9. Aardahl, C.L.; Rassat, S.D. Overview of Systems Considerations for On-Board Chemical Hydrogen Storage. *Int. J. Hydrogen Energy* **2009**, *34*, 6676–6683. [CrossRef]
10. Liu, Z.Q.; Marder, T.B. B–N versus C–C: How similar are they? *Angew. Chem. Int. Ed.* **2008**, *47*, 242–244. [CrossRef] [PubMed]
11. Marder, T.B. Will we soon be fueling our automobiles with ammonia–borane? *Angew. Chem. Int. Ed.* **2007**, *46*, 8116–8118. [CrossRef] [PubMed]
12. Kong, V.C.Y.; Kirk, D.W.; Foulkes, F.R.; Hinatsu, J.T. Development of hydrogen storage for fuel cell generators. II: Utilization of calcium hydride and lithium hydride. *Int. J. Hydrogen Energy* **2003**, *28*, 205–214. [CrossRef]
13. Kim, K.J.; Montoya, B.; Razani, A.; Lee, K.H. Metal Hydride Compacts of Improved Thermal Conductivity. *Int. J. Hydrogen Energy* **2001**, *26*, 609–613. [CrossRef]
14. Visaria, M.; Mudawar, I.; Pourpoint, T.; Kumar, S. Study of heat transfer and kinetics parameters influencing the design of heat exchangers for hydrogen storage in high-pressure metal hydrides. *Int. J. Heat Mass Transf.* **2010**, *53*, 2229–2239. [CrossRef]
15. Walker, G. *Solid-State Hydrogen Storage: Materials and Chemistry*; Woodhead Publishing: Cambridge, UK, 2008.
16. Mori, D.; Hirose, K. Recent challenges of hydrogen storage technologies for fuel cell vehicles. *Int. J. Hydrogen Energy* **2009**, *34*, 4569–4575. [CrossRef]
17. Schlapbach, L.; Zuttel, A. Hydrogen-Storage Materials for Mobile Applications. *Nature* **2001**, *414*, 353–358. [CrossRef] [PubMed]
18. Poirier, E.; Dailly, A. Thermodynamic study of the adsorbed hydrogen phase in Cu-based metal-organic frameworks at cryogenic temperatures. *J. Phys. Chem. C* **2008**, *112*, 13047–13052. [CrossRef]
19. Mellouli, S.; Askri, F.; Dhaou, H.; Jemni, A.; Nasrallah, S.B. Parametric Studies on a Metal-Hydride Cooling System. *Int. J. Hydrogen Energy* **2009**, *34*, 3945–3952. [CrossRef]
20. Kim, K.J.; Feldman, K.T.; Lloyd, G.; Razani, A.; Shanahan, K.L. Performance of High Power Metal Hydride Reactors. *Int. J. Hydrogen Energy* **1998**, *23*, 355–362. [CrossRef]
21. Suda, S.; Komazaki, Y.; Kobayashi, N. Effective thermal conductivity of metal hydride beds. *J. Less Common Met.* **1983**, *89*, 317–324. [CrossRef]

22. Yang, F.S.; Wang, G.X.; Zhang, Z.X.; Meng, X.Y.; Rudolph, V. Design of the Metal Hydride Reactors—A Review on the Key Technical Issues. *Int. J. Hydrogen Energy* **2010**, *35*, 3832–3840. [CrossRef]

23. Mellouli, S.; Askri, F.; Dhaou, H.; Jemni, A.; Nasrallah, S.B. Numerical Study of Heat Exchanger Effects on Charge/Discharge Times of Metal-Hydrogen Storage Vessel. *Int. J. Hydrogen Energy* **2009**, *34*, 3005–3017. [CrossRef]

24. Kaplan, Y.; Veziroglu, T.N. Mathematical Modelling of Hydrogen Storage in a LaNi$_5$ Hydride Bed. *Int. J. Energy Res.* **2003**, *27*, 1027–1038. [CrossRef]

25. Cipiti, F.F.; Cacciola, G. Finite Element-Based Simulation of a Metal Hydride-Based Hydrogen Storage Tank. *Int. J. Hydrogen Energy* **2009**, *34*, 8574–8582.

26. Ghafir, M.F.A.; Batcha, M.F.M.; Raghavan, V.R. Prediction of the Thermal Conductivity of Metal Hydrides—The Inverse Problem. *Int. J. Hydrogen Energy* **2009**, *34*, 7125–7130. [CrossRef]

27. Bershadsky, E.; cJosephy, Y.; Ron, M. Permeability and Thermal Conductivity of Porous Metallic Matrix Hydride Compacts. *J. Less Common Met.* **1989**, *153*, 65–78. [CrossRef]

28. Ron, M.; Bershadsky, E.; Josephy, Y. The Thermal Conductivity of Porous Metal Matrix Hydride Compacts. *J. Less Common Met.* **1991**, *172–174*, 1138–1146. [CrossRef]

29. Sanchez, R.; Klein, H.P.; Groll, M. Expanded Graphite as Heat Transfer Matrix in Metal Hydride Beds. *Int. J. Hydrogen Energy* **2003**, *28*, 515–527.

30. Klein, H.P.; Groll, M. Heat Transfer Characteristics of Expanded Graphite Matrices in Metal Hydride Beds. *Int. J. Hydrogen Energy* **2004**, *29*, 1503–1511. [CrossRef]

31. Lee, M.; Kim, K.J.; Hopkins, R.R.; Gawlik, K. Thermal Conductivity Measurements of Copper-Coated Metal Hydrides (LaNi$_5$, Ca$_{0.6}$Mm$_{0.4}$Ni$_5$, and LaNi$_{4.75}$Al$_{0.25}$) for Use in Metal Hydride Hydrogen Compression Systems. *Int. J. Hydrogen Energy* **2009**, *34*, 3185–3190. [CrossRef]

32. Asano, K.; Yamazaki, Y.; Iijima, Y. Hydrogenation and Dehydrogenation Behavior of LaNi$_{5-x}$Co$_x$ (x = 0, 0.25, 2) Alloys Studied by Pressure Differential Scanning Calorimetry. *Mater. Trans.* **2002**, *43*, 1095–1099. [CrossRef]

33. Dhaou, H.; Askri, F.; Salah, M.B.; Jemni, A.; Nasrallah, S.B.; Lamloumi, J. Measurement and Modelling of Kinetics of Hydrogen Sorption by LaNi$_5$ and Two Related Pseudobinary Compounds. *Int. J. Hydrogen Energy* **2007**, *32*, 576–587. [CrossRef]

34. Pukazhselvan, D.; Kumar, V.; Singh, S.K. High Capacity Hydrogen Storage: Basic Aspects, New Developments and Milestones. *Nano Energy* **2012**, *1*, 566–589. [CrossRef]

35. Mosher, D.A.; Tang, X.; Brown, R.J.; Arsenault, S.; Saitta, S.; Laube, B.L.; Dold, R.H.; Anton, D.L. *High Density Hydrogen Storage System Demonstration Using NaAlH$_4$ Based Complex Compound Hydrides*; United Technologies Research Center: East Hartford, CT, USA, 2007.

36. Bahadori, R.; Gutierrez, H.; Manikonda, S.; Meinke, R. Monte Carlo method simulation for two-dimensional heat transfer in homogenous medium and proposed application to quench propagation simulation. *IEEE Trans. Appl. Supercond.* **2017**, *27*, 1–5. [CrossRef]

37. Shafiee, S.; McCay, M.H. Different Reactor and Heat Exchanger Configurations for Metal Hydride Hydrogen Storage Systems—A Review. *Int. J. Hydrogen Energy* **2016**, *41*, 9462–9470. [CrossRef]

38. Shafiee, S. A Method to Enhance the Heat Transfer Mechanism in Metal Particle Porous Beds with Applicability to Paramagnetic Materials for Hydrogen Storage. Ph.D. Thesis, Florida Institute of Technology, Melbourne, FL, USA, 2016.

39. Shafiee, S.; McCay, M.H.; Kuravi, S. Effect of Magnetic Fields on Thermal Conductivity in a Ferromagnetic Packed Bed. *Exp. Therm. Fluid Sci.* **2017**, *86*, 160–167. [CrossRef]

40. Zhang, Z.T.; Guo, Q.T.; Yu, F.Y.; Li, J.; Zhang, J.; Li, T.J. Motion Behavior of Non-Metallic Particles under High Frequency Magnetic Field. *Trans. Nonferr. Met. Soc. China* **2009**, *19*, 674–680. [CrossRef]

41. Mei, M.R.; Klausner, J.F.; Rahmatian, N. Interaction Forces Between Soft Magnetic Particles in Uniform and Non-Uniform Magnetic Fields. *Acta Mech. Sin.* **2010**, *26*, 921–929.

42. Stucki, F.; Schlapbach, L. Magnetic Properties of LaNi$_5$, FeTi, Mg$_2$Ni and Their Hydrides. *J. Less Common Met.* **1980**, *74*, 143–151. [CrossRef]

43. Kim, G.H.; Chun, C.H.; Lee, S.G.; Lee, J.Y. A Studey on the Microstructural Change of Surface of the Intermetallic Compound LaNi$_5$ by Hydrogen Absorption. *Scr. Metall. Mater.* **1993**, *29*, 485–490. [CrossRef]

44. Liu, N.; Ma, Q.M.; Xie, Z.; Liu, Y.; Li, Y.C. Structures, Stabilities and Magnetic Moments of Small Lanthanum-Nickel Clusters. *Chem. Phys. Lett.* **2007**, *436*, 184–188. [CrossRef]

45. Alam, F.A.; Matar, S.F.; Nakhl, M.; Ouaini, N. Investigation of Changes in Crystal and Electronic Structures by Hydrogen within LaNi5 from First-Principles. *Solid State Sci.* **2009**, *11*, 1098–1106. [CrossRef]
46. Albertini, F.; Canepa, F.; Cirafici, S.; Franceschi, E.A.; Napoletano, M.; Paoluzi, A.; Pareti, L.; Solzi, M. Composition Dependence of Magnetic and Magnetothermal Properties of Ni-Mn-Ga Shape Memory Alloys. *J. Magn. Magn. Mater.* **2004**, *272–276*, 2111–2112. [CrossRef]
47. Marcos, J.; Manosa, L.; Planes, A.; Casanova, F.; Batlle, X.; Labarta, A.; Martinez, B. Magnetic Field Induced Entropy Change and Magnetoelasticity in Ni-Mn-Ga Alloys. *J. Magn. Magn. Mater.* **2004**, *272–276*, 224413. [CrossRef]
48. Grechnev, G.E.; Logosha, A.V.; Panfilov, A.S.; Kuchin, A.G.; Vasijev, A.N. Effect of Pressure on the Magnetic Properties of YNi5, LaNi5, and CeNi5. *Low Temp. Phys.* **2011**, *37*, 138. [CrossRef]
49. *LaNi5*; SDS No. 685933; Sigma-Aldrich: Saint Louis, MO, USA, 16 November 2017.

Article

Investigation of Catalytic Effects and Compositional Variations in Desorption Characteristics of LiNH$_2$-nanoMgH$_2$

Sesha S. Srinivasan [1,*], **Dervis Emre Demirocak** [2], **Yogi Goswami** [3] **and Elias Stefanakos** [3]

[1] Department of Physics, Florida Polytechnic University, 4700 Research Way, Lakeland, FL 33805, USA
[2] Department of Mechanical & Industrial Engineering, Texas A&M University-Kingsville, Kingsville, TX 78363, USA; dervis.demirocak@tamuk.edu
[3] Clean Energy Research Center, College of Engineering, University of South Florida, Tampa, FL 33620, USA; goswami@usf.edu (Y.G.); estefana@usf.edu (E.S.)
* Correspondence: ssrinivasan@flpoly.org; Tel.: +1-813-451-1876

Received: 9 June 2017; Accepted: 4 July 2017; Published: 7 July 2017

Abstract: LiNH$_2$ and a pre-processed nanoMgH$_2$ with 1:1 and 2:1 molar ratios were mechano-chemically milled in a high-energy planetary ball mill under inert atmosphere, and at room temperature and atmospheric pressure. Based on the thermogravimetric analysis (TGA) experiments, 2LiNH$_2$-nanoMgH$_2$ demonstrated superior desorption characteristics when compared to the LiNH$_2$-nanoMgH$_2$. The TGA studies also revealed that doping 2LiNH$_2$-nanoMgH$_2$ base material with 2 wt. % nanoNi catalyst enhances the sorption kinetics at lower temperatures. Additional investigation of different catalysts showed improved reaction kinetics (weight percentage of H$_2$ released per minute) of the order TiF$_3$ > nanoNi > nanoTi > nanoCo > nanoFe > multiwall carbon nanotube (MWCNT), and reduction in the on-set decomposition temperatures of the order nanoCo > TiF$_3$ > nanoTi > nanoFe > nanoNi > MWCNT for the base material 2LiNH$_2$-nanoMgH$_2$. Pristine and catalyst-doped 2LiNH$_2$-nanoMgH$_2$ samples were further probed by X-ray diffraction, Fourier transform infrared spectroscopy, transmission and scanning electron microscopies, thermal programmed desorption and pressure-composition-temperature measurements to better understand the improved performance of the catalyst-doped samples, and the results are discussed.

Keywords: hydrogen storage; complex hydrides; nanocatalyst; LiNH$_2$; MgH$_2$; ball milling

---

1. Introduction

The depletion of fossil fuels, especially oil in the near future, rising environmental concerns due to global warming, and the necessity of a secure energy supply have created a worldwide interest in the renewable energy technologies during the last decade. Among many forms of alternative energy options, hydrogen has attracted much attention as an energy carrier due to its potential for the replacement of oil in stationary and mobile applications. However, viable hydrogen storage technology remains the biggest challenge in the utilization of the hydrogen despite intensive research efforts throughout the world. As of now, there is no single material that is capable of attaining the desired set of targets designated by the US Department of Energy (DOE) and FreedomCAR industrial partners. Solid-state hydrogen storage can be broadly classified into two groups considering the mechanisms involved, namely, physisorption, as in carbon-nanotubes (CNT)/metal organic frameworks (MOFs), and chemisorption, as in metal/complex hydrides. The complex metal hydrides, which have been extensively studied recently, have high volumetric and gravimetric densities, but suffer from high desorption temperatures, reversibility, and sluggish kinetics [1]. Therefore, improving the desorption kinetics, reversibility and desorption temperatures of the complex metal hydrides remains a challenge.

Among many complex hydrides investigated, some of the amides (i.e., $LiNH_2/Mg(NH_2)_2$) had shown favorable storage capacity and reversibility.

The studies on the interaction of lithium with hydrogen and nitrogen as early as 1910 led to the discovery of the $LiNH_2$-LiH system [2,3]. However, a detailed investigation of this system as a potential hydrogen storage material was not carried out until 2002 when Chen et al. [4] first reported the promising results for the lithium nitride (Li_3N) system, in which reaction consists of two steps and given as:

$$Li_3N + 2H_2 \leftrightarrow Li_2NH + LiH + H_2 \leftrightarrow LiNH_2 + LiH \tag{1}$$

The theoretical capacity of Li_3N system is 10.4 wt. %; however, only the second step of the reaction path given in (1) is practical for reversible hydrogen storage since the first reaction step has a very low equilibrium pressure (~0.07 bar) [4]. Further investigation of the second step of the reaction (1), which has a theoretical capacity of 6.5 wt. % and favorable thermodynamics, revealed the role of the lithium hydride (LiH). The elementary steps are given by the reaction steps (2) and (3) as follows [5].

$$2LiNH_2 \rightarrow Li_2NH + NH_3 \tag{2}$$

$$NH_3 + LiH \rightarrow LiNH_2 + H_2 \tag{3}$$

The reaction (3) was found to be ultrafast (~25 ms) and responsible for the capture of NH_3 which is detrimental for the fuel cells [5]. After these prolific works, the research on hydrogen storage in amides focused on compositional changes, replacement of Li with other light or more electronegative alkali/earth alkaline metals, mechanical activation, effects of catalysts, reaction mechanism, reversibility issues, and the mitigation of NH_3 emission.

The high desorption temperature (~285 °C at 1 bar) of the $LiNH_2$ + LiH system, which is not feasible for mobile applications, led to studies on the destabilization of the system. One of the ways involves replacing Li with more electronegative metals such as Mg, Na, and Ca. Nakomori et al., partially replaced Li in $LiNH_2$ with 90 at. % Li and 10 at. % Mg, and obtained a 50 K reduction in the desorption temperature [6]. Later, Luo developed a new hydrogen storage material by completely replacing LiH with MgH_2 in the reaction (1). The resulting compound had a 4.5 wt. % gravimetric capacity with a plateau pressure of 30 bar at 200 °C [7]. Further studies on the $LiNH_2$-MgH_2 system concentrated on the reaction mechanism, kinetics, structural characterization, and thermodynamics of the system [8–17]. It was shown that the $2LiNH_2$-MgH_2 system transforms into the $Mg(NH_2)_2$-2LiH system after the first desorption/absorption cycle, and the proposed reaction mechanism is [8,9]:

$$MgH_2 + 2LiNH_2 \rightarrow Li_2Mg(NH)_2 + 2H_2 \leftrightarrow [Mg(NH_2)_2] + 2LiH \tag{4}$$

The different compositions of the Li-Mg-N-H system were also investigated. Leng et al. obtained ~7 wt. % capacity with a $3Mg(NH_2)_2$-8LiH compound, and Nakomori et al. investigated a $3Mg(NH_2)_2$-12LiH compound with a capacity of 9.1 wt. % [18–20]. Despite the higher gravimetric capacities of these systems, desorption temperatures were relatively higher than the $Mg(NH_2)_2$-2LiH compound. Xiong et al. examined the LiH-$Mg(NH_2)_2$ with molar ratios of 1:1, 2:1 and 3:1, and showed that the lower the Li content the higher the NH_3 emission, and the higher the Li content the higher the desorption temperature [21]. A $LiNH_2$-MgH_2 (1:1) compound was also investigated by several researchers, and the results showed considerable NH_3 emission, and the revealed reaction mechanism was quite different than the $LiNH_2$-MgH_2 (2:1) compound [13,22,23].

Besides the Li-N-H and Li-Mg-N-H systems, other (i.e., Li-Ca-N-H) metal amide-metal hydride systems were also investigated. Some researchers focused on the decomposition of metal amides alone to discover new working pairs, and to address the NH_3 emission problem [18–26], whereas some other authors examined the Li-Ca-N-H system, but the results were not superior compared to Li-N-H and Li-Mg-N-H systems [27,28].

Apart from the type of the complex hydride system, the preparation procedure also plays a significant role in the hydrogen storage characteristics of the complex hydrides. The favorable effects of mechanical activation (MA) via ball milling, such as reducing the onset temperature of desorption, enhancing the reaction kinetics, and lowering the activation energy, have been known for a while, and these favorable effects are associated with the creation of nanocrystallites, smaller particle sizes, and increased surface area [29–31]. Regarding the amide systems, MA is especially important in preventing the release of NH_3 by enhancing the homogeneous mixing of the constituents. Since the conversion of amide to imide is an ammonia-mediated process as explained in Equations (2) and (3), a metal hydride compound can effectively capture NH_3 only if the M-amide–M-imide interface could be created at the nanoscale. It was shown that increasing the milling time decreases the grain size monotonically and increases the surface area up to around 5 h milling, as well as enhances the desorption kinetics considerably [30,32–34]. Xie et al. further confirmed that reducing the particle size enhances the reaction kinetics, and reduces the NH_3 emission [35]. Osborn et al., investigated the low temperature milling, and the results showed that desorption kinetics is faster for the sample milled at $-196\,^\circ C$ compared to the samples milled at $-40\,^\circ C$ and $20\,^\circ C$ [36].

Ammonia emission, even in trace levels, is undesirable in the amide systems because it poisons the fuel cells and causes loss of hydrogen, which in return results in the loss of gravimetric capacity [37]. The effect of NH_3 emission on cyclic behavior and capacity loss is further elaborated, and the results showed that NH_3 emission can be mitigated by ball milling. Ammonia emission is more pronounced for the Li-Mg-N-H system compared to the Li-N-H system since the reaction rate of MgH_2 with NH_3 is slower than the reaction rate between LiH and NH_3 [38–41].

Another important strategy in improving the performance of the complex hydrides is the utilization of catalysts. The seminal study of Bogdanovic on Ti-doped $NaAlH_4$ paved the way for further investigations of the favorable effects of various catalysts in complex hydrides [42]. The favorable effects of catalysts in enhancing the reactions kinetics, lowering the desorption temperature, and alleviating NH_3 emission in the amide systems have been studied by many researchers [43–49]. Ickikawa et al. was the first to investigate the effects of $TiCl_3$, Ni, Co and Fe on NH_3 emission, reaction kinetics, and the desorption temperature of the Li-N-H system, and the results showed that $TiCl_3$ is superior compared to Ni, Co and Fe [39]. Isobe et al. focused on Ti^{nano}, Ti^{micro}, $TiCl_3$, TiO_2^{nano}, and TiO_2^{micro} in the Li-N-H system, and proved the importance of the particle size of the catalysts. The nanocatalysts were superior compared to micro catalysts [44]. Yao et al. investigated Mn, V, MnO_2, and V_2O_5 in the Li-N-H system, and showed that Mn, V, MnO_2, and V_2O_5 has no effect in hydrogen desorption, but enhances the ammonia emission [45]. The catalytic studies on the Li-Mg-N-H system revealed that as-prepared single-wall carbon nanotube (SWCNT) considerably improves the reactions kinetics compared to purified SWCNT/ multiwall carbon nanotube (MWCNT), graphite and activated carbon [46]. Janot et al. [47] showed that Nb_2O_5, $TiCl_3$ and Pd have insignificant effect on kinetics of the Li-Mg-N-H system, and these results were further confirmed elsewhere [48]. Wang et al. reported no improvements on kinetics of the Li-Mg-N-H system using Ti, Fe, Co, Ni, Pd, Pt and their oxides, but they did not disclose the details. However, they showed enhanced kinetics using a potassium-modified $Mg(NH_2)_2$-2LiH compound [49]. Utilization of the transition metal nitrides (TaN, TiN) were also shown to enhance the kinetics [50].

To the best of our knowledge, the effect of various nanocatalysts (i.e., nanoCo, nanoTi, nanoFe, nanoNi, TiF_3 and MWCNT) on the performance of the $LiNH_2$-nanoMgH$_2$ complex hydride has not been systematically investigated to date. Additionally, we have utilized a preprocessed MgH_2 (nanoMgH$_2$) in preparation of the $LiNH_2$-nanoMgH$_2$ complex hydride to better understand the benefits of reduced particle size [51–53]. In this study, the effects of compositional changes on hydrogen storage characteristics of $LiNH_2$-nanoMgH$_2$ compound has been investigated using (1:1) and (2:1) molar ratios, respectively. The effects of nanoNi catalyst concentration and various other catalysts (i.e., TiF_3, nanoCo, nanoTi, nanoFe and MWCNT) on desorption kinetics, the on-set decomposition temperature, and the gravimetric capacity of the $LiNH_2$-nanoMgH$_2$ compound are reported.

2. Materials and Methods

LiNH$_2$ was purchased from Sigma-Aldrich, St. Louis, MO, USA, with purity not less than 95%, and MgH$_2$ was procured from Alfa Aesar (Ward Hill, MA, USA) with a purity of 98%. All the materials were kept and handled in an inert nitrogen atmosphere in a glove box. LiNH$_2$ was used as received, whereas MgH$_2$ was preprocessed for 15 h in ball mill under Ar/H$_2$ medium to obtain nanoMgH$_2$ with finer particle/grain size (<10 nm) compared to as-received MgH$_2$. The catalyst TiF$_3$ was purchased from Sigma-Aldrich with 99.9% purity, nanoTi, nanoCo, nanoNi and nanoFe (purity of 90%, particle size range of 3–20 nm, and average surface area of 35–130 m^2/g) were purchased from QuantumSphere Inc. (Santa Ana, CA, USA) and MWCNT (purity of at least 60%) was purchased from Sigma-Aldrich. All the catalysts were used without further purification. The samples were ball milled using an 80 mL stainless steel bowl, which had a custom-built lid to facilitate evacuating/purging with hydrogen/argon (5%/95%) before starting and after every 2 h of ball milling. High-energy ball milling was carried out by Fritsch Pulverisette P6 planetary mill, the ball milling parameters such as the ball-to-powder ratio, milling speed and milling time were to 20:1, 300 RPM and 1–5 h, respectively. The base materials, xLiNH$_2$-MgH$_2$, were synthesized by milling xLiNH$_2$ (x = 1 or 2) and MgH$_2$ for 5 h, and then the desired catalysts were added. The resulting compound was ball milled again for 15 min to make sure a thorough dispersion of catalysts in the base materials was obtained. After the ball milling operation, all the samples were kept in a glove box until further characterization as explained below.

The microstructural and chemical analyses were carried out to confirm the morphology and composition of chemical elements by scanning electron microscopy (SEM) and energy dispersive X-rays (EDX). The powder X-ray diffraction (XRD) of the samples was carried out by Philips X'pert diffractometer with CuKα radiation of λ = 1.54060 Å. The as-milled samples were prepared inside the glove box, and sealed with Parafilm® tape, which shows peaks at 2θ angles of 21° and 23°. The diffraction data was analyzed using PANalytical X'pert Highscore software version 1.0f.

The Perkin-Elmer Spectrum One Fourier transform infrared (FTIR) spectrometer was utilized to measure the bond stretches of the complex hydride compound, and the instrument's working range was between 370–7800 cm^{-1} with a resolution of 0.5 cm^{-1}.

The gravimetric weight loss experiments were conducted by TA (New Castle, DE, USA) instrument's SDT-Q600 equipment, which is the combination of thermogravimetric analysis (TGA) and differential scanning calorimetry (DSC). The as-prepared samples and their catalyst doped versions were heated at a rate of 5 °C/min, and the data was analyzed with TA Universal Analysis 2000 software.

The thermal volumetric sorption analyses of the pristine and catalysts loaded base materials were carried out by Setaram (Caluire FRANCE) HyEnergy's PCTPro 2000 (i.e., a fully automated Sievert's type apparatus) and Quantachrome's (Boynton Beach, FL, USA) Autosorb 1C thermal programmed desorption (TPD) equipment.

3. Results and Discussion

3.1. Thermogravmetric Analysis (TGA) and Thermal Programmed Desorption (TPD)

TGA and TPD analyses facilitate the rapid screening of the complex hydride materials in a relatively short duration and gives invaluable information on the desorption characteristics such as gravimetric storage capacity and hydrogen decomposition temperature. To determine the optimal catalyst concentration, we focused on the nanoNi catalyst, which showed enhanced performance in various complex hydrides as discussed in the Introduction. It is highly desirable to use a minimum amount of catalyst to limit the cost of the complex hydride. As shown in Figure S1 (see supplementary information), 2 wt. % nanoNi showed the best desorption performance at temperatures up to 300 °C. Temperatures higher than 300 °C are not practical for the mobile applications; therefore, 2 wt. % nanoNi doping is the optimal catalyst loading for the 2LiNH$_2$-nanoMgH$_2$. The concentration of 2 wt. % nanoNi catalyst was also investigated for the LiNH$_2$-nanoMgH$_2$ system, and the TGA and TPD results are given in Figures 1 and 2, respectively.

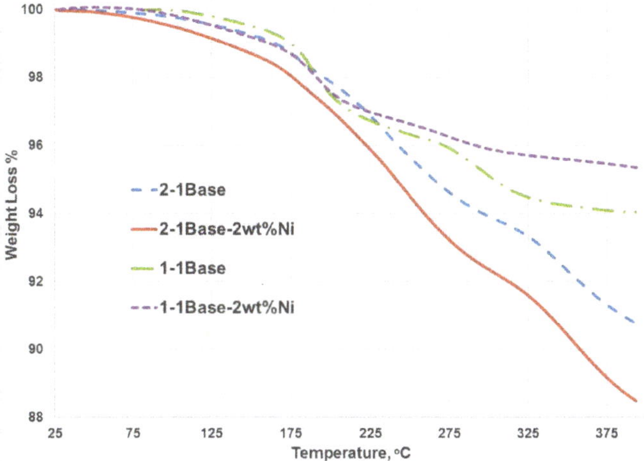

Figure 1. Thermogravimetric analysis (TGA) profiles of pristine and Ni-doped $LiNH_2$-nanoMgH_2 with 1:1 and 2:1 ratios.

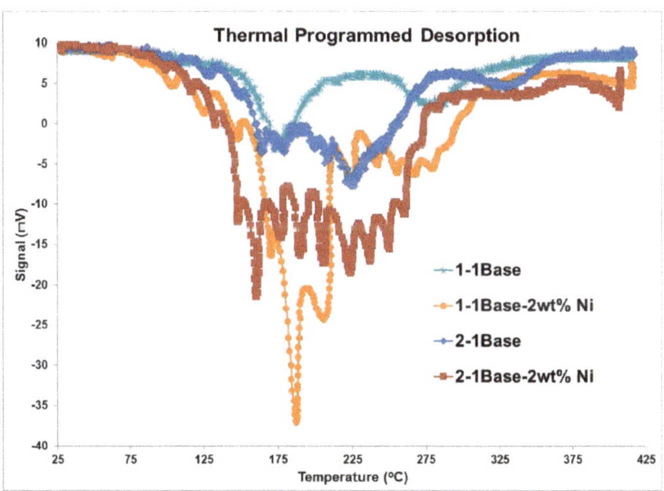

Figure 2. Thermal programmed desorption (TPD) profiles of pristine and 2 wt. % nanoNi-doped $LiNH_2$-nanoMgH_2 with 1:1 and 2:1 ratios.

Considering the TGA results of $LiNH_2$-nanoMgH_2 given in Figure 1, the addition of 2 wt. % nanoNi catalyst improved the hydrogen desorption up until around 200 °C, but hampered the ultimate gravimetric capacity (i.e., capacity at 375 °C) of the $LiNH_2$-nanoMgH_2 system. This result is in line with TPD studies. According to Figure 2, the hydrogen signal of the 2 wt. % nanoNi-doped $LiNH_2$-nanoMgH_2 increases up until 200 °C during the TPD experiments, then sharply decreases with increasing temperatures. From the practical application point of view, high temperature performance of the complex hydrides is not critical for the mobile applications; therefore, it is concluded that nanoNi doping is desirable for the $LiNH_2$-nanoMgH_2 system.

Regarding the TGA of $2LiNH_2$-nanoMgH_2 material given in Figure 1, the 2 wt. % nano-Ni doping enhances the gravimetric capacity of the base material at all temperatures studied. The TPD results given in Figure 2, however, show stronger hydrogen desorption signal for 2 wt. % nanoNi added material compared to the base material, which is further evidence for higher H_2 desorption from 2 wt. % nanoNi-doped material. Comparing gravimetric capacities of $LiNH_2$-MgH_2 with the molar ratios of 1:1 and 2:1 given in Figures 1 and 2, the $2LiNH_2$-$nanoMgH_2$ compound is superior to $LiNH_2$-$nanoMgH_2$, hence $2LiNH_2$-$nanoMgH_2$ is selected for further investigation.

A closer look into Figures 1 and 2 on the thermogravimetric and thermal programmed desorption profiles of 2 wt. % nanoNi-doped $LiNH_2$-$nanoMgH_2$ with 1:1 and 2:1 ratios are discussed here. The TGA of the 2 wt. % doped 1:1 sample shows the gaseous weight loss close to 5% with two desorption steps, whereas the 2 wt. % nanoNi-doped 2:1 sample shows double the weight loss capacity (~10%) with single major decomposition step and inflections at higher temperatures (~325 °C). This is confirmed from the TPD profile (Figure 2) of 2 wt. % doped 2:1 compound, where the total effective desorption or decomposition attributed in a single broader step when compared to TPD profiles of 2 wt. % doped 1:1 $LiNH_2$-$nanoMgH_2$ compound, where there are two sharp decomposition steps below 225 °C. We have also demonstrated TGA with other nickel concentrations, and found that lower nanoNi catalyst concentration of 2 wt. % is ideal to improve the gaseous hydrogen decomposition characteristics such storage capacity and the rate of desorption (see Supplementary Material Figure S1).

The TGA and TPD results of mixing 2 wt. % nanoNi, nanoCo, nanoFe, nanoTi and TiF_3 are given in Figures 3 and 4, respectively. Thus, these catalysts enhance the hydrogen desorption kinetics as well as the gravimetric capacity in comparison to the base material.

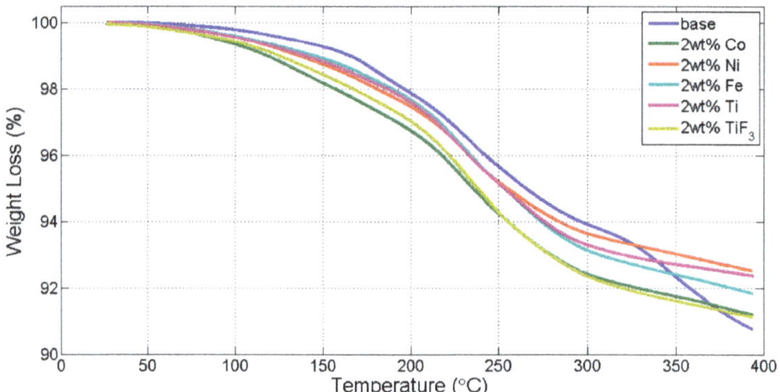

Figure 3. TGA analysis of the different catalysts on $LiNH_2$-$nanoMgH_2$ (2:1).

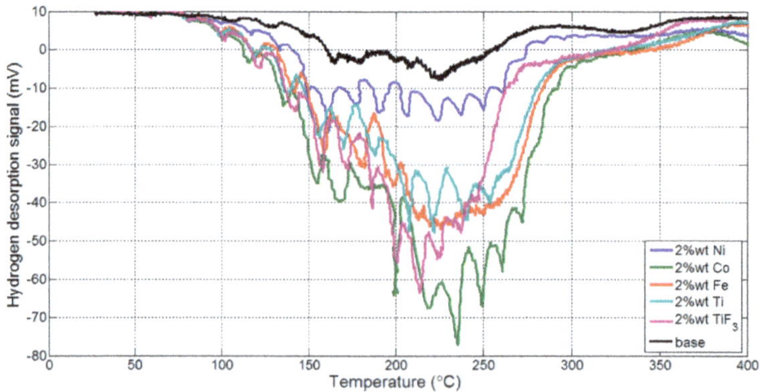

Figure 4. TPD characteristics of different catalysts on $LiNH_2$-$nanoMgH_2$ (2:1) from TPD analysis.

A closer analysis on the linear portion and start-up on-set point on the profiles of Figure 3, the reaction kinetics (wt. %/min) and on-set decomposition temperature were estimated for all the catalysts and are summarized in Table 1. It can been seen from Table 1 that TiF_3 enhances the reaction kinetics most with the order of TiF_3 > nanoNi > nanoTi > nanoCo > nanoFe, and the highest reduction in on-set desorption temperature was obtained from the nanoCo catalyst (nanoCo > TiF_3 > nanoTi > nanoFe > nanoNi).

Table 1. Reaction kinetics and on-set decomposition temperature of $2LiNH_2$-nanoMgH$_2$ with different catalysts.

Reaction Kinetics (wt. %/min)		On-Set Temperature (°C)	
TiF_3	0.5816	nanoCo	121.96
nanoNi	0.5330	TiF_3	125.99
nanoTi	0.5312	nanoTi	136.65
nanoCo	0.5255	nanoFe	142.86
nanoFe	0.5113	nanoNi	149.30

The thermal programed desorption profiles, unlike TGA, may only provide insights into the decomposition temperature along with strength of signal indicating the concentration of gaseous decomposition. Therefore, the trend in reaction kinetics as obtained from TGA in Figure 3 above may not be comparable to the TPD profiles of Figure 4. It seems, however, that the TPD of nanoCo excelled in the highest hydrogen concentration with the order of nanoCo > TiF_3 > nanoFe = nanoTi > nanoNi > base. By comparing the TGA profiles at an instant decomposition temperature say 225 °C, it is discernible again that nanoCo and TiF_3 outperformed with highest hydrogen release capacity and then continues with the order nanoCo > TiF_3 > nanoFe = nanoTi = nanoNi > base. Overall, the catalyst-doped $2LiNH_2$-MgH$_2$ thus enhances the hydrogen decomposition kinetics while maintaining available hydrogen content similar to that of the base hydride compound. The high reactivity of cobalt nanoparticles and the Ti^{3+} state of TiF_3 thus increases the reaction kinetics for the hydrogen absorption and reversible desorption. Further research is needed to understand why only Co among all transition metal nanoparticles (Fe, Ti, Ni) is superior in hydrogen decomposition at low temperatures.

3.2. Desorption Kinetics Using Sievert's Type Measurements

The ramping desorption kinetics (1 °C/min) of the base material $2LiNH_2$-MgH$_2$ and 2 wt. % Ni-added $2LiNH_2$-MgH$_2$ compound are given in Figure 5. As expected, Ni addition did not alter the overall desorption capacity of $2LiNH_2$-MgH$_2$ since the final desorption temperature was 325 °C. However, Ni addition showed enhanced kinetics and higher desorption capacity up to 250 °C.

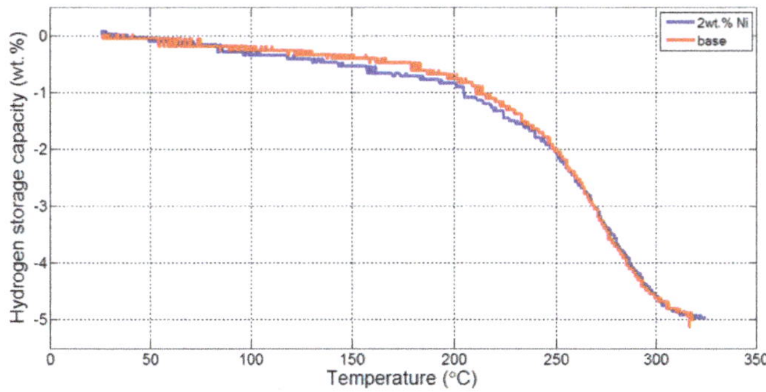

Figure 5. Ramping kinetics of $2LiNH_2$-nanoMgH$_2$ and $2LiNH_2$-nanoMgH$_2$ + 2 wt. % Ni.

The absorption kinetics of 2 wt. % Ni added 2LiNH$_2$-MgH$_2$ and the base material 2LiNH$_2$-MgH$_2$ at 180, 200, and 220 °C are given in Figure 6 The catalyst-added compound performed best in terms of kinetics at 200 °C, reaching 2.6 wt. % capacity in 60 min, whereas the base material can only reach a capacity of 1.75 wt. % in 60 min, as shown in Figure 6. The absorption kinetics of 2 wt. % Ni added 2LiNH$_2$-MgH$_2$ at 180, 200, and 220 °C up to 5 h is also given in Figure S2 (see Supplementary Material). The ultimate capacity of 2 wt. % Ni-added 2LiNH$_2$-MgH$_2$ is around 3.5 wt. % at 200 °C (Figure S2), which is lower than the theoretical and reported values earlier [41]. This discrepancy is due to differences in the material preparation procedure, slightly different compound ratios (LiNH$_2$ to MgH$_2$ ratio is 2:1 in this study as compared to 2:1.1 in Ref [41]), NH$_3$ emission, and the self-decomposition of Mg(NH$_2$)$_2$ to Mg$_3$N$_2$ at elevated temperatures. Among these, the self-decomposition of Mg(NH$_2$)$_2$ to Mg$_3$N$_2$ is considered to be the main reason, since every absorption cycle preceded by the evacuation of the 2LiNH$_2$-MgH$_2$ compound at 325 °C for 60 min to make sure for the complete desorption of the sample. Moreover, the self-decomposition of Mg(NH$_2$)$_2$ to Mg$_3$N$_2$ was also confirmed with the XRD measurements, as explained in Section 3.3.

On the other hand, Luo et al. showed a 25% capacity loss after 270 cycles where 7% of the capacity loss was attributed to NH$_3$ emission that increases with higher desorption temperatures [41]. Since the desorption temperature in this study was much higher (325 °C compared to 240 °C in Ref [41]), the main reasons for the apparent capacity loss are twofold: self-decomposition of Mg(NH$_2$)$_2$ to Mg$_3$N$_2$, and NH$_3$ emission. Therefore, high temperatures should be avoided in utilization of the 2LiNH$_2$-MgH$_2$ compound. To better understand the capacity loss due to self-decomposition of Mg(NH$_2$)$_2$ to Mg$_3$N$_2$ and NH$_3$ emission, further investigations are underway by utilizing gas chromatography and a residual gas analyzer coupled with a quadruple mass-spectrometer.

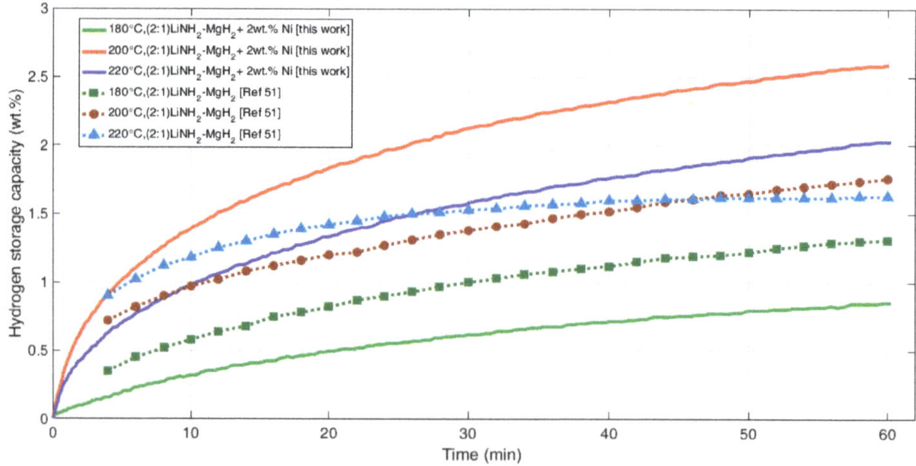

Figure 6. Absorption kinetics of 2LiNH$_2$-nanoMgH$_2$ + 2 wt. % Ni and the base material 2LiNH$_2$-MgH$_2$ at 180, 200, and 220 °C [54].

It is well known that the addition of transition metal catalysts to complex hydrides does not change the thermodynamic properties (i.e., enthalpy of formation) of the base material, but enhances the kinetic properties of the base material by lowering the activation energy [55]. The exact mechanism for the improved kinetics by Ni addition is not well understood currently. However, the enhanced kinetics due to Ni addition is considered to be associated with the increased heterogeneous sites for the nucleation of the complex hydride phases given in Equation (4). Additionally, Ni addition can facilitate the hydrogen diffusion through the complex hydride matrix [56].

3.3. X-ray Diffraction (XRD), Fourier Transform Infrared Spectroscopy (FTIR) Scanning Electron Microscopy (SEM), and Energy Dispersive X-rays (EDX)

The conversion or yield of the products have been determined by metrological characterization tools such as X-ray diffraction for phase identification, FTIR spectroscopic analysis for chemical bonding information, and SEM/EDX microscopic tools to evaluate the microstructure and nanoparticle size determination in addition to compositional (elemental) distributions.

X-ray diffraction and Fourier transform infrared spectroscopy were employed to characterize the $2LiNH_2$-MgH_2 compound after ball milling (BM), and after pressure composition temperature (PCT) measurements in hydrogenated/dehydrogenated conditions, to verify the validity of the reaction mechanism given in reaction (3). FTIR spectra of $LiNH_2$, $2LiNH_2$-MgH_2 after 5 h ball milling (BM) and after PCT measurements in hydrogenated/dehydrogenated conditions are given in Figure 7. The characteristics of N-H asymmetric and symmetric vibrations of $LiNH_2$ at 3313 and 3259 cm^{-1}, respectively, were observed in $2LiNH_2$-MgH_2 after 5 h BM. FTIR spectrum of hydrogenated $2LiNH_2$-MgH_2 showed the characteristic bands of $Mg(NH_2)_2$ at 3278 and 3325 cm^{-1} and Mg_3N_2 band at 3160 cm^{-1} [10,57]. On the other hand, the FTIR spectrum of the fully dehydrogenated $2LiNH_2$-MgH_2 showed the characteristic peaks of $Li_2Mg(NH)_2$ compound at 3163 and 3180 cm^{-1} [38].

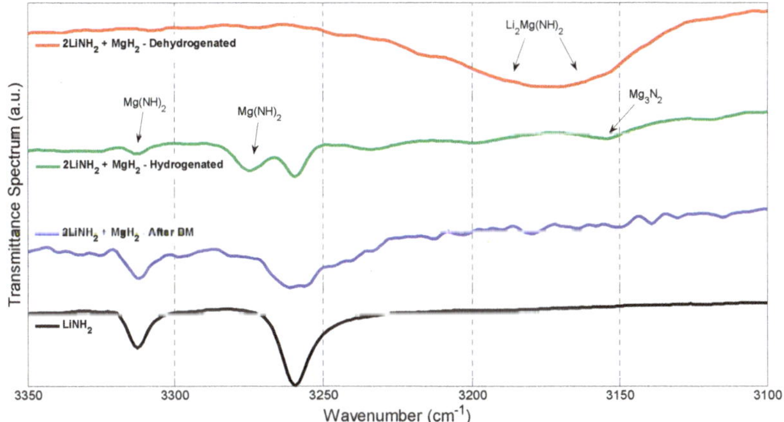

Figure 7. Fourier transform infrared spectroscopy (FTIR) spectra of $LiNH_2$-nanoMgH_2 (2:1) before and after pressure composition temperature (PCT).

The XRD profiles of the empty sample holder, $2LiNH_2$-MgH_2 after 5 h BM and after PCT measurements in hydrogenated/dehydrogenated conditions, are given in Figure 8. Self-decomposition of $Mg(NH_2)_2$ to Mg_3N_2 at elevated temperatures was discussed by Luo et al., and further proved unambiguously by XRD and FTIR measurements in this study [41]. However, the XRD of the dehydrogenated sample resulted in an unidentified peak around 27.5° which needs to be further investigated. This peak could be the result of an oxide formation at high desorption temperatures due to the impurities in the as-received raw materials. The SEM/EDX mapping and image of the of $2LiNH_2$-MgH_2 + 2 wt. % Ni is given in Figure 9, which clearly shows the uniform dispersion of nanoNi particles among the $2LiNH_2$-MgH_2 compound. Therefore, 15 min is a reasonable duration for mixing catalyst with the base material. Moreover, the EDX spectral analysis of these samples demonstrated the correct 2 wt. % fraction of elemental Ni catalyst on a base hydride matrix. Additionally, the morphology and particles size of the nanoNi (obtained from QuantumSphere Inc.) catalyst was determined by the transmission electron microscopy (TEM), and is shown in Figure 10. From the TEM microstructure, it is evidenced that we have used cluster sizes of 3–10 nm nickel nanoparticles admixed with the base hydride compound that has enhanced hydrogen absorption and desorption kinetics.

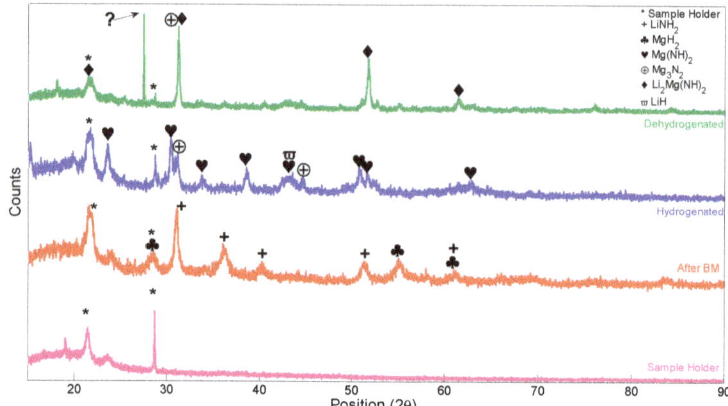

Figure 8. X-ray diffraction (XRD) spectra of LiNH$_2$-nanoMgH$_2$ after ball milling and after PCT measurements in hydrogenated/dehydrogenated states.

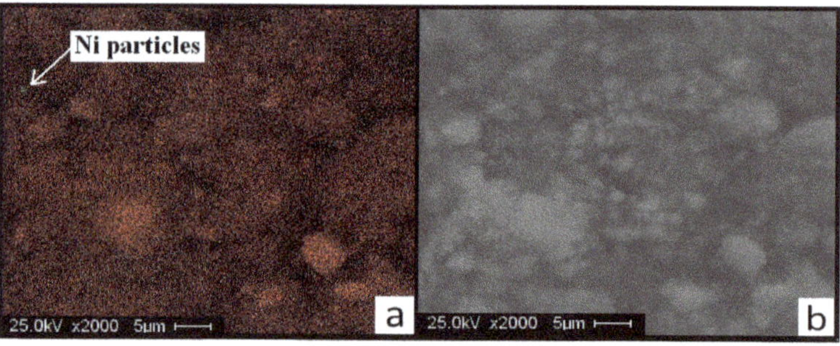

Figure 9. Scanning electron microscopy (SEM)/ energy dispersive X-rays (EDX) mapping (**a**) and image (**b**) of the 2LiNH$_2$-nanoMgH$_2$ + 2 wt. % Ni. Red and green color (labeled) represents Mg and Ni, respectively.

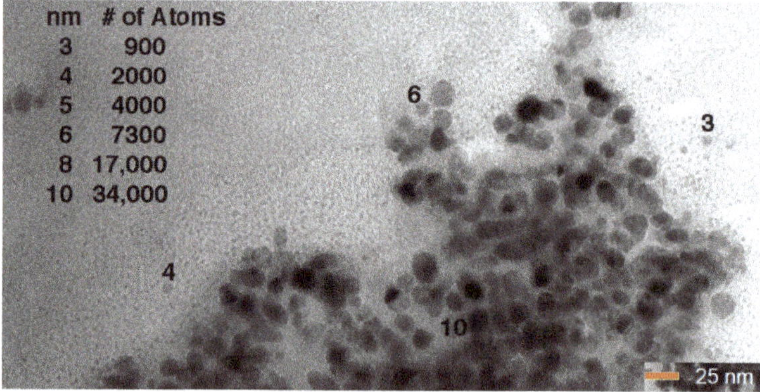

Figure 10. Transmission electron microscopy (TEM) image of nickel nanopartile manufactured by the QuatumSphere Inc. [58].

4. Conclusions

The complex hydride $LiNH_2$-nanoMgH$_2$ compound was systematically investigated considering different compositional variations and the effect of various catalysts. The thermogravimetric and thermal programmed desorption results revealed that $2LiNH_2$-nanoMgH$_2$ has higher hydrogen storage capacity and faster sorption kinetics when compared to $LiNH_2$-nanoMgH$_2$. Among the various concentrations of nanoNi additives on the base material of $2LiNH_2$-nanoMgH$_2$, 2 wt. % nanoNi showed enhancement in reaction kinetics. Additionally, the TiF_3 doping demonstrates greater reaction kinetics (0.5816 wt. %/min), whereas the nanoCo doping shows the lowest on-set decomposition temperature (121.96 °C) as obtained from the TGA results. The absorption kinetics of $2LiNH_2$-MgH$_2$ mixed with 2 wt. % nanoNi was rapid at 200 °C, and more than a twofold increase in kinetics when compared to the base material within the first 60 min of absorption. Structural, microstructural, and chemical investigations using metrological tools further supported that high temperature desorption is detrimental to the overall capacity of the $2LiNH_2$-MgH$_2$ compound.

Supplementary Materials: The following are available online at http://www.mdpi.com/2076-3417/7/7/701/s1, Figure S1: TGA analysis of Ni doped $LiNH_2$-nanoMgH$_2$ (2:1); Figure S2: Absorption kinetics of $2LiNH_2$-nanoMgH$_2$ + 2 wt. % Ni at 180, 200 and 220 °C.

Acknowledgments: The authors wish to acknowledge the Florida Energy Systems Consortium (FESC) and US Department of Energy for the project funding and support.

Author Contributions: Sesha S. Srinivasan and Dervis Emre Demirocak conceived and designed the experiments; Sesha S. Srinivasan and Dervis Emre Demirocak performed the experiments; Yogi Goswami and Elias Stefanakos analyzed the data; Sesha S. Srinivasan and Dervis Emre Demirocak wrote the paper. Elias Stefanakos and Yogi Goswami proofread the manuscript.

Conflicts of Interest: The authors declare no conflict of interest.

Abbreviations

BM	Ball Milling
CNT	Carbon Nanotube
DSC	Differential Scanning Calorimetry
DOE	Department of Energy
EDX	Energy Dispersive X-rays
FTIR	Fourier Transform Infrared Spectroscopy
MA	Mechanical Activation
MOF	Metal Organic Frameworks
MWCNT	Multiwall Carbon Nanotube
PCT	Pressure Composition Temperature
RPM	revolutions per minute
SDT	Simultaneous TGA and DSC
SEM	Scanning Electron Microscopy
SWCNT	Single Wall Carbon Nanotube
TCD	Thermal Conductivity Detector
TEM	Transmission Electron Microscopy
TGA	Thermogravimetric Analysis
TPD	Thermal Program Desorption
wt. %	weight percent
XRD	X-Ray Diffraction

References

1. Orimo, S.; Nakamori, Y.; Eliseo, J.R.; Zuettel, A.; Jensen, C.M. Complex hydrides for hydrogen storage. *Chem. Rev.* **2007**, *107*, 4111–4132. [CrossRef] [PubMed]

2. Dafert, F.; Miklauz, R. New compounds of nitrogen and hydrogen with lithium. *Monatsh. Chem.* **1910**, *31*, 981. [CrossRef]

3. Ruff, O.; Goeres, H. Li imide and some compounds of N, H and Li. *Ber. Dtsch. Chem. Ges.* **1910**, *44*, 502–506. [CrossRef]

4. Chen, P.; Xiong, Z.; Luo, J.; Lin, J.; Tan, K.L. Interaction of hydrogen with metal nitrides and imides. *Nature* **2002**, *420*, 302–304. [CrossRef] [PubMed]

5. Hu, Y.H.; Ruckenstein, E. Ultrafast reaction between LiH and NH_3 during H_2 storage in Li_3N. *J. Phys. Chem. A* **2003**, *107*, 9737–9739. [CrossRef]

6. Nakamori, Y.; Orimo, S. Destabilization of Li-based complex hydrides. *J. Alloys Compd.* **2004**, *370*, 271–275. [CrossRef]

7. Luo, W. ($LiNH_2$-MgH_2): A viable hydrogen storage system. *J. Alloys Compd.* **2004**, *381*, 284–287. [CrossRef]

8. Luo, W.; Rönnebro, E. Towards a viable hydrogen storage system for transportation application. *J. Alloys Compd.* **2005**, *404*, 392–395. [CrossRef]

9. Xiong, Z.; Wu, G.; Hu, J.; Chen, P. Ternary imides for hydrogen storage. *Adv. Mater.* **2004**, *16*, 1522–1525. [CrossRef]

10. Luo, W.; Sickafoose, S. Thermodynamic and structural characterization of the Mg-Li-NH hydrogen storage system. *J. Alloys Compd.* **2006**, *407*, 274–281. [CrossRef]

11. Lohstroh, W.; Fichtner, M. Reaction steps in the Li-Mg-NH hydrogen storage system. *J. Alloys Compd.* **2007**, *446*, 332–335. [CrossRef]

12. Chen, Y.; Wu, C.Z.; Wang, P.; Cheng, H.M. Structure and hydrogen storage property of ball-milled $LiNH_2$/MgH_2 mixture. *Int. J. Hydrog. Energy* **2006**, *31*, 1236–1240. [CrossRef]

13. Pottmaier, D.; Dolci, F.; Orlova, M.; Vaughan, G.; Fichtner, M.; Lohstroh, W.; Baricco, M. Hydrogen release and structural transformations in $LiNH_2$-MgH_2 systems. *J. Alloys Compd.* **2010**, *509*, 719–723. [CrossRef]

14. Chen, P.; Xiong, Z.; Yang, L.; Wu, G.; Luo, W. Mechanistic investigations on the heterogeneous solid-state reaction of magnesium amides and lithium hydrides. *J. Phys. Chem. B* **2006**, *110*, 14221–14225. [CrossRef] [PubMed]

15. Wang, J.; Li, H.; Wang, S.; Liu, X.; Li, Y.; Jiang, L. The desorption kinetics of the $Mg(NH_2)_2$ + LiH mixture. *Int. J. Hydrog. Energy* **2009**, *34*, 1411–1416. [CrossRef]

16. Yang, J.; Sudik, A.; Wolverton, C. Activation of hydrogen storage materials in the Li-Mg-NH system: Effect on storage properties. *J. Alloys Compd.* **2007**, *430*, 334–338. [CrossRef]

17. Xiong, Z.; Hu, J.; Wu, G.; Chen, P.; Luo, W.; Gross, K.; Wang, J. Thermodynamic and kinetic investigations of the hydrogen storage in the Li-Mg-NH system. *J. Alloys Compd.* **2005**, *398*, 235–239. [CrossRef]

18. Leng, H.; Ichikawa, T.; Hino, S.; Hanada, N.; Isobe, S.; Fujii, H. New metal-NH system composed of $Mg(NH_2)_2$ and LiH for hydrogen storage. *J. Phys. Chem. B* **2004**, *108*, 8763–8765. [CrossRef]

19. Nakamori, Y.; Kitahara, G.; Miwa, K.; Towata, S.; Orimo, S. Reversible hydrogen-storage functions for mixtures of Li_3N and Mg_3N_2. *Appl. Phys. A Mater. Sci. Process.* **2005**, *80*, 1–3. [CrossRef]

20. Leng, H.; Ichikawa, T.; Fujii, H. Hydrogen Storage Properties of Li-Mg-N-H Systems with Different Ratios of LiH/$Mg(NH_2)_2$. *J. Phys. Chem. B* **2006**, *110*, 12964–12968. [CrossRef] [PubMed]

21. Xiong, Z.; Wu, G.; Hu, J.; Chen, P.; Luo, W.; Wang, J. Investigations on hydrogen storage over Li-Mg-NH complex-the effect of compositional changes. *J. Alloys Compd.* **2006**, *417*, 190–194. [CrossRef]

22. Osborn, W.; Markmaitree, T.; Shaw, L.L. Evaluation of the hydrogen storage behavior of a $LiNH_2$ + MgH_2 system with 1:1 ratio. *J. Power Sources* **2007**, *172*, 376–378. [CrossRef]

23. Leng, H.; Ichikawa, T.; Hino, S.; Hanada, N.; Isobe, S.; Fujii, H. Synthesis and decomposition reactions of metal amides in metal-NH hydrogen storage system. *J. Power Sources* **2006**, *156*, 166–170. [CrossRef]

24. Lu, J.; Choi, Y.J.; Fang, Z.Z.; Sohn, H.Y. Effect of milling intensity on the formation of LiMgN from the dehydrogenation of $LiNH_2$-MgH_2 (1:1) mixture. *J. Power Sources* **2010**, *195*, 1992–1997. [CrossRef]

25. Leng, H.; Ichikawa, T.; Isobe, S.; Hino, S.; Hanada, N.; Fujii, H. Desorption behaviours from metal-NH systems synthesized by ball milling. *J. Alloys Compd.* **2005**, *404*, 443–447. [CrossRef]

26. Pinkerton, F. Decomposition kinetics of lithium amide for hydrogen storage materials. *J. Alloys Compd.* **2005**, *400*, 76–82. [CrossRef]

27. Tokoyoda, K.; Hino, S.; Ichikawa, T.; Okamoto, K.; Fujii, H. Hydrogen desorption/absorption properties of Li-Ca-NH system. *J. Alloys Compd.* **2007**, *439*, 337–341. [CrossRef]

28. Chu, H.; Xiong, Z.; Wu, G.; He, T.; Wu, C.; Chen, P. Hydrogen storage properties of Li-Ca-NH system with different molar ratios of $LiNH_2$/CaH_2. *Int. J. Hydrog. Energy* **2010**, *35*, 8317–8321. [CrossRef]

29. Zaluska, A.; Zaluski, L.; Ström-Olsen, J. Structure, catalysis and atomic reactions on the nano-scale: A systematic approach to metal hydrides for hydrogen storage. *Appl. Phys. A Mater. Sci. Process.* **2001**, *72*, 157–165. [CrossRef]

30. Markmaitree, T.; Ren, R.; Shaw, L.L. Enhancement of lithium amide to lithium imide transition via mechanical activation. *J. Phys. Chem. B* **2006**, *110*, 20710–20718. [CrossRef] [PubMed]

31. Liu, Y.; Zhong, K.; Luo, K.; Gao, M.; Pan, H.; Wang, Q. Size-Dependent Kinetic Enhancement in Hydrogen Absorption and Desorption of the Li-Mg-N-H System. *J. Am. Chem. Soc.* **2009**, *131*, 1862–1870. [CrossRef] [PubMed]

32. Varin, R.; Jang, M.; Polanski, M. The effects of ball milling and molar ratio of LiH on the hydrogen storage properties of nanocrystalline lithium amide and lithium hydride (LiNH$_2$ + LiH) system. *J. Alloys Compd.* **2010**, *491*, 658–667. [CrossRef]

33. Shaw, L.L.; Ren, R.; Markmaitree, T.; Osborn, W. Effects of mechanical activation on dehydrogenation of the lithium amide and lithium hydride system. *J. Alloys Compd.* **2008**, *448*, 263–271. [CrossRef]

34. Shahi, R.R.; Yadav, T.; Shaz, M.; Srivastava, O. Effects of mechanical milling on desorption kinetics and phase transformation of LiNH$_2$/MgH$_2$ mixture. *Int. J. Hydrog. Energy* **2008**, *33*, 6188–6194. [CrossRef]

35. Xie, L.; Liu, Y.; Li, G.; Li, X. Improving Hydrogen Sorption Kinetics of the Mg(NH$_2$)$_2$-LiH System by the Tuning Particle Size of the Amide. *J. Phys. Chem. C* **2009**, *113*, 14523–14527. [CrossRef]

36. Osborn, W.; Markmaitree, T.; Shaw, L.L.; Hu, J.Z.; Kwak, J.H.; Yang, Z. Low temperature milling of the LiNH$_2$ + LiH hydrogen storage system. *Int. J. Hydrog. Energy* **2009**, *34*, 4331–4339. [CrossRef]

37. Ikeda, S.; Kuriyama, N.; Kiyobayashi, T. Simultaneous determination of ammonia emission and hydrogen capacity variation during the cyclic testing for LiNH$_2$-LiH hydrogen storage system. *Int. J. Hydrog. Energy* **2008**, *33*, 6201–6204. [CrossRef]

38. Markmaitree, T.; Osborn, W.; Shaw, L.L. Comparisons between MgH$_2$- and LiH-containing systems for hydrogen storage applications. *Int. J. Hydrog. Energy* **2008**, *33*, 3915–3924. [CrossRef]

39. Markmaitree, T.; Osborn, W.; Shaw, L.L. Comparative studies of reaction rates of NH$_3$ with MgH$_2$ and LiH. *J. Power Sources* **2008**, *180*, 535–538. [CrossRef]

40. Luo, W.; Stewart, K. Characterization of NH$_3$ formation in desorption of Li-Mg-NH storage system. *J. Alloys Compd.* **2007**, *440*, 357–361. [CrossRef]

41. Luo, W.; Wang, J.; Stewart, K.; Clift, M.; Gross, K. Li-Mg-NH: Recent investigations and development. *J. Alloys Compd.* **2007**, *446*, 336–341. [CrossRef]

42. Bogdanovic, B.; Schwickardi, M. Ti-doped alkali metal aluminium hydrides as potential novel reversible hydrogen storage materials. *J. Alloys Compd.* **1997**, *253*, 1–9. [CrossRef]

43. Ichikawa, T.; Isobe, S.; Hanada, N.; Fujii, H. Lithium nitride for reversible hydrogen storage. *J. Alloys Compd.* **2004**, *365*, 271–276. [CrossRef]

44. Isobe, S.; Ichikawa, T.; Hanada, N.; Leng, H.; Fichtner, M.; Fuhr, O.; Fujii, H. Effect of Ti catalyst with different chemical form on Li-NH hydrogen storage properties. *J. Alloys Compd.* **2005**, *404*, 439–442. [CrossRef]

45. Yao, J.; Shang, C.; Aguey-Zinsou, K.; Guo, Z. Desorption characteristics of mechanically and chemically modified LiNH$_2$ and (LiNH$_2$ + LiH). *J. Alloys Compd.* **2007**, *432*, 277–282. [CrossRef]

46. Chen, Y.; Wang, P.; Liu, C.; Cheng, H.M. Improved hydrogen storage performance of Li-Mg-NH materials by optimizing composition and adding single-walled carbon nanotubes. *Int. J. Hydrog. Energy* **2007**, *32*, 1262–1268. [CrossRef]

47. Janot, R.; Eymery, J.B.; Tarascon, J.M. Investigation of the processes for reversible hydrogen storage in the Li-Mg-NH system. *J. Power Sources* **2007**, *164*, 496–502. [CrossRef]

48. Barison, S.; Agresti, F.; Lo Russo, S.; Maddalena, A.; Palade, P.; Principi, G.; Torzo, G. A study of the LiNH$_2$-MgH$_2$ system for solid state hydrogen storage. *J. Alloys Compd.* **2008**, *459*, 343–347. [CrossRef]

49. Wang, J.; Liu, T.; Wu, G.; Li, W.; Liu, Y.; Araújo, C.M.; Scheicher, R.H.; Blomqvist, A.; Ahuja, R.; Xiong, Z. Potassium Modified Mg(NH$_2$)$_2$/2LiH System for Hydrogen Storage. *Angew. Chem. Int. Ed.* **2009**, *48*, 5828–5832. [CrossRef] [PubMed]

50. Ma, L.P.; Wang, P.; Dai, H.B.; Cheng, H.M. Catalytically enhanced dehydrogenation of Li-Mg-NH hydrogen storage material by transition metal nitrides. *J. Alloys Compd.* **2009**, *468*, L21–L24. [CrossRef]

51. Srinivasan, S.S.; Niemann, M.U.; Hattrick-Simpers, J.R.; McGrath, K.; Sharma, P.C.; Goswami, D.Y.; Stefanakos, E.K. Effects of nano additives on hydrogen storage behavior of the complex hydride LiBH$_4$/LiNH$_2$/MgH$_2$. *Int. J. Hydrog. Energy* **2010**, *35*, 9646–9652. [CrossRef]

52. Vittetoe, A.W.; Niemann, M.U.; Srinivasan, S.; McGrath, K.; Kumar, A.; Goswami, D.Y.; Stefanakos, E.K.; Thomas, S. Destabilization of LiAlH$_4$ by nanocrystalline MgH$_2$. *Int. J. Hydrog. Energy* **2009**, *4*, 2333–2339. [CrossRef]
53. Niemann, M.U.; Srinivasan, S.S.; Kumar, A.; Stefanakos, E.K.; Goswami, D.Y.; McGrath, K. Processing analysis of the ternary LiNH$_2$-MgH$_2$-LiBH$_4$ system for hydrogen storage. *Int. J. Hydrog. Energy* **2009**, *34*, 8086–8093. [CrossRef]
54. Markmaitree, T.; Shaw, L.L. Synthesis and hydriding properties of Li$_2$Mg(NH)$_2$. *J. Power Sources* **2010**, *195*, 1984–1991. [CrossRef]
55. Varin, R.A.; Zbroniec, L.; Polanski, M.; Bystrzycki, J. A review of recent advances on the effects of microstructural refinement and nano-catalytic additives on the hydrogen storage properties of metal and complex hydrides. *Energies* **2010**, *4*, 1–25. [CrossRef]
56. Bazzanella, N.; Checchetto, R.; Miotello, A. Atoms and nanoparticles of transition metals as catalysts for hydrogen desorption from magnesium hydride. *J. Nanomater.* **2011**. [CrossRef]
57. Kojima, Y.; Kawai, Y.; Ohba, N. Hydrogen storage of metal nitrides by a mechanochemical reaction. *J. Power Sources* **2006**, *159*, 81–87. [CrossRef]
58. Maloney, K.; Dopp, R. *Advanced Materials for Clean Energy Applications*; Online Webinar Event, 26 June 2007; Elsevier: Amsterdam, The Netherlands, 2007.

Article

Simulation Investigation on Combustion Characteristics in a Four-Point Lean Direct Injection Combustor with Hydrogen/Air

Jianzhong Li [1,*], Li Yuan [2] and Hukam C. Mongia [3]

[1]   Key Laboratory of Aero-engine Thermal Environment and Structure, Ministry of Industry and Information Technology, Nanjing University of Aeronautics and Astronautics, Nanjing 210016, China
[2]   School of National Defense Engineering, PLA University of Science and Technology, 88 Biaoying Rd., Nanjing 210007, China; 80njyuanli@163.com
[3]   School of Mechanical Engineering, Purdue University, West Lafayette, IN 47907-2088, USA; hmongia43@hotmail.com
*   Correspondence: ljzh0629@nuaa.edu.cn; Tel.: +86-025-84895927

Academic Editors: Elias K. Stefanakos and Sesha S. Srinivasan
Received: 5 May 2017; Accepted: 12 June 2017; Published: 14 June 2017

Abstract: To investigate the combustion characteristics in multi-point lean direct injection (LDI) combustors with hydrogen/air, two swirl–venturi 2 × 2 array four-point LDI combustors were designed. The four-point LDI combustor consists of injector assembly, swirl–venturi array and combustion chamber. The injector, swirler and venturi together govern the rapid mixing of hydrogen and air to form the mixture for combustion. Using clockwise swirlers and anticlockwise swirlers, the co-swirling and count-swirling swirler arrays LDI combustors were achieved. Using Reynolds-Averaged Navier–Stokes (RANS) code for steady-state reacting flow computations, the four-point LDI combustors with hydrogen/air were simulated with an 11 species and 23 lumped reaction steps H_2/Air reaction mechanism. The axial velocity, turbulence kinetic energy, total pressure drop coefficient, outlet temperature, mass fraction of OH and emission of pollutant NO of four-point LDI combustors, with different equivalence ratios, are here presented and discussed. As the equivalence ratios increased, the total pressure drop coefficient became higher because of increasing heat loss. Increasing equivalence ratios also corresponded with the rise in outlet temperature of the four-point LDI combustors, as well as an increase in the emission index of NO EI_{NO} in the four-point LDI combustors. Along the axial distance, the EI_{NO} always increased and was at maximum at the exit of the dump. Along the chamber, the EI_{NO} gradually increased, maximizing at the exit of chamber. The total temperature of four-point LDI combustors with different equivalence ratios was identical to the theoretical equilibrium temperature. The EI_{NO} was an exponential function of the equivalence ratio.

Keywords: gas turbine engine; lean direct injection; four-point; low emissions combustion

---

1. Introduction

The atmospheric environment is becoming worse, and the environmental protection consciousness of people is stronger than ever. To reduce the effects of the misuse of fossil fuels and their destructive impacts on nature, extensive efforts has been applied to develop hydrogen fuel as an alternative to hydrocarbon fuels [1–4]. The pollutants produced by the combustion of hydrogen are lower and less damaging than those produced by the combustion of hydrocarbon fuels, therefore, hydrogen could be used as a novel, renewable and environmentally-friendly fuel. The hydrogen fuelled combustion system in operation would mean that there would be no ozone layer depletion, and greenhouse gases and acid rain would be minimised. Hydrogen fuelled combustion of gas turbine engines has huge environmental advantages for the current system, because only water and nitrogen oxides (NO_x) are the emissions of hydrogen combustion [5,6]. The hydrogen-enrichment of natural gas could be used to solve several drawbacks encountered for turbulent premixed combustion [7–9]. Using the CHEMKIN PREMIX code based on the GRI (Gas Research Institute) kinetic mechanism, the laminar burning velocities in a combustor with hydrogen–methane/air were calculated [10]. The methane reactivity could be enhanced slightly in lean mixtures conditions. Using Time-Resolved Particle Image Velocimetry, Di Sarli [11] investigated the transient interactions of premixed flames and toroidal vortex in combustors with hydrogen-enriched methane/air. The hydrogen presence would increase the main toroidal vortex velocity to affect the flow field and generate different sub-vortices. While the hydrogen substitutes for the methane, the vortex would wrinkle the flame front which induces the small flame pockets separates from the main front. Using a Large Eddy Simulation (LES) method, the unsteady propagation of the premixed flames around toroidal vortices in a combustor with hydrogen-enriched methane/air was simulated [12]. With the hydrogen mole fraction varying from 0 to 0.5 for lean mixtures, good predictions of experimental data could be obtained. The flame shape and structure affected by the vortex were presented. The hydrogen diffusion time was higher than that of the flame roll-up round the vortex.

Hydrogen will be a great potential surrogate for fuel fossil fuels for gas turbine engines in the future, but currently, it is not widely used for gas turbine engines. Previous research on the application of hydrogen focused on internal combustion engines. For example, hydrogen has been seen as suitable for spark ignition engines as fuel, and compression ignition engines are also now in the process of modification to operate with hydrogen [2]. There are only sporadic studies of hydrogen applications for gas turbine engines. Daniel [13] investigated a low emission combustor using hydrogen lean direct injections. At an inlet temperature of 588 to 811 K and inlet pressure of 0.4 to 1.4 MPa, the NO_x emissions and combustion performance were measured, and the equivalence ratios were up to 0.48. With multiple injection points and quick mixing, all the injectors were installed to achieve lean direct injection (LDI) technology.

The combination of nitrogen dioxide (NO_2) and nitric oxide (NO) is usually called as NO_x. The formation of photochemical smog, acid rain and harmful ozone is caused by such NO_x emissions as from aircraft engines [14,15]. For fuelled hydrogen gas turbine engines, low NO_x emissions of combustion are usually required in order to meet the demand of environmental emissions regulations. The temperature, mixedness, residence time, and engine operation pressure would play an important role in NO_x emission production. For designing advanced gas turbine combustors in the future, locally leaner combustion should be developed to achieve low NO_x emissions [16–18].

Because there is an exponential correlation between the NO_x emissions and local combustion temperature, the local temperature peaks and NO_x emissions could be reduced through decreasing the burning zone equivalence ratio (below stoichiometric values). There are several low emission combustion technologies, such as lean prevaporized premixed (LPP) combustion [19–21], rich quench lean (RQL) combustion [22–26], and lean direct injection (LDI) combustion [27]. Compared with traditional designs, the LDI combustion concept has received considerable attention as an extremely promising technology that utilizes rapid mixing of fuel-air to produce lean combustible mixtures, as well as reduce NO_x emissions from the gas turbine engine. LPP combustion technologies could reduce the NO_x emission from gas turbine engines, but there are some shortcomings of autoignition and flashback problems for LPP combustor, so that the ignition delay times of the kerosene fuels is shorter when the kerosene is at high pressure and in high temperature conditions [28]. Using multiple smaller assemblies to produce an injector/swirler assembly, rapid mixing could be achieved. Then, multiple small burning zones would be produced. To promote reduced NO_x formation, a rapid mixing process, a uniform fuel air mixture, and a short residence time in high temperature zones, are generally best [29]. The evolution of LDI technologies for combustor design was summarized by the authors of [30,31]. To define and refine the next-generation and look for a tool to provide guidance, the combustion characteristics in multi-lean direct injection combustors were investigated with the National Combustion Code (NCC). The different vane angles swirlers were used to form a 3×3 element array LDI combustor. Using experimental data, the predictions of CO emission index, NO_x emissions index and lean blowout were verified for two different geometry configurations [32]. To gather experience for the next-generation LDI-2, the non-reacting and reacting flow computation method was used to simulate the LDI-2 combustor, which consists of thirteen-element injectors. The NCC includes an approach of spray-modeling, mesh-refinement, kinetics-modeling and ignition. The emissions predictions of EI_{NOx}, the emission index of CO (EI_{CO}) and the unburned hydrocarbon (UHC) were evaluated for LDI-2 [33]. A parameterized model of swirl–venturi LDI was used to explore a design method [34]. The Reynolds-averaged Navier–Stokes equation for steady-state reacting computations was used to simulate the 20 three-element LDI combustor. The 18-step Jet-A reduced reaction mechanism was used to directly solve the species concentrations of combustion in the LDI combustor. The axial flow field would be vulnerable to the geometric perturbations. The turbulent kinetic energy, axial velocity, fuel distribution, static temperature and species mass fractions were analyzed. At inlet temperatures between 835 and 865 K and inlet pressure of 1034 kPa, the emission of 9-point swirl–venturi LDI (SV-LDI) was measured [35]. The two swirler blade angles were 45° and 60°, respectively. The NO_x emissions of LDI with 45° swirler was lower than that of the LDI with 60° swirler. The swirling flow field characteristics of the multipoint LDI combustor with nine fuel injectors were also investigated [36]. Using an injector, an axial swirler and a convergent–divergent venturi, the fuel/air rapid mixing and a recirculation zone for a stabilized flame were achieved. The vane angle of each swirler was 60° and the swirl number was 1.0. The transient swirling flow field structure, the component of velocity and Reynolds stresses were measured by a 3D Particle Image Velocimetry (PIV) system. If the recessed center swirler arrays of LDI combustor are different, the flow structures are significantly different. For both co-swirler LDI combustor and counter-swirler combustor, the short strong central recirculation zones are produced. For the baseline arrays LDI combustor, the turbulent activity near the swirler exit was highest. In a self-sustaining combustion oscillations combustor, the thermoacoustic coupling was observed [37]. The oscillation amplitude was highest when the equivalence ratio was highest. The oscillation strength was affected slightly by the different fuel distributions, whereas the overall equivalence ratio was constant. The proper orthogonal decomposition (POD) was used to present the energetic spatial components that could represent OH* distribution and periodic variation. The fluctuation location and magnitude could be visualized by the POD modes. There were similar phases and locations between OH* variations and periodic changes. A multipoint lean direct injection combustor, which includes 36 fuel injectors and fuel-air mixers was investigated [38]. The construction consists of the injectors, swirlers and fuel distributer, where Jet-A was used as the fuel. The effects of inlet total temperature, inlet total pressure, and equivalence ratio on the NO_x emission were investigated when the inlet temperatures and inlet pressures could reach up to 866 K and 4825 kPa. The spray combustion characteristics of a multi-point LDI combustor was investigated [39]. The RANS code was used to simulate the flow and combustion characteristics in the combustor. Using several spray sub-models for the liquid spray modeling, the spray properties were analyzed and discussed. The short flames from individual injectors, a uniform temperature profile at the chamber exit and the uniformly low temperature distribution inside the combustor were observed. The flow field at the injector exits becomes highly strained when the air flow velocity is increasing. The injector structure, rapid mixing, reacting, and operating cycle conditions would affect the performance, operability, and emission of the combustor.

The swirl–venturi four-point lean direct injection combustor with hydrogen/air is explored and the influence of key geometric, swirl–venturi array and equivalence ratio on reacting flow and emission production characteristics is discussed. A swirl–venturi 2×2 array four-point LDI combustor has been designed. The injector, swirler and venturi together govern the rapid mixing of hydrogen and air to form a combustible mixture. Using the clockwise swirler and anticlockwise swirler, the co-swirling and count-swirling swirler arrays of LDI combustors were achieved. With the Reynolds-averaged

Navier–Stokes code for steady-state reacting computations, the four-point LDI combustors with hydrogen/air were simulated with a reduced 23-step reaction mechanism. The axial velocity, swirl number, velocity angle, effective area, total pressure drop coefficient, total temperature, mass fraction of OH and emission of pollutant NO of a hydrogen fuelled four-point LDI combustor, with different equivalence ratios, were achieved and discussed.

2. Problem Formulation

2.1. Parametric Geometry Definition of Four-Point LDI

Using a parametric modeling scheme, a swirl–venturi 2×2 array LDI combustor was produced and is shown in Figure 1. The scheme consists of inlet, injector, swirl–venturi array, combustion chamber. The swirlers include the CW swirler with clockwise vanes and ACW swirler with anticlockwise vanes. There are two swirlers arrangements in LDI combustors, which are the co-swirling array combustor, and the counter-swirling array combustor, as shown in Figure 2. All swirlers of the co-swirling array are in the same swirling direction. The adjacent swirlers of the counter-swirling array have an alternating swirl direction. The dimensions of swirler and venturi modules are shown by the authors of [40]. The width of the four-point LDI combustor has been changed to 82.18 mm, and the length of the four-point LDI combustor is also 300 mm.

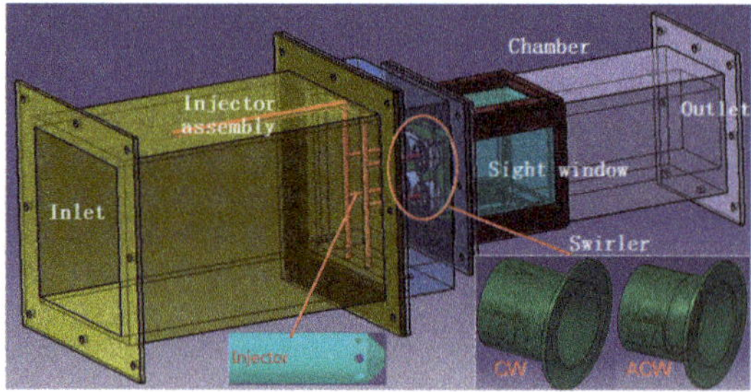

Figure 1. Geometric definition of four-point lean direct injection (LDI) combustor.

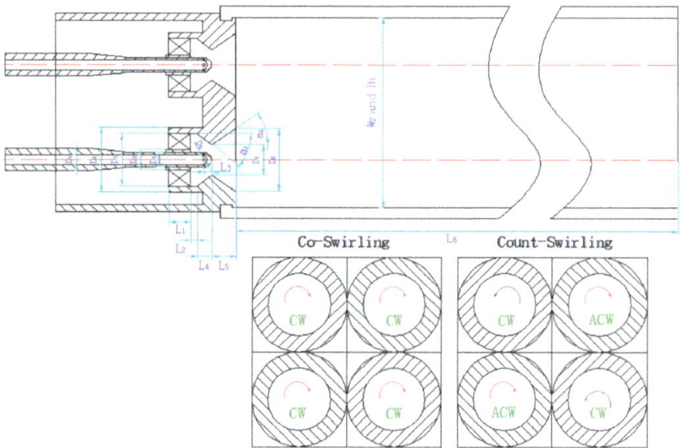

Figure 2. Schematic diagram of the co-swirling and count-swirling swirler arrays of four-point LDI combustors.

2.2. Computational Approach and Modeling

Reynolds-averaged Navier–Stokes code for steady-state reacting computations was used to investigate the flow and combustion characteristics of the four-point LDI combustor [40–42]. The conservation equations of mass, momentum, energy and species mass are expressed for steady flow as:

$$div(\rho\overline{u}) = S_m \tag{1}$$

$$\frac{\partial}{\partial x_j}(\rho\overline{u_i u_j}) = -\frac{\partial P}{\partial x_i} + \frac{\partial}{\partial x_j}(\mu\frac{\partial u}{\partial x} - \rho\overline{u_i u_j}) + \overline{F_i} \tag{2}$$

$$\nabla\overline{u}[(\rho E + p)] = \nabla(k_{eff}\nabla T - \sum_j h_j j_j + \overline{\overline{\tau_{eff}}} \cdot \overline{u}) + S_h \tag{3}$$

$$\nabla(\rho\overline{u}Y_i) = -\overline{\nabla J_i} + R_i + S_i \tag{4}$$

where,ρ, $\overline{u}$ and S_m represent density, fluid velocity vector and flow mass through control volume, P and $\overline{F_i}$ represent static pressure and all the power on the elemental volume; k_{eff}, j_j and S_h represent effective coefficients of heat conduction, diffusion mass of species j and reaction heat and other volume heat source; Y_i, J_i, R_i and S_i represent mass fraction, diffusion flux, net generation rate and extra generation rate by source defined by dispersed phase and user for species i, respectively.

In this paper, the realizable $\kappa - \varepsilon$ closure turbulence model is considered to solve the turbulence problem of the four-point LDI combustor. The equations of turbulent kinetic energy K and turbulent dissipation ε could be written as

$$\frac{\partial}{\partial x_i}(\rho k u_i) = \frac{\partial}{\partial x_j}[(\mu + \frac{\mu_\tau}{\sigma_k})\frac{\partial k}{\partial x_j}] + G_k + G_b - \rho\varepsilon - Y_M + S_k \tag{5}$$

$$\frac{\partial}{\partial x_i}(\rho\varepsilon u_i) = \frac{\partial}{\partial x_j}[(\mu + \frac{\mu_\tau}{\sigma_\varepsilon})\frac{\partial\varepsilon}{\partial x_j}] + \rho C_1 E\varepsilon - \rho C_2\frac{\varepsilon^2}{k + \sqrt{v\varepsilon}} + G_{1\varepsilon}\frac{\varepsilon}{k}C_{3\varepsilon}G_b + S_\varepsilon \tag{6}$$

where, G_k is turbulent kinetic energy generated by the average velocity gradient, G_b is turbulent kinetic energy generated by buoyancy. The G_k and G_b could be written as

$$G_k = \mu_\tau(\frac{\partial u_i}{\partial x_j} + \frac{\partial u_j}{\partial x_i})\frac{\partial u_i}{\partial x_j} \tag{7}$$

$$G_b = \beta g_i\frac{\mu_\tau}{Pr_\tau}\frac{\partial T}{\partial x_i} \tag{8}$$

where, $C_{1\varepsilon}$, C_2, σ_k and $\sigma_\varepsilon$are the constants of 1.44, 1.9, 1.0 and 1.2, respectively.

By solving the RANS equations coupled with the chemistry, the numerical simulations of the combustion process in the four-point LDI combustor induced by the flame for the H_2/Air mixture were performed. The eddy dissipation concept (EDC) combustion model [40,43] was used to solve the interaction of the turbulence and chemistry. The reaction consists of two processes. One of these processes is that the reaction is induced by collision in turbulence micro-scale structure which is controlled by chemical kinetics. The other process is that the mixing time is more than the reaction time in the vortex clouds and the reaction rate is governed by the mixing rate. The H_2/Air reaction mechanism is a reduced mechanism which consists of 11 species (H_2, O_2, OH, H_2O_2, HO_2, H_2O, H, N, NO, O, N_2) and 23 lumped reaction steps [40,44–47]. A splitting operator method was used to separately treat the aerodynamic process and the chemical process in the LDI combustion simulation.

2.3. Mesh Generation and Boundary Condition

Simplified computational domains for co-swirling and count-swirling four-point LDI combustors were modeled respectively. Using the commercial gambit software, the unstructured tetrahedral volume grids were generated and are shown in Figure 3. Using a multi-block approach, the meshes of the flow domain for the inlet passages, seven vanes passages of swirlers, the venturi and the dump were refined to ensure uniform grid quality control. The hexahedral cells are used for each vane passage and jet holes of the injector and the mesh size was identical. To reduce computing time and memory spaces of workstation, the appropriate minimum grid quantity was pursued. Therefore, the grid-sensitivities for single and four-point LDI combustors with non-reaction flow were investigated, as shown in Table 1. Compared with the effective areas of simulation and experiment results, the error of all the cases are the range of ±3.6%, which indicates that the computation grids for single and four point LDI combustors are feasible. For the single LDI, the computation grids with 2.2 million cells were an appropriate minimum grid quantity. Based on the computation grids of single LDI, the computation grids with 6.266 million cells and 6.265 million cells were acceptable meshes for co-swirling and count-swirling four-point LDI combustors, respectively.

Table 1. Grid-sensitivities for single and four-point LDI combustors with non-reaction flow.

Models / Parameter	Number of Mesh (M)	M_{in} (g/s)	ΔP (%)	A_{cd} (mm^2)	ρ_{air} (kg/m^3)	A_{cd} (Expt) (mm^2)	Error (%)
Single-LDI-CW	2.25	8.495		97.644		97.5	−0.15
	4.0	8.636		99.25			−1.8
Single-LDI-ACW	2.197	8.232	3.0	94.621	1.245	96.5	1.95
	4.0	8.572		98.529			−2.1
Four-point LDI-Co	6.266	33.22		381.84		395.6	3.48
Four-point LDI-Count	6.265	33.42		384.14		398.4	3.58

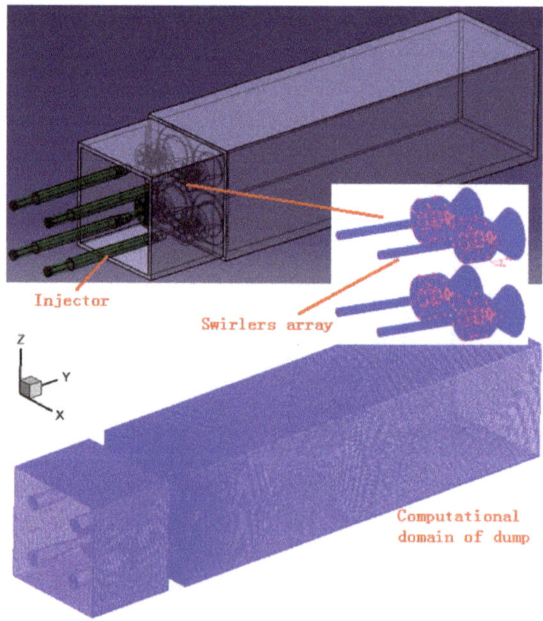

Figure 3. Mesh for a four-point LDI combustor.

The inflow boundary condition of the four-point LDI combustors includes an airflow mass-flow rate (m_{in}), fuel mass-flow rate (m_f) and inlet temperature (T_{t3}). An outflow boundary was set for the exit. Table 2 lists the boundary conditions for the four-point LDI combustors computation, in which $m_{in\text{-}1}$

and $m_{in\text{-}2}$ represent the airflow mass-flow rate of the single-element LDI combustor and four-point LDI combustor, respectively. The inlet temperature and operating pressure (P_3) are constant for valid comparison. The airflow mass-flow rates of the four-point LDI combustors are four times that of the single-element LDI combustor. The fuel mass-flow rates vary with different equivalence ratio.

Table 2. Operating conditions of four-point LDI combustors.

Parameter	Value
$m_{in\text{-}1}$ (kg/s)	0.00857
$m_{in\text{-}2}$ (kg/s)	0.03428
T_{t3} (K)	295.4
P_3 (Pa)	101,325

3. Results and Discussion

The co-swirling and count-swirling four-point LDI combustors with different equivalence ratios (0.3–1.0) were simulated with H_2/Air skeletal and reduced reaction mechanism. The flow and combustion characteristics of the four-point LDI combustors are now discussed. Figures 4 and 5 show axial velocity profiles in line, crossed with two planes, which are the center plane through the axis and the axial planes located at 2.54 mm, 20.32 mm, 81.28 mm, namely $Z = -20.32$ mm plane, $Z = 20.32$ mm plane, $Y = 2.54$ mm plane, $Y = 20.32$ mm plane, $Y = 81.28$ mm plane. Axial velocity profiles and contours present that the flow field of each swirler centerline is slightly asymmetrical. There are two asymmetrical recirculation zones near the two side walls of the combustion chambers. There are four interaction vortexes between four array swirlers. Axial velocity profiles are double-peak shaped which looks like a hump. Through reducing the fuel mass-flow rate to decrease the equivalence ratio, the axial velocities along the centerline are decreased. The axial velocity distribution is irregular and presents in a three-dimensional shape. The axial velocities are highest along the centerline of the combustion chamber. The fuel and airflow would be in the center of the combustion chamber which is induced by the velocity distribution. The distribution would result in the flame and high temperature zone being lengthened and moved backward.

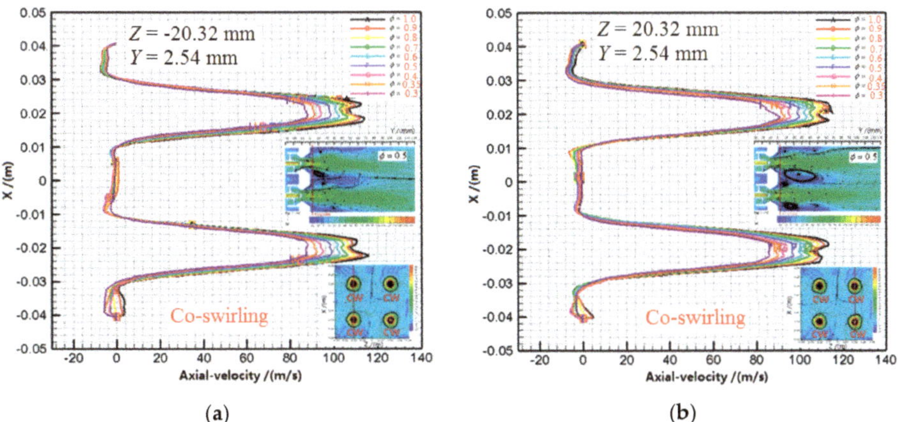

Figure 4. *Cont.*

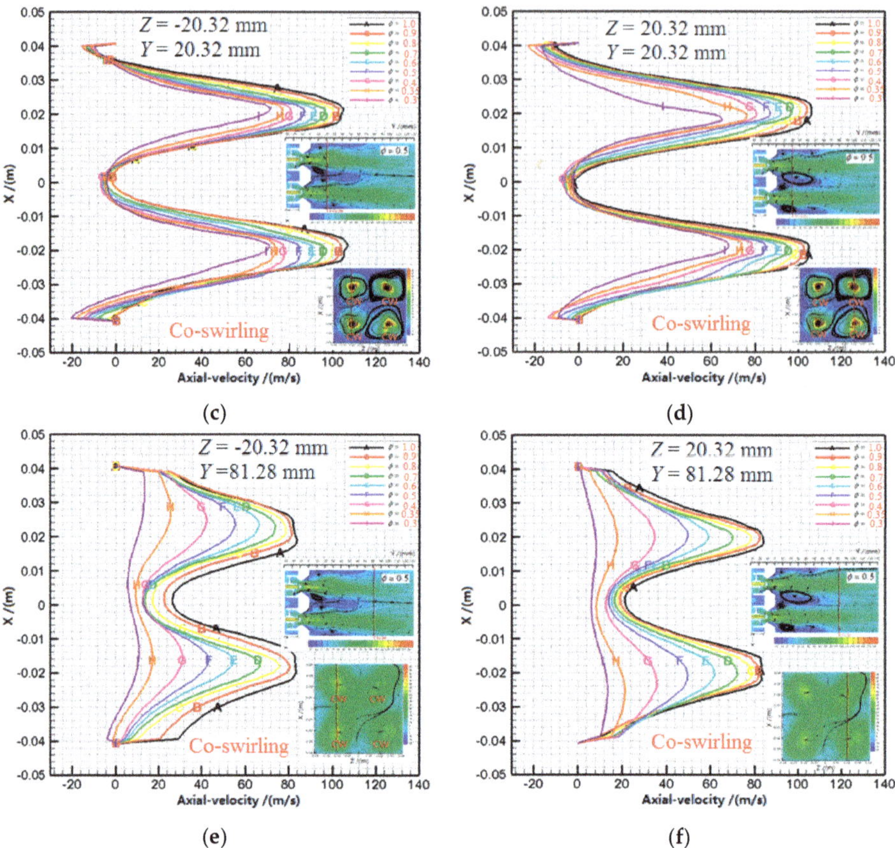

Figure 4. Axial velocity profiles and contours compared for co-swirling four-point LDI combustors with different equivalence ratios.

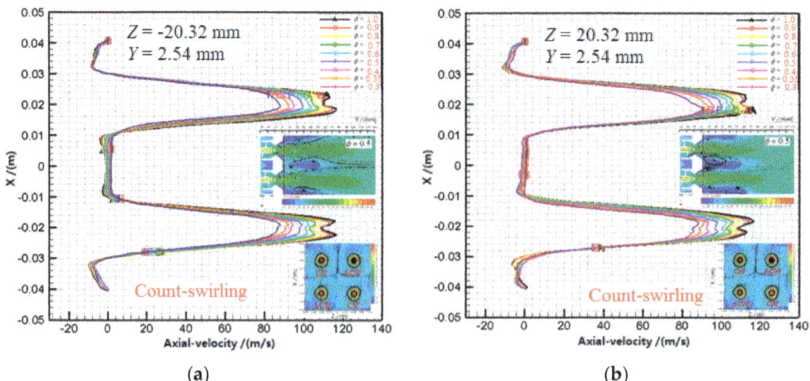

Figure 5. *Cont.*

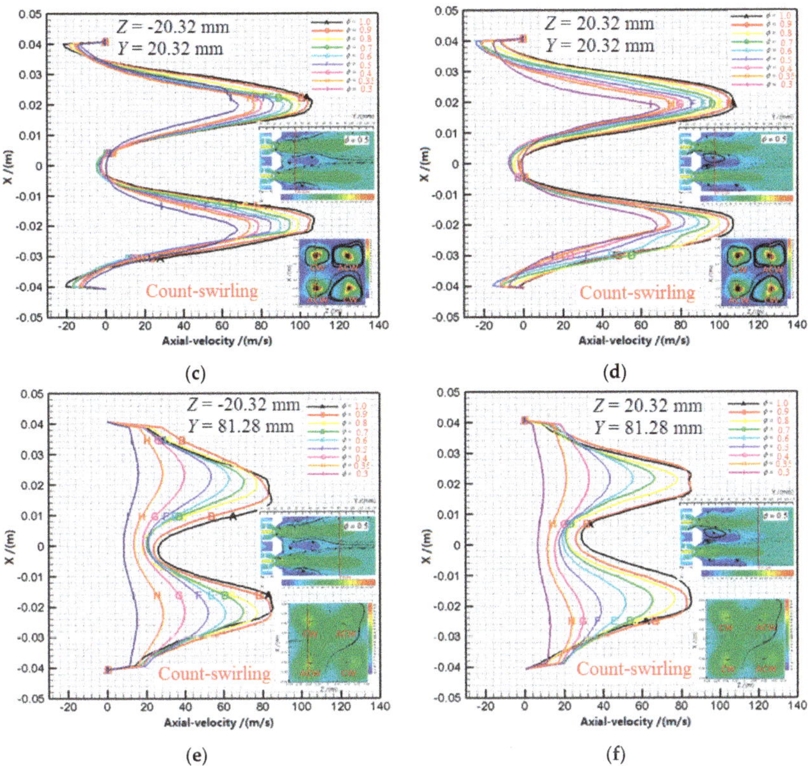

Figure 5. Axial velocity profiles and contours compared for count-swirling four-point LDI combustors with different equivalence ratios.

Figure 6 shows the turbulence kinetic energy contours of the co-swirling and counter-swirling four-point LDI combustors with an equivalence ratio of 0.5. There is no interaction between swirler–venturi at $Y = 2.54$ mm plane of co-swirling and count-swirling four-point LDI combustors. The individual flow field of each swirler–venturi is presented, respectively. The turbulence kinetic energy is highest at swirling flow fan of four-point LDI combustors. While at $Y = 10.16$ mm plane of co-swirling and count-swirling four-point LDI combustors, the interaction of turbulence kinetic energy between the four swirlers begins to emerge and the intensity of the count-swirling LDI combustor is slightly more than that of the co-swirling LDI combustor. For co-swirling and count-swirling four-point LDI combustors, the interaction of flow field between the four swirlers could be neglected before $Y = 10.16$ mm plane, only there are interaction effect in downstream flow field. While at $Y = 20.32$ mm plane of co-swirling and count-swirling four-point LDI combustors, the interaction of flow field between the four swirlers increases. The interaction of co-swirling LDI combustor is obviously less than that of the count-swirling LDI combustor. With the swirling flow running downstream, there is acute interaction effect between the swirlers of the four-point LDI combustor at $Y = 40.64$ mm and the turbulence kinetic energy is highest at the interface of the swirlers. The turbulence kinetic energy of count-swirling four-point LDI combustor is higher than that of the co-swirling four-point LDI combustor. This indicates that the count-swirling is available to improve the rapid mixing for combustion. When at $Y = 60.96$ mm of four-point LDI combustors, the highest turbulence kinetic energy locates at the center of the LDI combustor; this is because the spinning air from swirlers are colliding in the center of the count-swirling four-point LDI combustor. When at

$Y = 81.28$ mm of four-point LDI combustors, the turbulence kinetic energy contours of the co-swirling four-point LDI combustor and the count-swirling four-point LDI combustor begin to be uniform and the turbulence kinetic energy also decreases. When at $Y = 101.6$ mm of four-point LDI combustors, the turbulence kinetic energy is further weakened and tending toward uniformity. The turbulence kinetic energies of the co-swirling four-point LDI combustor and the count-swirling four-point LDI combustor are completely uniform. The different swirler arrangements for the four-point LDI combustors would result in there being different distributions of turbulence kinetic energy. This would affect the fuel-air distributions of the primary combustion zone to induce the temperature difference, which is a disadvantage for reducing the NO emission.

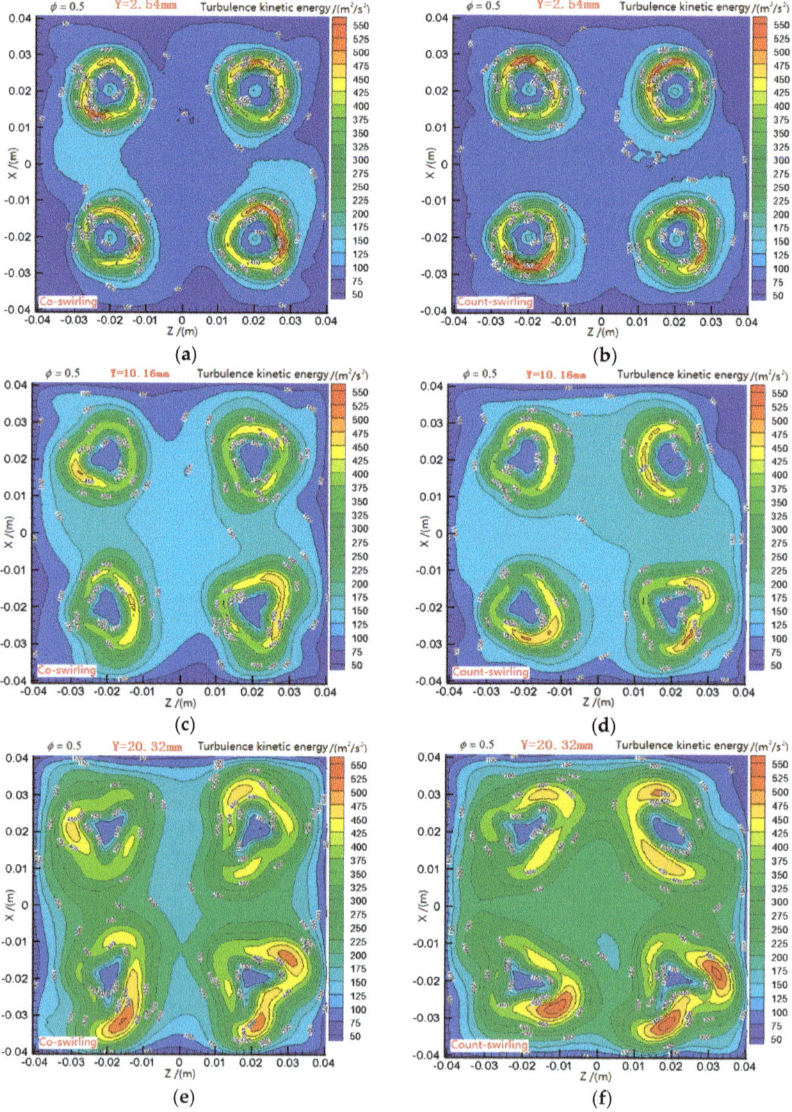

Figure 6. *Cont.*

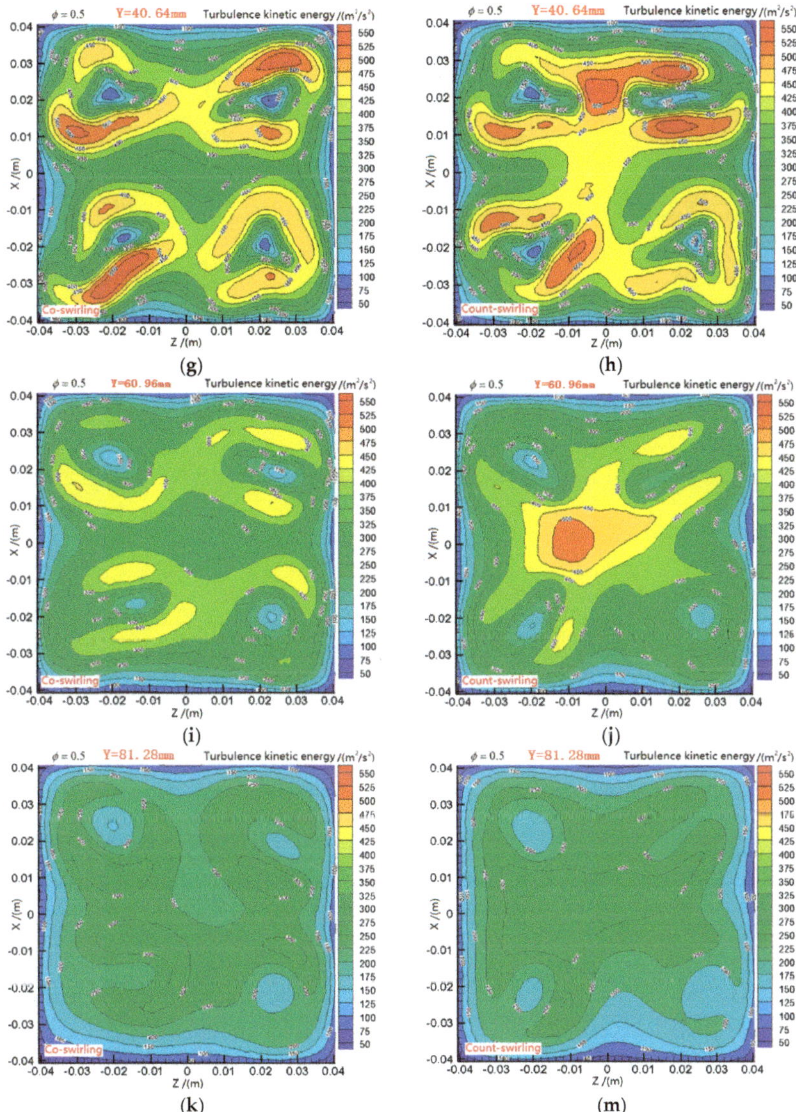

Figure 6. *Cont.*

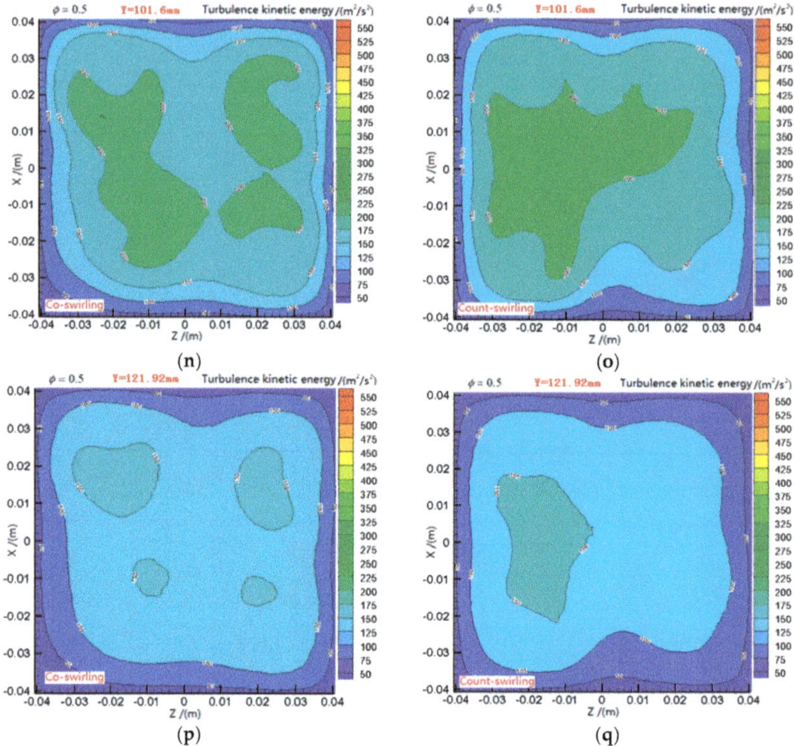

Figure 6. Turbulence kinetic energy contours compared for count-swirling four-point LDI combustors with different equivalence ratios.

Figure 7 shows the pressure drop coefficients of co-swirling and count-swirling four-point LDI combustors with different equivalence ratios. With the equivalence ratio increasing, the static temperature in the combustion chamber would also be increased, which results in increased velocity, which in turn, increases the total pressure loss. In other words, the pressure drop coefficients of co-swirling and count-swirling four-point LDI combustors are different with different equivalence ratio.

The total temperature profiles of co-swirling and count-swirling four-point LDI combustors with different equivalence ratios are shown in Figure 8. The total temperatures of co-swirling and count-swirling four-point LDI combustors with different equivalence ratios are identical to the theoretical equilibrium temperature. This tendency verified the validation of the numerical methods and codes. When the equivalence ratio is from 0.3 to 1.0, the total temperature increases in the co-swirling and count-swirling four-point LDI combustors, with the increasing of the equivalence ratio. The temperatures of co-swirling and count-swirling four-point LDI combustors increase at approximately same rate. When the equivalence ratio is from 0.3 to 1.0, the temperature rise is from 1150 K to 2050 K. This indicates that hydrogen could serve as fuel for high temperature rise combustors, if the security of storage can be solved easily.

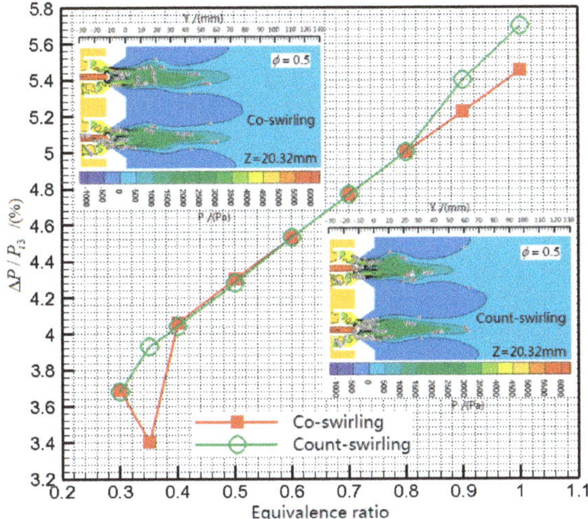

Figure 7. Total pressure drop coefficient profiles and contours compared between co-swirling and count-swirling four-point LDI combustors with different equivalence ratios.

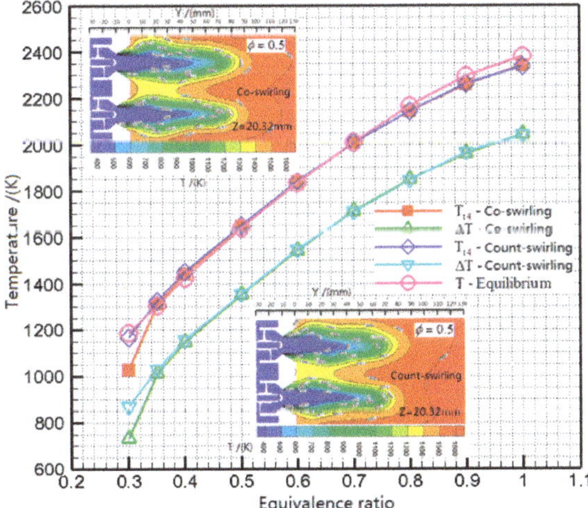

Figure 8. Total temperature profiles and contours compared between co-swirling and count-swirling four-point LDI combustors with different equivalence ratios.

Figure 9 shows the EI_{NO} profiles of co-swirling and count-swirling four-point LDI combustors with different equivalence ratios. With the equivalence ratio decreasing, the EI_{NO} are reduced sharply in the four-point LDI combustors with the equivalence ratio from 0.6 to 1.0. Corresponding to equivalent ratio of 0.6, the total temperature is approximately 1800 K. Therefore, the tendency of EI_{NO} is in line with the formation mechanism of thermal NO [42,43] which is directly related to temperature when the combustion temperature is more than 1800 K. This verifies the validation of the formation model of NO for numerical simulation. While the equivalence ratio is less than 0.6, the EI_{NO} value is very low and the mass fraction of NO is very small.

Figures 8 and 9 show the total temperature and EI_{NO} contours in the co-swirling and count-swirling four-point LDI combustors with the equivalence ratio of 0.5. For co-swirling and count-swirling four-point LDI combustors, the high temperature zones are all at the rear of combustion chamber, and the formation of NO occurred there, not near the flame at the middle of the combustion dump. This indicates that the reaction rate of H_2/Air is more than the formation rate of NO. Along the chamber, the EI_{NO} increases gradually, maximizing at the exit of chamber.

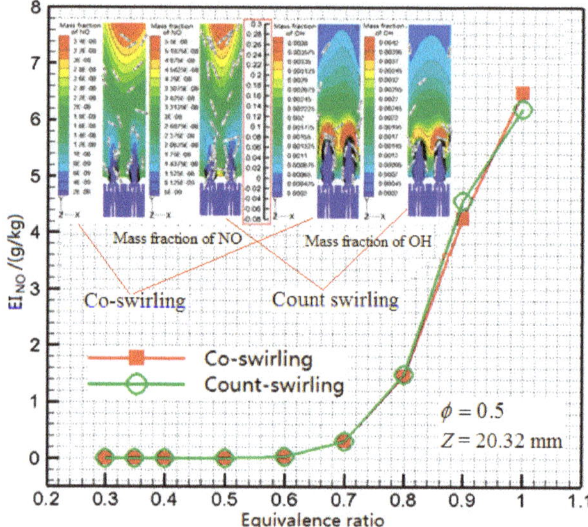

Figure 9. EI_{NO} profiles compared between co-swirling and count-swirling four-point LDI combustors with different equivalence ratios.

4. Conclusions

In this computational study, co-swirling and count-swirling four-point LDI combustors were produced and combustion characteristics in *t* LDI combustors with hydrogen/air were explored. Using Reynolds-averaged Navier-Stokes code for steady-state reacting computations, co-swirling and count-swirling four-point LDI combustors were simulated with a reduced 23-step hydrogen/air reaction mechanism. The axial velocity, total pressure drop coefficient, total temperature and emission of pollutant NO were presented and discussed. Pressure drop coefficients of co-swirling and count-swirling four-point LDI combustors with different equivalence ratio were observed. The flow field indicates that there existed some degree of flow asymmetry along each swirler centerline and the asymmetrical recirculation zones were produced near the two side walls of the combustion dump. The shape, location and length of the two asymmetrical recirculation zones changed with the decrease of equivalence ratio. The total temperature increased rapidly at the axial distance, and then increased slowly to the theoretical equilibrium temperature. The high temperature zones were all at the rear of the combustion chamber where the formation of NO took place. Along the chamber, the EI_{NO} increased gradually and maximized at the exit of the chamber. The total temperatures of co-swirling and count-swirling four-point LDI combustors with different equivalence ratios were identical to the theoretical equilibrium temperature. As the equivalence ratios increase, the total temperatures and temperature rises of co-swirling and count-swirling four-point LDI combustors also increase at approximately the same rate. When the equivalence ratio decreased from 1.0 to 0.6, the EI_{NO} reduced too, with the equivalence ratio decreasing and declining sharply.

Acknowledgments: This work was supported by "the Fundamental Research Funds for the Central Universities", No. NS2014019.

Author Contributions: Hukam C. Mongia led the study. Jianzhong Li designed the simulation and wrote the paper. Li Yuan analyzed the data and checked the paper.

Conflicts of Interest: The authors declare no conflict of interest.

References

1. Hamada, K.I.; Rahman, M.M.; Aziz, A.R.A. Time-averaged heat transfer correlation for direct injection hydrogen fueled engine. *Int. J. Hydrog. Energy* **2012**, *37*, 19146–19157. [CrossRef]
2. Das, L.M. Hydrogen engine: Research and development (R&D) programmes in Indian Institute of Technology (IIT), Delhi. *Int. J. Hydrog. Energy* **2002**, *27*, 953–965.
3. Kamil, M.; Rahman, M.M.; Bakar, R.A. Performance evaluation of external mixture formulation strategy in hydrogen fuelled engine. *J. Mech. Eng. Sci.* **2011**, *1*, 87–98. [CrossRef]
4. Kamil, M.; Rahman, M.M.; Bakar, R.A. Modeling of SI engine forduel fuels of hydrogen, gasoline and methane with port injection feeding system. *Technology* **2012**, *29*, 1399–1416.
5. Marek, C.; Smith, T.; Kundu, K. Low emission hydrogen combustors for gas turbines using lean direct injection. In Proceedings of the 41st AIAA/ASME/SAE/ASEE Joint Propulsion Conference & Exhibit, Tuscon, AZ, USA, 10–13 July 2005.
6. Ugur, K. Aircraft emissions at Turkish airports. *Energy* **2006**, *31*, 372–384.
7. Schefer, R.W.; Wicksall, D.M.; Agrawal, A.K. Combustion of hydrogen-enriched methane in a lean premixed swirl-stabilized burner. *Proc. Combust. Inst.* **2002**, *29*, 843–851. [CrossRef]
8. Bauer, C.G.; Forest, T.W. Effect of hydrogen addition on the performance of methane-fueled vehicles. Part I: Effect on S.I. engine performance. *Int. J. Hydrog. Energy* **2001**, *26*, 55–70. [CrossRef]
9. Ángel, M.; Mendoza, G.; Alzatepiedrahíta, M.V. La infancia contemporánea. *Int. J. Chem. React. Eng.* **2014**, *12*, 77–89.
10. Sarli, V.D.; Benedetto, A.D. Laminar burning velocity of hydrogen—Methane/air premixed flames. *Int. J. Hydrog. Energy* **2007**, *32*, 637–646. [CrossRef]
11. Sarli, V.D.; Benedetto, A.D.; Long, E.J.; Hargrave, G.K. Time-Resolved Particle Image Velocimetry of dynamic interactions between hydrogen-enriched methane/air premixed flames and toroidal vortex structures. *Int. J. Hydrog. Energy* **2012**, *37*, 16201–16213. [CrossRef]
12. Sarli, V.D.; Benedetto, A.D. Effects of non-equidiffusion on unsteady propagation of hydrogen-enriched methane/air premixed flames. *Int. J. Hydrog. Energy* **2013**, *38*, 7510–7518. [CrossRef]
13. Daniel, C.; Robert, I. Investigation of low emission combustors using hydrogen lean direct injection. *Incas Bull.* **2011**, *3*, 45–52. [CrossRef]
14. Schumann, U. Effects of aircraft emissions on ozone, cirrus clouds, and environmental climate. *Air Space Eur.* **2000**, *2*, 29–33. [CrossRef]
15. Berntsen, T.; Gauss, M.; Grewe, V.; Hauglustaine, D.; Isaksen, I.S.A. Sources of NO_X at cruise altitudes: Implications for predictions of ozone and methane perturbations due to NO_X from aircraft. In Proceedings of the International Conference on European Conference on Aviation, Atmosphere and Climate (AAC), Friedrichshafen, Germany, 30 June–3 July 2003; pp. 190–196.
16. Cabot, G.; Vauchelles, D.; Taupin, B.; Boukhalfa, A. Experimental study of lean premixed turbulent combustion in a scale gas turbine chamber. *Exp. Therm. Fluid Sci.* **2004**, *28*, 683–690. [CrossRef]
17. Ying, H.; Vigor, Y. Dynamics and stability of lean-premixed swirl-stabilized combustion. *Prog. Energy Combust. Sci.* **2009**, *35*, 293–364.
18. Nanduri, J.R.; Parsons, D.R.; Yilmaz, S.L.; Celik, I.B.; Strakey, P.A. Assessment of RANS-based turbulent combustion models for prediction of emissions from lean premixed combustion of methane. *Combust. Sci. Technol.* **2010**, *182*, 794–821. [CrossRef]
19. Shehata, M. Emissions and wall temperatures for lean prevaporized premixed gas turbine combustor. *Fuel* **2009**, *88*, 446–455. [CrossRef]
20. Allouis, C.; Beretta, F.; Amoresano, A. Experimental study of lean premixed prevaporized combustion fluctuations in a gas turbine burner. *Combust. Sci. Technol.* **2008**, *180*, 900–909. [CrossRef]
21. Bernier, D.; Lacas, F.; Candel, S. Instability mechanisms in a premixed prevaporized combustor. *J. Propuls. Power* **2004**, *20*, 648–656. [CrossRef]

22. Randal, M.; Domingo, S.; William, S.; Albert, C. The Pratt & Whitney Talon X low emissions combustor: revolutionary results with evolutionary technology. In Proceedings of the AIAA Aerospace Sciences Meeting and Exhibit, Reno, NV, USA, 8–11 January 2007.

23. Sebastian, G.; Marc, F.; Gilles, B.; Bernhard, B.; Katharina, G.; Oliver, K.; Sebastian, S.; Steffen, T.; Christian, O.P. Influence of steam dilution on the combustion of natural gas and hydrogen in premixed and rich-quench-lean combustors. *Fuel Process. Technol.* **2013**, *107*, 14–22.

24. Burger, V.; Yates, A.; Mosbach, T.; Gunasekaran, B. Fuel influence on targeted gas turbine combustion properties part II: Detailed results. In Proceedings of the ASME Turbo Expo 2014 Turbine Technical Conference and Exposition, Düsseldorf, Germany, 16–20 June 2014.

25. Makida, M.; Yamada, H.; Shimodaira, K. Detailed research on rich-lean type single sector combustor for small aircraft engine tested under practical conditions up to 3 MPa. In Proceedings of the ASME Turbo Expo 2012 Turbine Technical Conference and Exposition, Copenhagen, Denmark, 11–15 June 2012.

26. Makida, M.; Kurosawa, Y.; Yamada, H. Influence of injection ratio of dual-injection type air-blast fuel nozzle on emission characteristics applied to rectangular single-sector combustor under atmospheric condition. In Proceedings of the ASME Turbo Expo 2014 Turbine Technical Conference and Exposition, Düsseldorf, Germany, 16–20 June 2014.

27. Lee, C.-M.; Kathleen, M.T.; Changlie, W. ISABA. In Proceedings of the High Pressure Low NO$_X$ Emission Research: Recent Progress at NASA Glenn Research Center, Beijing, China, 2–7 September 2007.

28. Yoon, C.; Huang, C.; Gejji, R.; Anderson, W. Computational investigation of combustion instabilities in a laboratory-scale LDI gas turbine engine. In Proceedings of the AIAA/ASME/SAE/ASEE Joint Propulsion Conference, San Jose, CA, USA, 14–17 July 2013.

29. Robert, R.T.; Changlie, W.; Kyung, J.C. Flame tube NO$_X$ emissions using a lean-direct-wall-injection combustor concept. In Proceedings of the AIAA 37th Joint Propulsion Conference and Exhibit, Sali Lake City, UT, USA, 8–11 July 2001.

30. Lee, C.-M.; Tacina, K.M.; Wey, C. High pressure low NO$_X$ emissions research: Recent progress at NASA glenn research center. In Proceedings of the International Society for Air Breathing Engines (ISABE), Beijing, China, 2–7 September 2007.

31. Tacina, K.M.; Wey, C. *NASA Glenn High Pressure Low NO$_X$ Emissions Research*; NASA: Cleveland, OH, USA, 2008.

32. Kumud, A.; Hukam, C.M.; Phil, L. Evaluation of CFD Best practices for combustor design: Part I—Non-reacting flows. In Proceedings of the 51st AIAA Aerospace Meeting Including the New Horizons Forum and Exposition, Grapevine, TX, USA, 2–7 January 2013.

33. Kumud, A.; Hukam, C.M.; Phil, L. CFD Computations of Emissions for LDI-2 Combustors with Simplex and Airblast Injectors. In Proceedings of the 50th AIAA/ASME/SAE/ASEE Joint Propulsion Conference, Cleveland, OH, USA, 28–30 July 2014.

34. Christopher, M.H. Characterization of Swirl-Venturi Lean Direct Injection Designs for Aviation Gas Turbine Combustion. *J. Propuls. Power* **2014**, *30*, 1334–1356.

35. He, Z.J.; Kathleen, M.T.; Lee, C.-M.; Robert, R.T.; Phil, L. *Effects of Spent Cooling and Swirler Angle on a 9-Point Swirl-Venturi Injector*; NASA: Park City, UT, USA, 2014.

36. Fu, Y.; Jeng, S.-M. Experimental Investigation of Swirling Air Flows in a Multipoint LDI Combustor. In Proceedings of the 43rd AIAA/ASME/SAE/ASEE Joint Propulsion Conference & Exhibit, Cincinnati, OH, USA, 8–11 July 2007.

37. Dolan, B.; Villalva, R.; Munday, D.; Zink, G.; Pack, S.; Gutmark, E. Flame Dynamics in a Multi-Nozzle Staged Combustor during High Power Operation. In Proceedings of the ASME Turbo Expo 2014 Turbine Technical Conference and Exposition, Düsseldorf, Germany, 16–20 June 2014.

38. Tacina, R.; Wey, C.; Laing, P.; Mansour, A. Sector tests of a Low-NO$_X$, lean-direct-injection, multipoint integrated module combustor concept. In Proceedings of the ASME Turbo Expo 2002 Power for Land, Sea, and Air, Amsterdam, The Netherlands, 3–6 June 2002.

39. Dewanji, D.; Rao, A.G.; Pourquie, M.; van Buijtenen, J. Simulation of Reacting Spray in a Multi-Point Lean Direct Injection Combustor. In Proceedings of the 48th AIAA/ASME/SAE/ASEE Joint Propulsion Conference & Exhibit, Atlanta, Georgia, 30 July–1 August 2012.

40. Li, J.; Yuan, L.; Hukam, C.M. Simulation of combustion characteristics in a hydrogen fuelled lean single-element direct injection combustor. *Int. J. Hydrog. Energy* **2017**, *42*, 3536–3548. [CrossRef]

41. Jiang, Z.; Chang, L.; Zhang, F. Dynamic characteristics of spherically converging detonation waves. *Shock Waves* **2007**, *16*, 257–267. [CrossRef]

42. Kim, S.E.; Choudhury, D.; Patel, B. Computations of complex turbulent flows using the commercial code fluent. In *Modeling Complex Turbulent Flows*; Springer: Berlin, Germany, 1999; pp. 259–276.

43. Magnussen, B.F. Modeling of pollutant formation in gas turbine combustors based on the eddy dissipation concept. In Proceedings of the 18th International Congress on Combustion Engines, International Council on Combustion Engines, Tianjin, China, 1989.

44. Baum, M.; Poinsot, T.J.; Haworth, D.C.; Darabiha, N. Direct numerical simulation of $H_2/O_2/N_2$ flames with complex chemistry in two-dimensional turbulent flows. *J. Fluid Mech.* **1994**, *281*, 1–32. [CrossRef]

45. Zeldovich, Y. The oxidation of nitrogen in combustion and explosions. *Acta Physicochim. USSR* **1947**, *21*, 577–628.

46. Connelly, B.C.; Long, M.B.; Smooke, M.D.; Hall, R.J.; Colket, M.B. Computational and experimental investigation of the interaction of soot and NO in coflow diffusion flames. *Proc. Combust. Inst.* **2009**, *32*, 777–784. [CrossRef]

47. Sheen, H.J.; Chen, W.J.; Jeng, S.Y.; Huang, T.L. Correlation of swirl number for a radial-type swirl generator. *Exp. Therm. Fluid Sci.* **1996**, *12*, 444–445. [CrossRef]

Article

Synthetic Rock Analogue for Permeability Studies of Rock Salt with Mudstone

Hongwu Yin [1,2], Hongling Ma [1,2,*], Xiangsheng Chen [1,2], Xilin Shi [1,2,*], Chunhe Yang [1,2], Maurice B. Dusseault [3] and Yuhao Zhang [1,2]

[1] State Key Laboratory of Geomechanics and Geotechnical Engineering, Institute of Rock and Soil Mechanics, Chinese Academy of Sciences, Wuhan 430071, China; hong5yin@163.com (H.Y.); bienangua@outlook.com (X.C.); chyang@whrsm.ac.cn (C.Y.); sdkjdxzyh@163.com (Y.Z.)
[2] University of Chinese Academy of Sciences, Beijing 100049, China
[3] Department of Earth & Environmental Sciences, University of Waterloo, Waterloo, ON N2L 3G1, Canada; mauriced@uwaterloo.ca
[*] Correspondence: HonglingMa@outlook.com (H.M.); xlshi@whrsm.ac.cn (X.S.); Tel.: +86-027-8719-8721 (H.M.); +86-027-8719-8210 (X.S.)

Received: 28 July 2017; Accepted: 11 September 2017; Published: 14 September 2017

Abstract: Knowledge about the permeability of surrounding rock (salt rock and mudstone interlayer) is an important topic, which acts as a key parameter to characterize the tightness of gas storage. The goal of experiments that test the permeability of gas storage facilities in rock salt is to develop a synthetic analogue to use as a permeability model. To address the permeability of a mudstone/salt layered and mixed rock mass in Jintan, Jiangsu Province, synthetic mixed and layered specimens using the mudstone and the salt were fabricated for permeability testing. Because of the gas "slippage effect", test results are corrected by the Klinkenberg method, and the permeability of specimens is obtained by regression fitting. The results show that the permeability of synthetic pure rock salt is 6.9×10^{-20} m^2, and its porosity is 3.8%. The permeability of synthetic mudstone rock is 2.97×10^{-18} m^2, with a porosity 17.8%. These results are close to those obtained from intact natural specimens. We also find that with the same mudstone content, the permeability of mixed specimens is about 40% higher than for the layered specimens, and with an increase in the mudstone content, the Klinkenberg permeability increases for both types of specimens. The permeability and mudstone content have a strong exponential relationship. When the mudstone content is below 40%, the permeability increases only slightly with mudstone content, whereas above this threshold, the permeability increases rapidly with mudstone content. The results of the study are of use in the assessment of the tightness of natural gas storage facilities in mudstone-rich rock salt formations in China.

Keywords: gas storage; material science; rock permeability; synthetic rock salt testing; Klinkenberg method

---

1. Introduction

Rock salt possesses characteristics of low porosity, low permeability, reasonable short-term mechanical strength and stiffness and a propensity to creep stably under deviatoric stresses. Rock salt is soluble in water (~1:7 volume ratio for salt and saturated brine), allowing caverns to be dissolved, and salt caverns possess good safety characteristics (environmental and physical security). Studies in the last seven decades have led to commissioning of dissolved salt caverns for the storage of liquids, gases and even solid wastes [1–6]. In several countries (USA, Germany, etc.), salt mines are used to store radioactive wastes. In addition to high safety, dissolved caverns with adequate borehole connections may have large volumetric capacity, fast injection and withdrawal speeds and low operating costs [7].

Underground energy storage (oil, gas, compressed air) has been implemented in very thick salt or salt domes [8]. The USA, Germany, France, Canada and other countries have established underground oil or gas storage, used for commercial short-term or seasonal storage, or as national strategic energy reserves (as in the USA Strategic Petroleum Reserve). Because of the demand for commercial and strategic energy storage, the implementation of large-volume salt cavern underground storage in China has begun.

The key to salt cavern storage security is to ensure the extremely low permeability of the rock salt so as to effectively block the leakage of oil and gas. This is straightforward in thick, clean, deep deposits. However, the rock salt in China we are working with has several characteristics, such as shallow depth, low salt thickness, high impurity content (hard gypsum mudstone, gray mudstone, salt mudstone, sandy mudstone), and so on [9,10]. The physical and mechanical characteristics of these deposits are complex and challenging to measure. Therefore, it is an important design aspect to carefully study the layered rock salt, especially the permeability characteristics, and to develop an understanding using field data and laboratory tests to develop a model.

The permeability characteristics of rock salt have been extensively studied. Field tests show that rock salt permeability is generally less than 10^{-17} m^2 [11,12]. Beauheim and Roberts [13] created a conceptual model for far-field Salado hydrology involved permeability in anhydrite layers and at least some impure halite layers. They note that the pure and most impure salt have negligible permeability because of low porosity and a lack of porosity inter-connection. Their research indicated that in the near-field of an opening in salt rock (excavation damaged zone (EDZ)), dilation, creep and shear can increase the permeability (this should be most severe at layer interfaces between salt and non-salt rocks because of deformation incompatibility). They report typical average permeability values for anhydrite of approximately $10^{-18} \sim 10^{-20}$ m^2, and for pure halite less than 10^{-20} m^2 [13]. Popp [14] combined gas permeability and P and S wave velocity measurements under hydrostatic and triaxial loading conditions on rock salt specimens from the Gorleben salt dome and the Morsleben salt mine. Isotropic loading markedly decreased permeability, tending toward the in situ matrix permeability ($<10^{-20}$ m^2), with a concomitant wave velocity increase because of progressive closure of grain boundary cracks. The experiments show that permeability change is not only a function of dilatancy, but also of microcrack linkage [14]. Allemandou and Dusseault [15], using before-and-after CAT-scans on 100-mm cores, showed explicit evidence of damage as a thick external annulus of slightly higher porosity (microcracks along grain boundaries), explaining why permeability is so sensitive to isotropic stress in the laboratory. Their results also showed large effects of increased stiffness (>50%) and unconfined compressive strength (>15%) in specimens that had been re-stressed to their in situ stress (annealed) for 72 h. Indeed, sampling damage and slow grain boundary annealing likely account for a substantial amount of the experimental scatter in laboratory measurements of permeability and transient creep in salt and can be taken as evidence that properties are altered in the EDZ in the ground.

In summary, during the investigation of the physical and mechanical properties of interlayers in bedded salt deposits and their effects on storage caverns, it is found that the porosity and permeability of salt and shale are minuscule and of the same order, and values for anhydrite may be somewhat greater. Specimen damage is a significant issue and must be recognized during test programs. In bedded salt storage caverns, anhydrite interlayers may present a greater risk of being a leakage path than shale interlayers for several reasons—creep incompatibility, stiffness and higher intrinsic permeability—so evaluation and testing are needed.

Stormont and Daemen [16] used a pressure pulse method for rock salt with a permeability below 10^{-17} m^2 and found that the permeability within the EDZ is $10^{-16} \sim 10^{-20}$ m^2, whereas the permeability of intact salt is less than 10^{-21} m^2. Wu et al. [17] tested the permeability of rock salt under different osmotic pressures and compared results using the Klinkenberg effect and the quasi-static pressure method. Yan et al. [18] analysed carbonate strata storage permeability and established relative models for excavation radius, permeability and porosity. Chen et al. [19] used the equivalent boundary gas percolation model to study the gas pressure distribution in the surrounding rock within five years under different injection pressure conditions in a salt cavern natural gas storage and found that the permeability of bedding surfaces between rock salt and non-salt interlayers has an important influence on the reservoir pressure distribution.

Other similar cases involving materials from salt cavern gas storage core holes have been studied in China to generate models that can allow some generalization of the results. Taking the rock salt underground oil and gas storage facility in Jintan, Jiangsu Province, as a prototype, Zhang et al. [20] developed a reservoir medium geomechanical model for permeability. Ren et al. [21] developed similar materials for cavity experiments, and based on similarity, Jiang et al. [22] developed an artificial model material of rock salt with interlayers.

There are few studies on the permeability characteristics of mudstone-rock salt mixes in the laboratory or in the field, so it is difficult to provide reliable guidance for the construction and maintenance of the pressure integrity for caverns in salt strata with non-salt interlayers. A representative stratigraphy of salt caverns in China is shown in Figure 1.

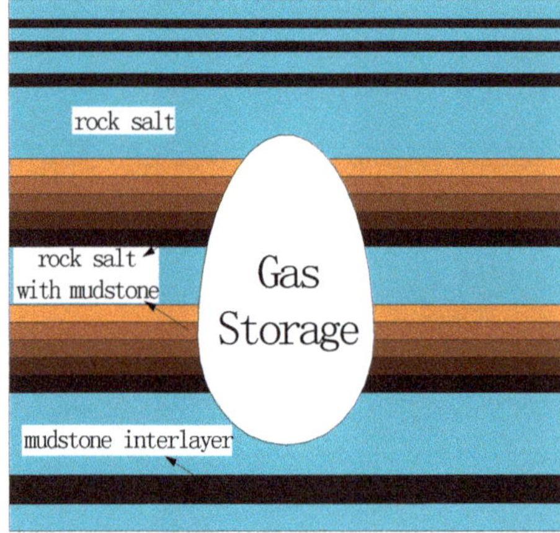

Figure 1. Schematic diagram of a salt cavern gas storage facility.

In this paper, we report on the development of a model material of rock salt with mud and then study and analyze its permeability. We took a storage cavern mudstone interlayer from Jintan and pure rock salt as our basic ingredients and made two kinds of synthetic specimens—mixed and layered—and then carried out permeability tests. These synthetic specimens can be made with any mudstone content, and once a specimen is made, it is straightforward to study the relationships between permeability and mudstone content. Indeed, the permeability testing reveals a reasonably regular relationship between the permeability of rock salt with mudstone and mudstone content, and we then explained the experimental phenomena and results. These results may help provide some methods and guidance for the study of the tightness of layered rock salt gas storage facilities in China.

2. Specimen Preparation

2.1. Material Selection and Specimen Preparation

In order to study the permeation behavior of rock salt with mudstone, we use natural mudstone from the gas storage facility in Jintan from a depth of approximately 934 m–935 m. Relatively pure rock salt is used to develop mud rock/salt specimens with different mudstone content. The intact rock salt has grains that are small and evenly distributed, and the grain boundaries are not strongly apparent. The salt is pale red and has a content of soluble matter over 96.3%.

The weakly consolidated clay forming a strongly consolidated rock through moderate epigenetic effects (such as compaction, dehydration, recrystallization and cementation) is called mudstone. The natural mudstone used in this paper is from the gas storage facility in Jintan from a depth of approximately 934 m–935 m. The mudstone we use is hard, brittle and relatively dense and has some secondary structures such as porous structure, pinhole structure and honeycomb structure in locality. Some of the internal cracks of mudstone are filled by glauberite in a later stage and are stellate distributed. X-ray diffraction shows that quartz, dolomite and illite have the largest proportions in the mudstone, and the average quartz content is about 32%. The mudstone has a small amount of interstitial salt, and the main soluble mineral in the mudstone is glauberite, $Na_2Ca(SO_4)_2$. The XRD result of mudstone is shown in Figure 2.

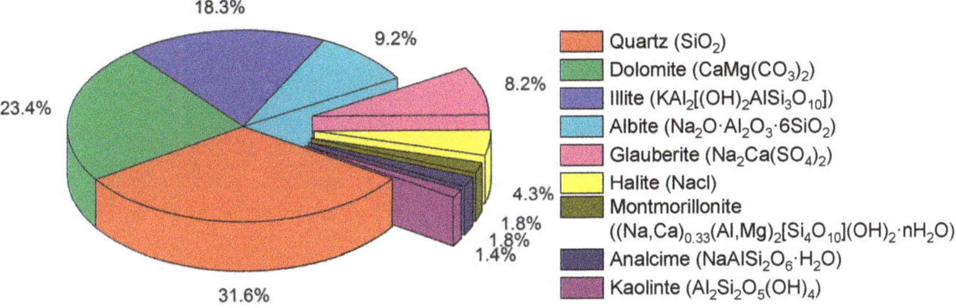

Figure 2. XRD results of mudstone.

The pure rock salt and mudstone (Figure 3) were pulverized into 0.5-mm and 2-mm particles respectively, uniformly mixed with a small amount of cement and near-saturated brine (~1.18 g/cm³). The pulverized rock salt particles are then placed in a drying box for 24 h, and the mudstone particles are put in a sealed bag.

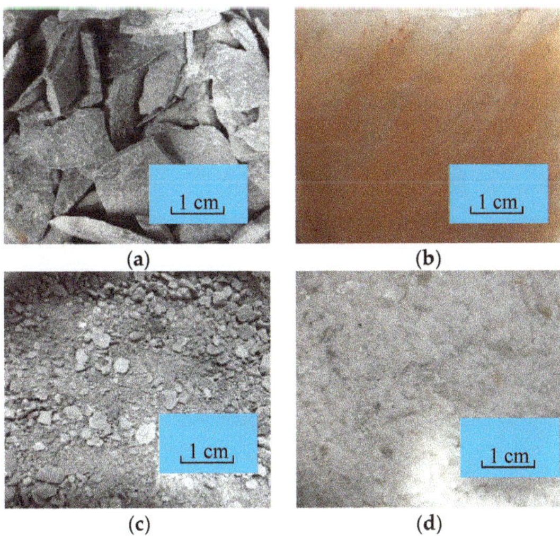

Figure 3. Material selection. (**a**) Mudstone; (**b**) rock salt; (**c**) the particles of mudstone; (**d**) the particles of rock salt.

We make specimens in accordance with the prescribed method [23] with a diameter of 25 mm and an H/D ratio of about two. The rock salt particles and mudstone particles are pressed and formed by a hydraulic pressure tester, with material proportioning shown in Table 1. Specimens are prepared as follows:

A.   Mixed specimens of rock salt with mudstone:

    1.   Weigh specified amounts of rock salt particles and mudstone particles, as well as cement and brine, then put them in a beaker and stir until homogeneous.

    2.   Add the mixed material to the pressing mold.

    3.   Using 120 MPa of compression stress, hold for 2 h.

B.   Layered specimens of rock salt with mudstone:

    4.   Weigh a specified amount of particles of rock salt and mudstone.

    5.   Place prepared material in the mold in the order of rock salt-mudstone-rock salt.

    6.   Using 120 MPa of compression stress, hold for 2 h.

Table 1. Material proportions.

Rock Salt with Mudstone	Material Proportion Y:N:C:W
Rock salt	1:0:0:0
Mudstone	0:1:0.12:0.02
Binder	0:0:6:1

Note: Y is the mass of the rock salt; N is the mass of the mudstone; C is the mass of cement; W is the mass of brine.

Note that in nature, the prolonged geological process of compaction takes place under higher temperature, but lower stress for millions of years. In the laboratory, we use ultra-high pressure to yield an artificial specimen with similar porosity to the natural specimens, but of course, there is no time for recrystallization through internal mass transfer.

The specimens are placed in a drying box for 24 h; the physical parameters are obtained, and then, the permeability test is performed. The density of synthetic pure salt is 2.125 g·cm^{-3}, and the density of synthetic pure mudstone is 2.283 g·cm^{-3}, values similar to those from the research of Jiang et al. [22]. The porosity of synthetic pure rock salt is 3.8% with a permeability of 6.93×10^{-20} m^2, and the porosity of synthetic pure mudstone is 17.8% with a permeability of 2.97×10^{-18} m^2. These values are close to those of natural specimens. The two kinds of synthetic specimens are shown in Figure 4, and the physical parameters are listed in Tables 2 and 3.

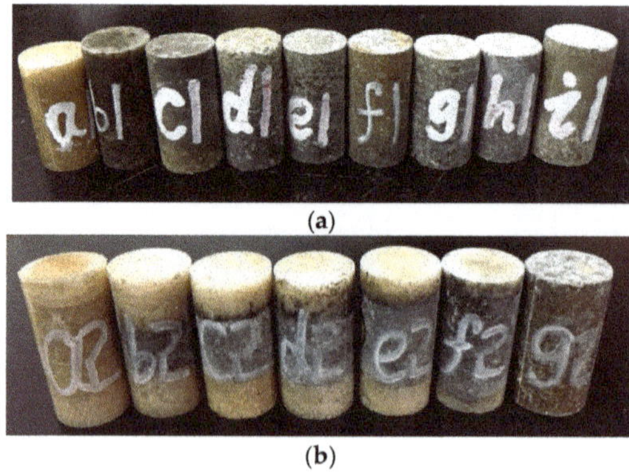

(a)

(b)

Figure 4. Synthetic specimens. (**a**) Mixed rock salt with mudstone specimens; (**b**) specimens of layered rock salt with mudstone.

Table 2. Physical parameters of mixed specimens.

No.	Diameter (mm)	Height (mm)	Mass (g)	Density (g·cm^{-3})	Mudstone Content (%)
a1	25.47	47.67	51.1576	2.106	0
b1	25.49	52.63	57.1631	2.128	0
c1	25.61	51.17	57.1781	2.169	20
d1	25.57	50.48	57.1842	2.206	40
e1	25.61	49.89	56.9402	2.216	50
f1	25.57	49.23	57.0120	2.255	60
g1	25.69	48.47	56.9275	2.266	80
h1	25.65	47.19	56.9406	2.335	100
i1	25.39	50.01	57.2017	2.259	100

Note: a1 is made of particles of pure rock salt; b1 is synthetic pure rock salt with a small amount of cement and brine; h1 is synthetic pure mudstone with a small amount of cement and brine; i1 is made of particles of pure mudstone.

Table 3. Physical parameters of layered rock salt.

No.	Diameter (mm)	Height (mm)	Mass (g)	Density (g·cm^{-3})	Mudstone Content (%)
a2	25.47	51.49	56.2170	2.143	0
b2	25.63	51.79	57.3667	2.147	20
c2	25.33	50.77	57.4733	2.246	40
d2	25.43	50.62	57.6456	2.242	50
e2	25.39	51.64	59.3861	2.271	60
f2	25.49	49.81	58.5075	2.302	80
g2	25.61	50.08	59.5015	2.307	100

Note: a2 is made of particles of pure rock salt; g2 is made of particles of pure mudstone.

In the preparation process, we find that the compressibility of mudstone is higher than that of rock salt. In Table 3, the mass of a2 is higher than that of g2, but the height relationship is opposite, a reasonable result as the height of the specimen is less with an increase in the proportion of mudstone. The relationship between the specimen height and the mudstone content of b1~h1 is shown in Figure 5.

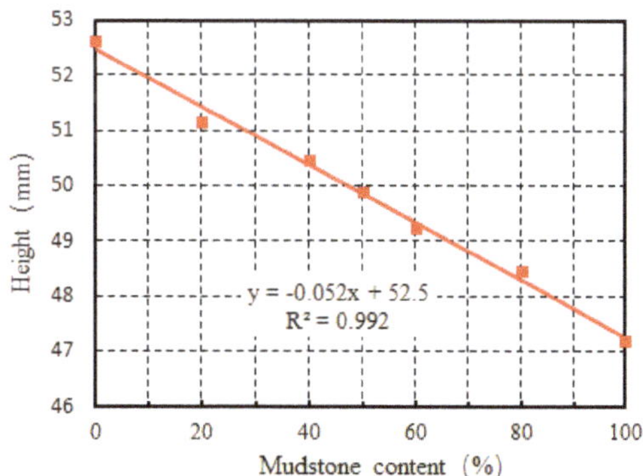

Figure 5. The relationship between the height and the mudstone content of b1~ h1.

2.2. Compressibility Analysis of Mudstone and Rock Salt

From Figure 5, the specimen height variation is linear with the mudstone content under the same mass and stress conditions, and the height of the specimen decreases with the increase of the mudstone content because the mudstone is more compressible than rock salt. The mudstone is porous and composed of clay minerals and other finely divided minerals, so its compressibility is higher than rock salt. Now, we compare the compressibility of mudstone and rock salt from the theoretical point of view.

Hall [24] defined a rock effective compression factor:

$$c_p = \frac{dV_P}{V_P \cdot dP} \tag{1}$$

where c_p is rock compression factor, V_P is the rock pore volume and P is pressure.

The commonly-used Hall chart curve [25] is shown in Figure 6, and the empirical formula for $c_p - \phi$ is:

$$c_p = \frac{2.59 \times 10^{-4}}{\phi^{0.44}} \tag{2}$$

where ϕ is the porosity of rock.

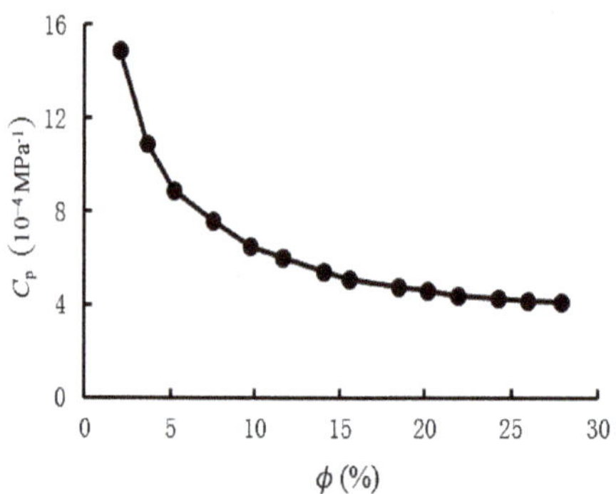

Figure 6. The compressibility $\times$ porosity chart from Hall [25].

When the ordinary single-phase solid material is compressed, the compression factor can be defined by the following equation [26]:

$$c_s = \frac{dV_s}{V_s \cdot d\sigma_s} \tag{3}$$

Here, c_s is the compression coefficient of rock solid matter in MPa^{-1}, and σ_s is the average of the three principal stresses of the solid matter in MPa.

In the case of elastic deformation, the compression coefficient for solid matter can be calculated by the following equation [27]:

$$c_s = \frac{3(1 - 2\nu)}{E_s} \tag{4}$$

where ν is Poisson's ratio and E_s is the Young's modulus. Li [26] derived the following relationship for compressibility and porosity:

$$c_p = \frac{\phi}{1 - \phi} c_s \tag{5}$$

For a rock of the same material, c_s is roughly a constant, so the compression coefficient of the rock increases as the porosity increases. At the same time, the mineral properties affect the compression coefficient, and the compression coefficient of the rock decreases as the stiffness of the mineral increases, which is partly why mudstone is easier to compress than salt rock.

3. Permeability Calculation Method

3.1. Klinkenberg Effect

When gas flows in a low permeability porous medium, in contrast to liquid laminar flow, there is a non-zero velocity at the contact with the solid phase, known as gas slippage or the Klinkenberg effect. Gas slippage only occurs when the mean-free-path of the gas molecules and the diameter of the flow channel are comparable, and the lower the rock permeability, the more pronounced the Klinkenberg effect [12,24]. Apparently [12], the mean-free-path of the gas molecules is equivalent to the pore size ($10^{-7} \sim 10^{-9}$ m) of the salt rock when the gas pressure is 0.06~6.00 MPa, and the Klinkenberg effect is significant at this condition, so calibration is needed. Li et al. [28] found that the slippage effect has a great influence on gas seepage when the permeability of a reservoir sandstone is less than 1 mD (millidarcy).

Klinkenberg [29] observed through experiments that in single-phase gas flow through a porous medium, the permeability is:

$$K_g = K_\infty\left(1 + \frac{b}{\overline{p}}\right) \tag{6}$$

where K_g is the gas permeability of the rock in millidarcy (mD, 1 mD = 10^{-3} μm^2); K_∞ is the absolute permeability (or Klinkenberg permeability) in mD; b is the slip factor (the strength of the slippage effect in the porous medium in MPa); and $\overline{p}$ is the average gas pressure in MPa.

Klinkenberg gave this expression for the slip factor b:

$$b = \frac{4C_1\overline{\lambda}p}{r} \tag{7}$$

where C_1 is a constant (close to one); r is the average radius of the pore in μm; $\overline{\lambda}$ is the mean-free path of gas molecules under average gas pressure, which is given by [30]:

$$\overline{\lambda} = \frac{C_2\mu\left(\frac{RT}{M}\right)^{\frac{1}{2}}}{p} \tag{8}$$

where C_2 is a constant (close to two); μ is the gas viscosity in mPa·s (10^{-3} Pa·s); R is the universal gas constant (0.86367 L·MPa/(g·mol)); T is the absolute temperature, K; M is the relative molecular weight of the gas.

Substituting Equation (8) into Equation (7):

$$b = \frac{a\mu\left(\frac{T}{M}\right)^{\frac{1}{2}}}{r} \tag{9}$$

where $a = 4C_1C_2R^{\frac{1}{2}}$.

Clearly, the smaller the pore throat, the higher the value of b [31] and the lower the specimen permeability. The relationship between permeability, porosity and pore throat radius for a circular equivalent pore throat is given by [32,33]:

$$K = \frac{\phi r^2}{8} \tag{10}$$

where K is the liquid permeability in mD, comparable to the Klinkenberg permeability (K_∞); and ϕ is rock porosity.

3.2. Permeability Calculation

The pseudo-pressure method is used to calculate the specimen permeability. The gas seepage equation is based on the following assumptions: (1) the fluid is a one-component gas; (2) the flow is isothermal; (3) the medium is uniform and isotropic, and the porosity ϕ is constant; (4) gravity can be neglected; and (5) flow is laminar so that Darcy's law holds.

Based on the above assumptions and conditions and combining continuity, momentum and state equations, the general form of the partial differential equation for gas seepage is derived [24]:

$$\frac{K_g}{\phi} \nabla \cdot \left[\frac{p \nabla p}{\mu(p) Z(p)} \right] = \frac{\partial}{\partial t} \left[\frac{p}{Z(p)} \right] \tag{11}$$

Here, p is the gas pressure; Z is the deviation factor for correction of ideal gas; $Z(p)$ is a function of the deviation factor with respect to p; and $\mu(p)$ is the gas viscosity function with respect to p.

The pseudo-pressure is defined as:

$$m(p) = 2 \int_{p_m}^{p} \frac{p}{\mu Z} dp \tag{12}$$

where $m(p)$ is the pseudo-pressure; p_m is any reference pressure, which can be taken as 0 or 0.1 MPa.

Equations (11) and (12) are used to obtain a partial differential equation, which describes the true gas percolation represented by the pseudo-pressure m.

$$\nabla^2 m = \frac{\phi c_g(p) \mu(p)}{K_g} \cdot \frac{\partial m}{\partial t} \tag{13}$$

where $c_g = \frac{1}{p} - \frac{1}{Z(p)} \left(\frac{dZ(p)}{dp} \right)_T$ is the isothermal compression coefficient of the real gas.

For an isotropic homogeneous specimen of length L and cross-sectional area A with the pressures at the upper ($x = 0$) and the lower ($x = L$) extremes delineated as p_0 and p_L, the seepage equation and the boundary conditions are as follows.

$$\frac{d^2 m}{dx^2} = 0 \qquad\qquad (0 < x < L) \tag{14}$$

$$m = m_0 = \frac{p_0^2}{\mu_0 Z_0} \qquad\qquad (x = 0) \tag{15}$$

$$m = m_L = \frac{p_L^2}{\mu_L Z_L} \qquad\qquad (x = L) \tag{16}$$

where μ_0 is the gas viscosity at the top of the specimen of length L; μ_L is the gas viscosity at the bottom of the specimen; Z_0 is the deviation factor at the top of the specimen; Z_L is the deviation factor at the bottom of the specimen.

Linear interpolation of (14)–(16) gives:

$$m = m_0 - \frac{m_0 - m_L}{L} x \tag{17}$$

where x is the distance from the top of the specimen.

Taking the average, $\mu = \bar{\mu}$, $Z = \bar{Z}$, we get $m - m_0 = \frac{p^2 - p_0^2}{\mu \bar{Z}}$. Taking Equation (12) into Equation (17), we can write:

$$p^2 = p_0^2 - \left(\frac{\mu \bar{Z}}{\mu_0 Z_0} p_0^2 - \frac{\mu \bar{Z}}{\mu_L Z_L} p_L^2 \right) \frac{x}{L} \tag{18}$$

For gas seepage, the volume flow varies with pressure. Using the mass flow F, which is invariant, and assuming that the flow rate is not too large and follows Darcy's law, Kong [24] obtained the volume flow rate Q_{sc} as:

$$Q_{sc} = \frac{F}{\rho_{sc}} = \frac{AK_g T_{sc}}{2pL\mu\overline{Z}T}(p_0^2 - p_L^2) \tag{19}$$

where Q_{sc} is the volume flow rate, ρ_{sc} is the gas density, T_{sc} is the temperature, all under standard conditions, and p is atmospheric pressure.

Equation (19) can also be written as:

$$K_g = \frac{2Q_{sc}pL\mu\overline{Z}T}{AT_{sc}(p_0^2 - p_L^2)} \tag{20}$$

During the test, we keep the outlet pressure constant, change the inlet pressure and get the gas flow rate Q_{sc} at different pressure differences. The gas permeability K_g is obtained according to Equation (20), and the Klinkenberg permeability K_∞ and slip factor b can be determined by fitting the permeability of gas obtained by multipoint measurements through Equation (6).

4. Test Results

4.1. Porosity

4.1.1. Testing Equipment and Principle of Porosity Measurement

Our helium porosity measuring instrument is designed using the standards of the American Petroleum Institute (API) [34] and Geology and Mineral Resources of China to measure the particle volume and helium porosity of rock (Figure 7). The helium atom has the advantages of small volume (minimal viscosity), stable ideal gas behavior and low adsorbtivity on clay mineral surfaces; this gives easy and accurate measurements without altering the nature of the specimen, which is why He measurements are standard in such cases.

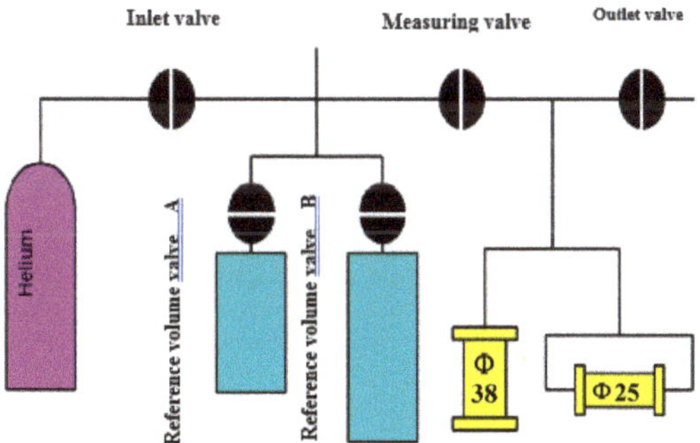

Figure 7. Helium porosity measurement instrument diagram.

Measurement of core porosity by the gas uses Boyle's law:

$$P_1 \cdot V_1 = P_2 \cdot V_2 \tag{21}$$

The porosity is calculated as follows:

$$\phi = \frac{V_P}{V_P + V_G} \times 100\% \tag{22}$$

where ϕ is porosity, V_P is the pore volume, and V_G is the particle volume. The formulae for measuring the particle volume and pore volume are:

$$P_1 \cdot V_R = P_2 \cdot (V_R + V_M - V_G) \tag{23}$$

$$P_1 \cdot V_R = P_2 \cdot (V_P + V_D + V_R) \tag{24}$$

where P_1 is the initial pressure, P_2 is the equilibrium pressure, V_R is the reference volume, V_M is the volume of the media cup, V_D is the volume of the closed system, V_G is the solids volume of the specimen and V_P is the pore volume of the specimen.

Taking Equations (22) and (23) into Equation (21), the porosity is determined. Table 4 displays the porosity test results.

Table 4. The results of the porosity test.

No.	Type	Mudstone Content (%)	Particle Volume (cm^3)	Pore Volume (cm^3)	Porosity (%)
a1	Mixed	0	23.35	1.21	4.9
b1	Mixed	0	26.07	0.79	2.9
c1	Mixed	20	24.66	1.69	6.4
d1	Mixed	40	23.75	3.22	11.9
e1	Mixed	50	23.02	2.67	10.4
f1	Mixed	60	23.51	2.43	9.4
g1	Mixed	80	21.69	3.47	13.8
h1	Mixed	100	20.53	3.95	16.1
i1	Mixed	100	21.39	5.11	19.3
a2	Layered	0	26.11	0.70	2.6
b2	Layered	20	24.89	1.77	6.6
c2	Layered	40	23.83	1.71	6.7
d2	Layered	50	23.65	1.99	7.8
e2	Layered	60	23.79	2.34	9.0
f2	Layered	80	22.66	2.77	10.9
g2	Layered	100	21.56	4.23	16.4

4.1.2. Porosity Test Results

From Table 4, the porosity of synthetic pure rock salt is 3.9%, and the porosity of synthetic pure mudstone is 17.8%. The test results are consistent with the range of natural rock porosity, indicating that our method of using such materials to make similar specimens is reasonable, under the circumstances (no analogue is perfect in geomaterials). The porosity and compressibility of the specimen increase with the mudstone content, which is consistent with the results described in Section 2.2.

4.2. Permeability Test

4.2.1. Test Equipment and Principles

The instrument used for the test is a low permeability measurement instrument designed using the standards of the American Petroleum Institute (API) [34] and Geology and Mineral Resources of China (Figures 8 and 9). It measures the permeability of the rock specimen under steady nitrogen (N_2) gas flow. Changing the upstream pressure, the Klinkenberg permeability and slip factor b can be fitted by the permeability of gas obtained by multipoint measurements.

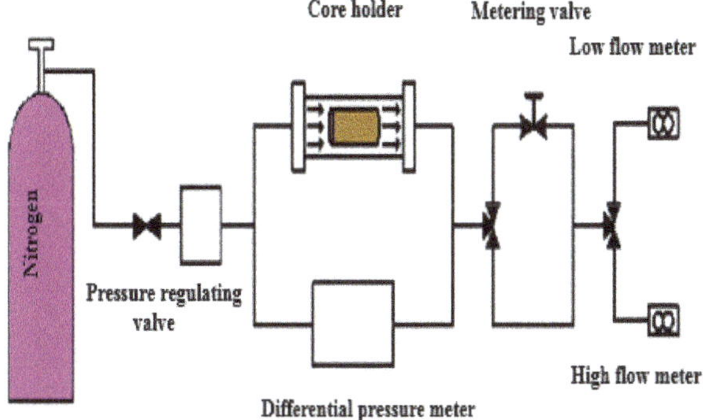

Figure 8. The diagram of the low permeability measurement instrument.

Figure 9. Low permeability measurement instrument.

4.2.2. Permeability Test Results

The permeability of the specimen at different inlet pressures can be calculated by Equation (20). In the experiment, p is the atmospheric pressure, and μ is the viscosity coefficient corresponding to the test temperature. The experimental process is isothermal, and the deviation factor Z is one [24] for N_2 at these conditions, so Equation (20) can be written as:

$$K_g = \frac{2QpL\mu}{A\left(p_0^2 - p_L^2\right)} \tag{25}$$

where Q is the test measured volume flow; p is the atmospheric pressure; μ is the viscosity coefficient corresponding to the test temperature; p_0 is the inlet pressure; p_L is the outlet pressure; L refers is the height of the specimen.

Gas flow is perpendicular to the direction of the mudstone salt interface for layered rock salt. During the test, the confining pressure is 1.38 MPa, the outlet pressure 0.10 MPa and the temperature 25 °C. Table 5 shows the results for synthetic pure rock salt Specimens a1, a2 and synthetic pure mudstone Specimens i1, g2.

Table 5. Permeability test results.

NO.	Inlet Pressure (MPa)	Reciprocal of Mean Pressure $1/p$ (MPa^{-1})	Volume Flow (mL·s^{-1})	K_g (10^{-21} m^2)
a1	0.333	4.74	0.008	2.65×10^2
a1	0.414	4.15	0.011	2.29×10^2
a1	0.475	3.62	0.014	2.18×10^2
a1	0.545	3.19	0.017	1.99×10^2
a2	0.305	4.92	0.004	1.75×10^2
a2	0.383	4.13	0.006	1.60×10^2
a2	0.450	3.63	0.008	1.51×10^2
a2	0.523	3.23	0.010	1.38×10^2
i1	0.311	4.84	0.109	4.43×10^3
i1	0.377	4.18	0.158	4.20×10^3
i1	0.451	3.62	0.226	4.11×10^3
i1	0.511	3.27	0.283	3.96×10^3
g2	0.322	4.77	0.093	3.49×10^3
g2	0.388	4.09	0.137	3.43×10^3
g2	0.457	3.58	0.189	3.34×10^3
g2	0.518	3.23	0.242	3.30×10^3

The volume flow rate increases with the increase of the inlet pressure, while the permeability decreases gradually. This indicates that the slippage effect is obvious and consistent with the results of Cosenza et al. [12].

5. Analysis of Test Results

5.1. Comparison of Porosity and Permeability between Synthetic Specimens and Natural Specimens

With Equation (20), we obtained the specimen gas permeability at four different inlet pressures, and the Klinkenberg permeability and slip factor b can be determined by the pseudo-pressure method. The results are shown in Figure 10; the indicators a1~i1 in the figure are mixed specimens of mudstone and rock salt, and Specimens a2~g2 are layered rock salt.

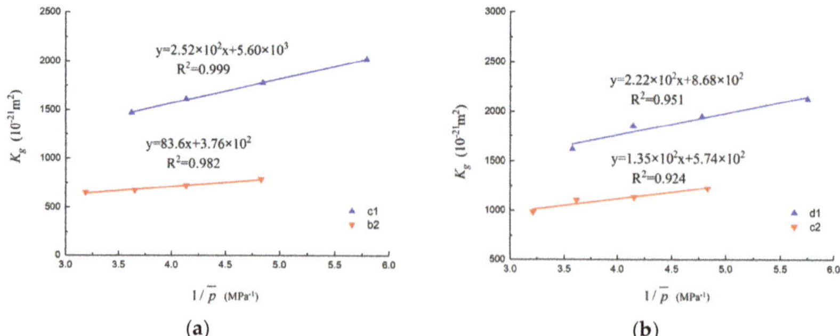

Figure 10. *Cont.*

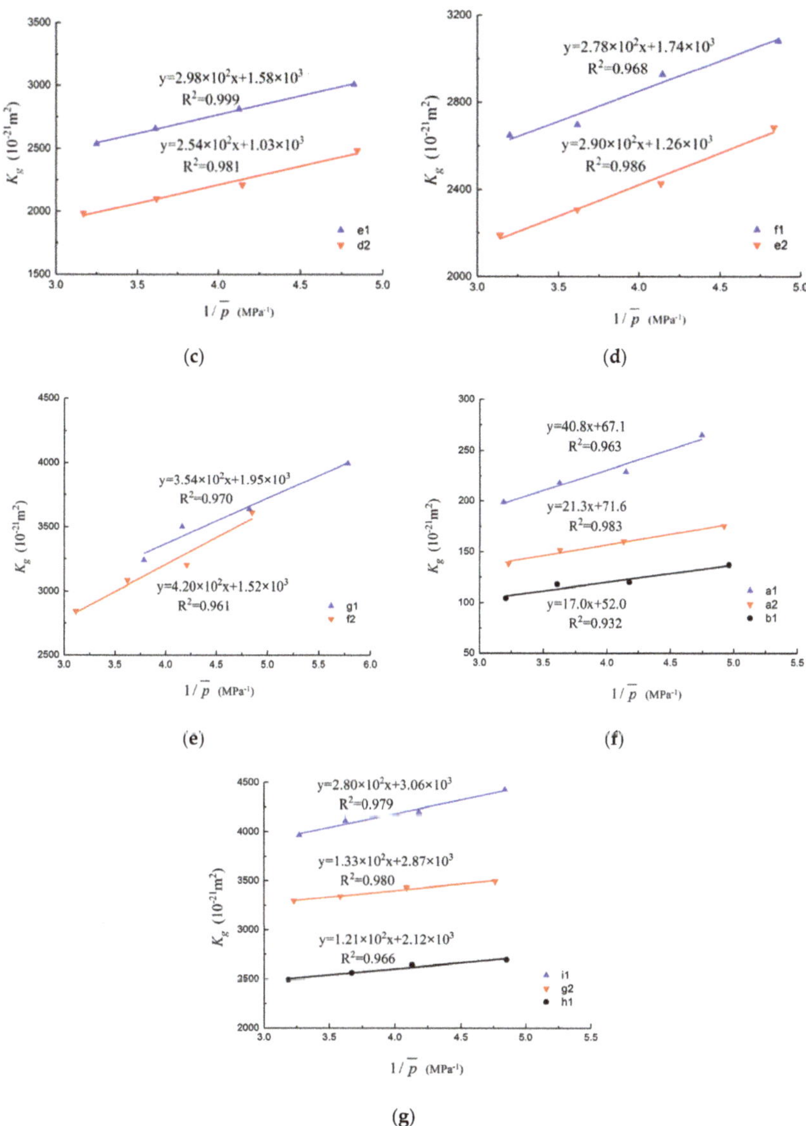

Figure 10. Fitting results of K_g and $1/p$. (**a**) Mudstone content is 20%; (**b**) mudstone content is 40%; (**c**) mudstone content is 50%; (**d**) mudstone content is 60%; (**e**) mudstone content is 80%; (**f**) synthetic pure rock salt and synthetic pure rock salt with binder; (**g**) synthetic pure mudstone and synthetic pure mudstone with binder.

From Figure 10, we can see that the correlation coefficients are close to one, so the results are consistent and reliable. From the data, the Klinkenberg permeability of synthetic pure rock salt specimens is 6.93×10^{-20} m^2 (the average values of a1 and a2), and that of synthetic pure mudstone specimens is 2.97×10^{-18} m^2 (the average values of i1 and g2). The test results are in agreement with the natural specimens, indicating that the model material is a reasonable analogue of the natural material. At the same time, the permeability of synthetic pure rock salt with a small amount of cement and brine (b1) is about 1/3 lower than that of synthetic pure rock salt, and the permeability of synthetic pure mudstone with a small amount of cement and brine (h1) is about 1/3 lower than that of synthetic pure mudstone. The cement and brine in the mudstone and rock salt particles play a role in blocking permeable channels.

According to the porosity in Table 3 and the Klinkenberg permeability in Table 6, the effective pore throat radius of the synthetic pure rock salt is 1.445×10^{-3} µm calculated by Equation (10) (the average values of a1 and a2), and the effective pore throat radius of the synthetic pure mudstone is 2.059×10^{-2} µm (the average values of i1 and g2). From the comparison of the two kinds of specimens in Figure 10, the results show that the permeability of the mixed specimens is about 40% higher than that of the layered mudstone/salt specimens. In a salt cavern gas facility, the permeability of the transition zone of rock salt and mudstone is likely to be higher than that of layered mudstone/salt interbed area. The permeability of the rock salt bands in situ is of course very low because the crystals are large, the grain arrangement is dense and the pore throat equivalent radius is small. Yang [35] found that the interface of rock salt and mudstone is not weak, and the rock salt crystals and mudstone particles form a zig-zag interface. Because the intact rock salt permeability is low, the interface of the rock salt with the mudstone plays a leading role in overall permeability. Note that the equivalent pore throat size of mixed specimens formed by two kinds of particles is larger than that of layered rock salt, so the permeability of the mixed specimens is slightly higher than the layered rock salt. The fitting results of Klinkenberg permeability are shown in Table 6.

Table 6. The fitting results.

No.	K_∞ (10^{-21} m^2)	b (MPa)	R^2
a1	67.1	0.61	0.963
b1	52.0	0.33	0.932
c1	5.60×10^2	0.45	0.999
d1	8.68×10^2	0.26	0.951
e1	1.58×10^3	0.19	0.999
f1	1.74×10^3	0.17	0.968
g1	1.95×10^3	0.19	0.970
h1	2.12×10^3	0.06	0.966
i1	3.06×10^3	0.09	0.979
a2	71.6	0.30	0.983
b2	3.76×10^2	0.22	0.982
c2	5.74×10^2	0.24	0.924
d2	1.03×10^3	0.29	0.981
e2	1.26×10^3	0.23	0.986
f2	1.52×10^3	0.28	0.961
g2	2.87×10^3	0.05	0.980

We tested the porosity and permeability of natural specimens in order to compare with synthetic specimens. The porosity of natural pure rock salt is 3.1%, and the porosity of synthetic pure mudstone is 16.9%. The porosity of synthetic pure rock salt is 3.8%, and the porosity of synthetic pure mudstone is 17.8%. The porosity results show that the synthetic specimen porosity is in good agreement with the natural specimen. The picture of the natural specimens and the permeability results can be seen in Figures 11 and 12.

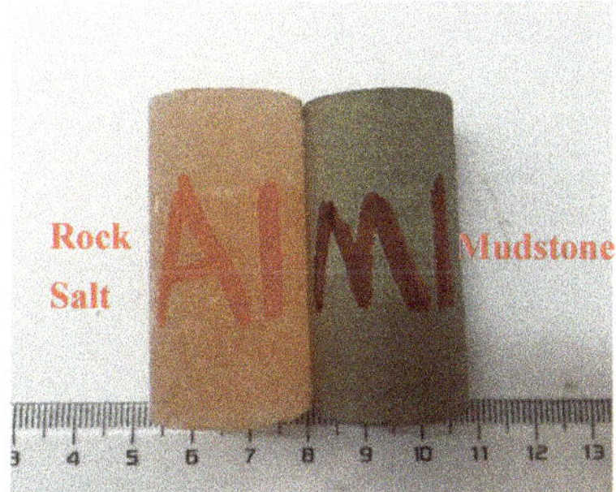

Figure 11. Natuural specimens.

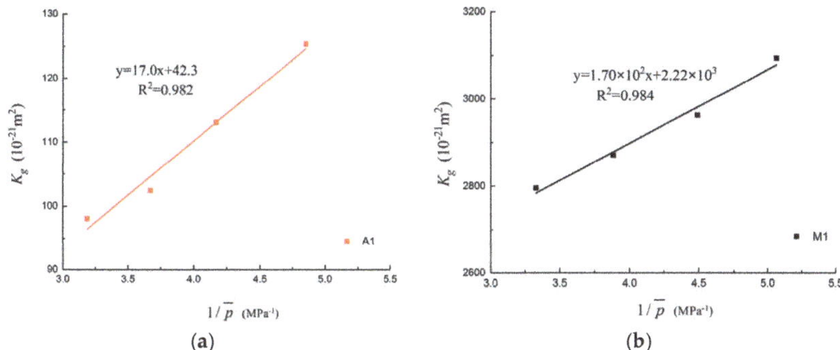

(**a**) (**b**)

Figure 12. Natural specimen permeability results. (**a**) Natural pure rock salt; (**b**) natural pure rock salt. Note: We define the salt rock with high salt content as pure salt rock (>95%, both for the natural rock salt and synthetic rock salt).

From Figure 12 and Table 7, we can see that the permeability of natural pure salt rock is 4.23×10^{-20} m^2, and the permeability of natural mudstone is 2.22×10^{-18} m^2. From Figure 10, we know that the permeability of synthetic pure salt rock is 6.93×10^{-20} m^2, and the permeability of synthetic mudstone is 2.97×10^{-18} m^2. The permeability results show that the synthetic specimens are in good agreement with the natural specimen.

Table 7. The porosity and permeability of the synthetic specimens and natural rock specimens.

No.	Specimen/Specimen Type	Porosity (%)	K_∞ $(10^{-21}$ m$^2)$
a1	Synthetic pure rock salt	4.9	67.1
a2	Synthetic pure rock salt	2.6	71.6
A1	Natural pure rock salt	3.1	42.3
i1	Synthetic pure mudstone	19.3	3.06×10^3
g2	Synthetic pure mudstone	16.4	2.87×10^3
M1	Natural pure mudstone	16.9	2.22×10^3

The test results of porosity and permeability of synthetic specimens are close to the natural specimens, which shows that the synthetic material model for permeability testing is a reasonable analogue for the ranges we addressed to emulate the natural cavern conditions.

5.2. Pore Size Distribution of Synthetic Specimens and Natural Specimens

In order to examine the difference between synthetic specimens and natural specimens, we carried out scanning-electron microscope assessment of the synthetic pure rock salt, synthetic pure mudstone, natural pure rock salt and natural pure mudstone. Some images are shown in Figure 13.

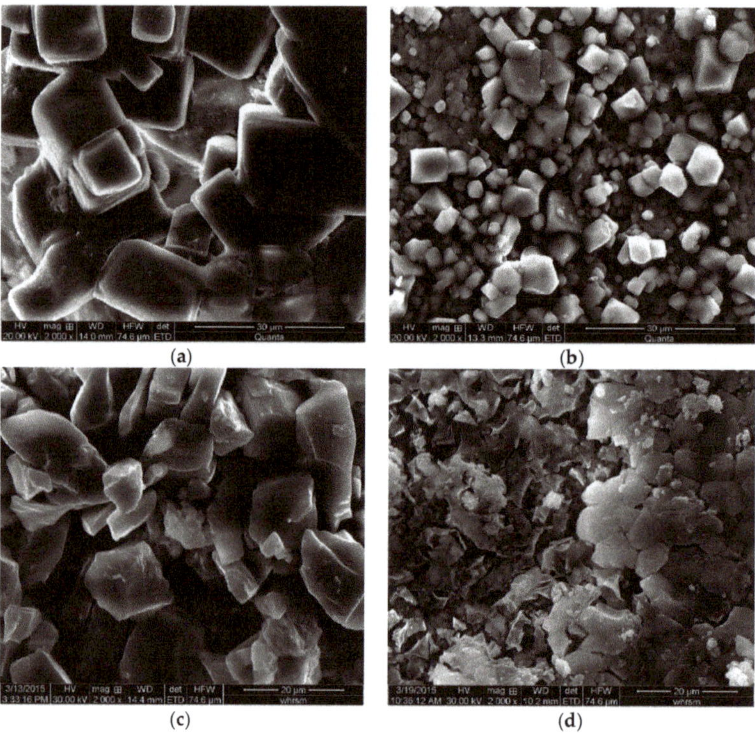

Figure 13. SEM pictures (2000 times). (**a**) Natural pure rock salt; (**b**) synthetic pure rock salt; (**c**) natural pure mudstone; (**d**) synthetic pure mudstone.

From Figure 13a,b, we can see that both the natural pure rock salt and the synthetic pure rock salt have a lattice crystal structure. The crystal lattice of the natural salt rock is larger than that of the synthetic mudstone, and the arrangement of natural pure rock salt is more disordered; this is because, in the process of comminution then remolding, the natural salt rock lattice is shattered and rearranged. From Figure 13c,d, we know that the synthetic mudstone is more closely arranged between particles.

We show the pore size distribution of synthetic pure rock salt, synthetic pure mudstone, natural pure rock salt and natural pure mudstone by mercury intrusion porosimetry (MIP) in Figure 14.

From Figure 14a,b, the intrusion and extrusion of mercury in the natural pure rock salt and the synthetic pure rock salt is mainly controlled by the 10-nm pore size, and the two are in good agreement. From Figure 14c,d, we can see that the intrusion and extrusion of mercury in the natural pure mudstone and the synthetic pure mudstone are mainly controlled by the 20-nm pore size, and the two are in good agreement.

The SEM and the pore size distribution of synthetic specimens and natural specimens are similar, which shows that synthetic material permeability model is a reasonable analogue to the natural specimens.

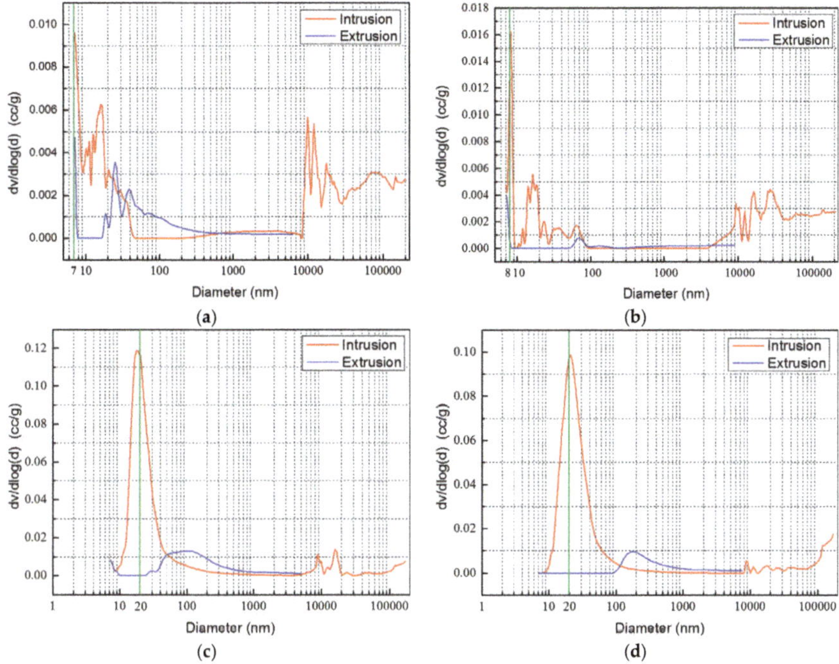

Figure 14. Pore size distribution. (**a**) Natural pure rock salt; (**b**) synthetic pure rock salt; (**c**) natural pure mudstone, (**d**) synthetic pure mudstone.

5.3. The Effect of Porosity on Permeability

As can be seen from Figure 15, the porosity affects the permeability greatly; there is a strong power function between permeability and porosity. The permeability increases with increasing porosity, and when the porosity is less than 10%, the increase is slower, while the permeability increases greatly when the porosity exceeds 10%. This indicates that the porosity has a large effect on the permeability and that the permeability depends on the internal pores and microcracks, as well as their interconnectivity (pore throat size distribution).

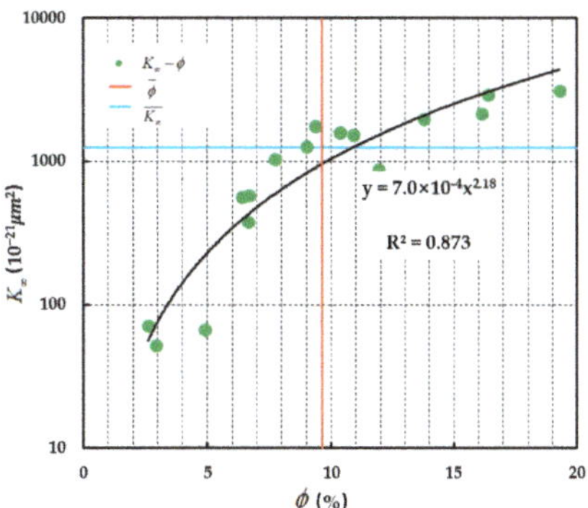

Figure 15. The relationship between K_∞ and ϕ.

5.4. The Influences of Mudstone Content on the Permeability of Specimens

It can be seen from Figure 16 that the Klinkenberg permeability increases with the increase of mudstone content for both kinds of specimens. There is a strong exponential relationship between Klinkenberg permeability and mudstone content of rock salt. We also find that for both kinds of specimens, when the mudstone content is below 40%, the permeability increases only slightly with mudstone content, whereas above this threshold, the permeability increases rapidly with mudstone content. When the mudstone content of rock salt is below 40%, the rock salt particles are closely arranged between each other, while the mudstone particles are relatively less, so the mudstone content has little impact on the Klinkenberg permeability. However, when the mudstone content ranges from 40%–100%, more particles of mudstone embed around rock salt particles, making the connection between the particles of rock salt less dense, so that some pore channels are formed and the pore throat radius is larger. Hence, with the increase of mudstone content, the Klinkenberg permeability increases rapidly when the mudstone content exceeds 40%, and when the mudstone content is 80%, the Klinkenberg permeability is close to the synthetic pure mudstone specimen value.

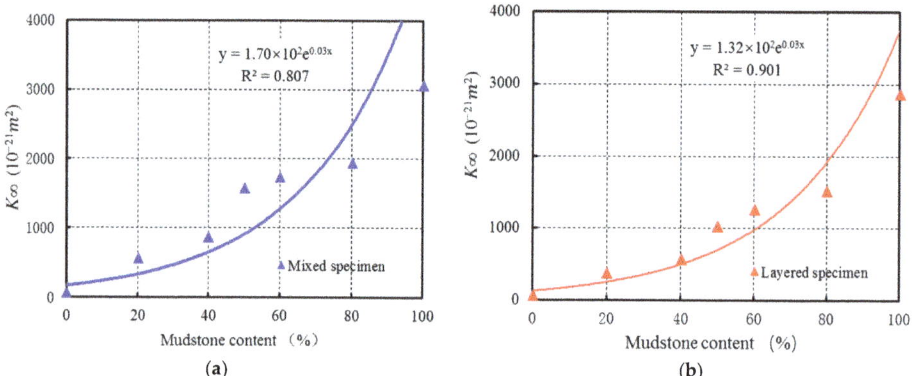

Figure 16. The relationship between K_∞ and mudstone content. (a) Mixed specimen; (b) layered specimen.

6. Conclusions

(1) During the experiment, it was found that the mudstone is more compressible than rock salt, and the compressibility of rock increased with the increase of porosity.

(2) The pseudo-pressure seepage equation has extensive application value, so the specimen permeability was calculated by the pseudo-pressure method. The Klinkenberg permeability of the synthetic specimens was obtained by curve-fitting. The permeability of synthetic pure rock salt was 6.93×10^{-20} m^2 and its porosity 3.8%, and the permeability of synthetic pure mudstone was 2.97×10^{-18} m^2, with a porosity 17.8%. The test results were close to the natural specimens, and it shows that this synthetic material permeability model is a reasonable analogue of use in engineering.

(3) Comparing permeability test results for the two kinds of synthetic specimens, it was found that the permeability of the mixed specimens was about 40% higher than that of the layered rock salt specimens at the same mudstone content. This suggests that in salt cavern gas storage cases, the layered rock salt is tight at the top and bottom of the cavity. The permeability of the transition zone of rock salt and mudstone around the cavity is higher, and the transition zone would be the permeable channel to be concerned with; although the permeabilities are still very low, and there would also be the beneficial effect of capillary blockage of the brine-filled channels in the field case.

(4) We also found that for both kinds of specimens, there was a strong exponential relationship between Klinkenberg permeability and mudstone content: when the mudstone content is below 40%,

the Klinkenberg permeability increases only slightly with mudstone content, whereas above this threshold, the Klinkenberg permeability increases significantly.

Note: The formation mechanisms of natural rock specimen and synthetic specimen were different. Natural rock was formed in the long-term diagenesis, while the synthetic specimens in this paper were formed in ultra-high pressure. The porosity and permeability of synthetic specimens were similar to the natural specimens, but the natural rock (mudstone/salt) had different pore space structures compared to the synthetic specimens. Therefore, the approach, respectively the outcome (permeability/porosity relationship), can only be a first attempt at investigation.

Acknowledgments: The authors wish to acknowledge the financial support of the National Natural Science Foundation of China (Grant Nos. 51774266, 51404241, 41602328), National Natural Science Foundation of China Innovative Research Team [Grant No. 51621006], and Natural Science Foundation for Innovation Group of Hubei Province, China [Grant No. 2016CFA014]. Moreover, the authors wish to thank the reviewers for constructive comments and suggestions that have helped us improve our manuscript.

Author Contributions: The paper was written by Hongwu Yin under the guidance of Chunhe Yang. The experiment scheme, data analysis and pre-literature research were carried out by Hongling Ma and Xilin Shi. Maurice B. Dusseault deepened the conclusions and did the English editing of the paper. The experiment was carried out by Hongwu Yin under the help of Xiangsheng Chen and Yuhao Zhang.

Conflicts of Interest: The authors declare no conflict of interest.

References

1. Li, Y.P.; Jiang, W.D.; Liu, J.; Chen, J.W.; Yang, C.H. Direct shear tests for layered salt rocks of yunying salt mine in hubei province. *Chin. J. Rock Mech. Eng.* **2007**, *26*, 001767–001772. (In Chinese)

2. Li, Y.P.; Yang, C.H.; Luo, C.W.; Qu, D.A. Study on sealability of underground energy storage in deep salt formation in yunying area, Hubei province. *Chin. J. Rock Mech. Eng.* **2007**, *26*, 2430–2436. (In Chinese)

3. Xu, F.; Yang, C.; Guo, Y.; Wang, T.; Wang, L.; Zhang, P. Effect of confining pressure on the mechanical properties of thermally treated sandstone. *Curr. Sci.* **2017**, *112*, 1101–1106

4. Liu, W.; Nawaz, M.; Li, Y.P.; Spiers, C.J.; Yang, C.H.; Ma, H.L. Experimental study of permeability of salt rock and its application to deep underground gas storage. *Chin. J. Rock Mech. Eng.* **2014**, *33*, 1953–1961. (In Chinese)

5. Wu, W.; Hou, Z.M.; Yang, C.H. Investigations on permeability of rock salt. *Chin. J. Geotech. Eng.* **2005**, *27*, 746–749. (In Chinese)

6. Zhou, H.W.; He, J.M.; Wu, Z.D. Permeability and meso-structure characteristics of bedded salt rock. *Chin. J. Rock Mech. Eng.* **2009**, *28*, 2068–2073. (In Chinese)

7. Xi, B.P.; Zhao, Y.S.; Zhao, Y.L.; Kang, Z.Q. Investigation on rheodestruction and permeability of surrounding rock for long-term running storage cavern in bedded rock salt. *Rock Soil Mech.* **2008**, *29*, 245–250. (In Chinese)

8. Petrasch, J.; Meier, F.; Friess, H.; Steinfeld, A. Tomography based determination of permeability, Dupuit–Forchheimer coefficient, and interfacial heat transfer coefficient in reticulate porous ceramics. *Int. J. Heat Fluid Flow* **2008**, *29*, 315–326. [CrossRef]

9. Yang, C.H.; Li, Y.P.; Qu, D.A.; Chen, F.; Yin, X.Y. Advances in researches of the mechanical behaviors of bedded salt rocks. *Adv. Mech.* **2008**, *38*, 484–494. (In Chinese)

10. Liu, W.; Li, Y.P.; Yang, C.H.; Ma, H.L.; Liu, J.X.; Wang, B.W.; Huang, X.L. Investigation on permeable characteristics and tightness evaluation of typical interlayers of energy storage caverns in bedded salt rock formations. *Chin. J. Rock Mech. Eng.* **2014**, *33*, 500–506. (In Chinese)

11. Berest, P.; Brouard, B.; Durup, J.G. Tightness Tests in Salt-Cavern Wells. *Oil Gas Sci. Technol.* **2002**, *56*, 451–469. [CrossRef]

12. Cosenza, P.; Ghoreychi, M.; Bazargan-Sabet, B.; Marsily, G.D. In situ rock salt permeability measurement for long term safety assessment of storage. *Int. J. Rock Mech. Min. Sci.* **1999**, *36*, 509–526. [CrossRef]

13. Beauheim, R.L.; Roberts, R.M. Hydrology and hydraulic properties of a bedded evaporite formation. *J. Hydrol.* **2002**, *259*, 66–88. [CrossRef]

14. Popp, T.; Kern, H.; Schulze, O. Evolution of dilatancy and permeability in rock salt during hydrostatic compaction and triaxial deformation. *J. Geophys. Res. Solid Earth* **2001**, *106*, 4061–4078. [CrossRef]

15. Allemandou, X.; Dusseault, M. Healing processes and transient creep of salt rock. *Geotech. Eng. Hard Soils-Soft Rocks.* **1993**, *1*, 3.

16. Stormont, J.C.; Daemen, J.J.K. Laboratory study of gas permeability changes in rock salt during deformation. *Int. J. Rock Mech. Min. Sci. Geomech. Abstr.* **1992**, *29*, 325–342. [CrossRef]

17. Wu, Z.D.; Zhou, H.W.; Ding, J.Y.; Ran, L.N.; Yi, H.Y. Research on permeability testing of rock salt under different permeability pressures. *Chin. J. Rock Mech. Eng.* **2012**, *31*, 3740–3746. (In Chinese)

18. Yan, Q.B.; Chen, M.L.; Wang, J.; Du, Y.; Wang, X.X. Correlation among permeability, porosity and pore throat radius of carbonate reservoirs. *Nat. Gas Ind.* **2015**, *35*, 30–36. (In Chinese)

19. Chen, W.Z.; Tan, X.J.; Wu, G.J.; Yang, J.P. Research on gas seepage law in laminated salt rock gas storage. *Chin. J. Rock Mech. Eng.* **2009**, *28*, 1297–1304. (In Chinese)

20. Zhang, Q.Y.; Liu, D.J.; Jia, C.; Shen, X.; Liu, J.; Duan, K. Development of geomechanical model similitude material for salt rock oil-gas storage medium. *Rock Soil Mech.* **2009**, *30*, 3581–3586. (In Chinese)

21. Ren, S.; Ren, Y.W.; Jiang, D.Y.; Chen, J.; Yang, C.H. Study of synthetic rock salt similar materials for solution mining test. *Chin. J. Rock Mech. Eng.* **2012**, *31*, 3716–3724. (In Chinese)

22. Jiang, D.Y.; Zhang, J.W.; Qu, D.A.; Chen, J.; Yang, C.H. Experimental study on a similar material of rock salt with interlayer. *J. China Coal Soc.* **2013**, *38*, 76–81. (In Chinese)

23. American Petroleum Institute. Practices for Core Analysis. In *SY/T5336–2006. API RP 40–1998 Recommended Practices for Core Analysis*, 2nd ed.; American Petroleum Institute: Washington, DC, USA, 2006.

24. Kong, X.Y. *Advanced Mechanics of Fluid in Porous Media*; Press of University of Science and Technology of China: Hefei, China, 2010. (In Chinese)

25. Hall, H.N. Compressibility of Reservoir Rocks. *J. Pet. Technol.* **1953**, *5*, 17–19. [CrossRef]

26. Li, C.L. The relationship between rock compressibility and porosity. *China Offshore Oil Gas (Geol.)* **2003**, *17*, 355–358. (In Chinese)

27. Liu, H.W. *Advanced Mechanics of Materials*; Higher Education Press: Beijing, China, 1985. (In Chinese)

28. Li, Q.; Gao, S.S.; Liu, H.X.; Ye, L.Y.; Gai, Z.H. Core permeability calculation methods and application scopes. *Nat. Gas Ind.* **2015**, *35*, 68–73. (In Chinese)

29. Klinkenberg, L.J. The Permeability of Porous Media to Liquids and Gases. *Socar Proc.* **1941**, *2*, 200–213. [CrossRef]

30. Scheidegger, A.E. The physics of flow through porous media. *Soil Sci.* **1958**, *86*, 355. [CrossRef]

31. Huang, J.Z.; Feng, J.M. Simplification of conventional methods for obtaining permeability. *Pet. Explor. Dev.* **1994**, *21*, 54–58. (In Chinese)

32. Gueguen, Y.; Palciauskas, V.; Jeanloz, R. Introduction to the Physics of Rocks. *Phys. Today* **1995**, *48*. [CrossRef]

33. Ge, J.L. *The Modern Mechanics of Fluids Flow in Oil Reservoir*; Petroleum Industry Press: Beijing, China, 2003. (In Chinese)

34. American Petroleum Institute. *Orifice Metering of Natural Gas and Other Related Hydrocarbon Fluids—Concentric, Square-Edged Orifice Meters—Part 1: General Equations and Uncertainty Guidelines*; API MPM CH14.3.1 Ed. 4; American Petroleum Institute: Washington, DC, USA, 2012.

35. Yang, C.H.; Li, Y.P.; Chen, F. *Mechanics Theory and Engineering of Bedded Rock Salt*; Science Press: Beijing, China, 2009. (In Chinese)

applied sciences

Article

Experimental Study on the Physical Simulation of Water Invasion in Carbonate Gas Reservoirs

Feifei Fang [1,2], Weijun Shen [1,3,*], Shusheng Gao [2,4], Huaxun Liu [2,4], Qingfu Wang [3] and Yang Li [4]

[1] School of Engineering Science, University of Chinese Academy of Sciences, Beijing 100049, China; fangfeifei13@mails.ucas.ac.cn

[2] Institute of Porous Flow and Fluid Mechanics, Chinese Academy of Sciences, Langfang 065007, Hebei, China; gaoshusheng69@petrochina.com.cn (S.G.); liuhuaxun@petrochina.com.cn (H.L.)

[3] Institute of Mechanics, Chinese Academy of Sciences, Beijing 100190, China; wangqingfu@imech.ac.cn

[4] PetroChina Research Institute of Petroleum Exploration and Development, Beijing 100083, China; liyang69@petrochina.com.cn

* Correspondence: wjshen763@imech.ac.cn; Tel.: +86-10-8254-4017

Received: 6 June 2017; Accepted: 29 June 2017; Published: 7 July 2017

Abstract: Water invasion in carbonate gas reservoirs often results in excessive water production, which limits the economic life of gas wells. This is influenced by reservoir properties and production parameters, such as aquifer, fracture, permeability and production rate. In this study, seven full diameter core samples with dissolved pores and fractures were designed and an experimental system of water invasion in gas reservoirs with edge and bottom aquifers was established to simulate the process of water invasion. Then the effects of the related reservoir properties and production parameters were investigated. The results show that the edge and bottom aquifers supply the energy for gas reservoirs with dissolved pores, which delays the decline of bottom-hole pressure. The high water aquifer defers the decline of water invasion in the early stage while the big gas production rate accelerates water influx in gas reservoirs. The existence of fractures increases the discharge area of gas reservoirs and the small water influx can result in a substantial decline in recovery factor. With the increase of permeability, gas production rate has less influence on recovery factor. These results can provide insights into a better understanding of water invasion and the effects of reservoir properties and production parameters so as to optimize the production in carbonate gas reservoirs.

Keywords: carbonate gas reservoirs; water invasion; recovery factor; aquifer size; production rate

1. Introduction

The production of water from gas producing wells is a common occurrence in gas fields. It is attributed to one or more of several reasons such as normal rise of gas-water contact, water coning or water fingering [1–4]. Water coning occurs in gas reservoirs with the edge-and-bottom aquifer drive, and water will flow into the wellbore from below and above of the perforations and consequently interfere with gas production [5]. Water coning is a common problem which increases the cost of producing operations and raises some environmental problems related to water disposal, and reduces the efficiency of the depletion mechanism and the overall recovery [6]. Thus, understanding the questions such as water invasion in gas reservoirs and its effects on recovery factor is significant for predicting gas productivity and optimizing extraction conditions.

Water invasion is a complex phenomenon observed in oil and gas reservoirs, which will occur due to pressure gradients close to the production well and the imbalance between the viscous and gravity forces around the completion interval [7,8]. There are three essential forces controlling the mechanism, which includes the capillary, viscous and gravity forces [6]. With the change of pressure drawdown, the viscous force pushes the gas phase into the production well, and the lift up of the gas-water interface will be caused by the dynamic pressure distribution near the production well [6,9–11]. In contrast, the gravity force will cause water to remain in the gas saturated resign of the reservoirs. When the dynamic force at the wellbore exceeds the gravity force, the pressure gradient will cause the local gas-water contact to rise upward, which will contribute to water invasion and interfere with gas production.

Some studies on water invasion in gas reservoirs and its impact on gas productivity have been considered in the past. Muskat and Wycokoff [12] studied the water coning phenomenon and demonstrated clearly that water coning would occur at high pressure drop. Hoyland [13] pointed out that the gravity force was dominant at initial reservoir conditions, and the viscous force would arise to part of the controlling mechanism when the pressure drawdown increased. Saad et al. [14] and Bahrami et al. [15] analyzed the problem of water coning in naturally fractured reservoirs using experimental work and field data, respectively. Cheng et al. [16] and Shen et al. [4] performed the effects of the reservoir properties on water coning with the numerical simulation. Hu et al. [17] and Xiong et al. [18] established a mathematical model to predict water properties and the performance of water invasion with the data of gas and water production. Perez et al. [19] used a coning radial model to evaluate the occurrence of coning in naturally fractured reservoirs. Hu et al. [20] and Shen et al. [21] used reservoir cores to analyze the effects of different aquifers on water invasion. Azim [7] proposed a fully coupled poroelastic multiphase fluidflow numerical model to understand the water coning phenomenon in naturally fractured reservoir under the effects of various rock and fluid properties. So far there has been a lot of research work carried out on the performance of water invasion in gas reservoirs. However, water invasion of gas reservoirs is not fully understood and many uncertainties still exist in the process. Hence, there is a necessity to understand water invasion and the effects of these reservoir and production parameters on the problem so as to optimize gas productivity in gas reservoirs.

In this study, the experimental system of water invasion in gas reservoirs with the edge-and-bottom aquifer was established, and the elastic expansion of water and gas was considered. Seven full diameter core samples with dissolved pores and fractures were designed and the physical simulations of water invasion in carbonate gas reservoirs were conducted. Then the effects of the related reservoir properties and production parameters, such as water invasion energy, aquifer, production rate, permeability and fracture, were studied and discussed. These results can provide a better understanding of water invasion for improving the gas recovery and development benefit in carbonate gas reservoirs.

2. Experimental Systems and Methods

2.1. Experimental Systems

In the physical simulation of water invasion in gas reservoirs, previous studies only considered the elastic expansion of formation water when simulating the edge-and-bottom aquifer [20,21]. However, the compressibility of formation water is between 3.4×10^{-4} and 5.0×10^{-4} MPa^{-1} in gas reservoirs with the finite edge-and-bottom aquifer [22]. Thus, the elastic expansion capacity of an aquifer is very weak and the aquifer expansion caused by the formation pressure drop is small. For example, when the pressure drop of formation water is 50 MPa, the aquifer expansion only accounts for about 2% of aquifer reserves. That is why water coning will occur at 20 times, 100 times and even infinite aquifer considering the aquifer expansion [21]. Therefore, there will be a false understanding that water coning will occur in strong aquifers if water invasion only considers the elastic expansion capacity of an aquifer, but this contradicts that water invasion in some gas reservoirs takes place without a strong aquifer.

In order to simulate water invasion of gas reservoirs with the edge-and-bottom aquifer accurately, the energy source of water invasion is not just the elastic expansion of aquifer, and the effects of rock pore compression and water soluble gas expansion should be considered. The solubility of natural gas in water is mainly affected by formation pressure, temperature, and the salinity of formation water [23,24]. With the increasing of formation pressure and temperature, the gas solubility in water rises gradually. The solubility declines gradually with the increasing of water salinity. According to the previous studies [23], there is a gas reservoir with origin formation temperature of 150 °C, formation pressure of 60 MPa, and water salinity of 19,029 mg/L. When the aquifer is 5 times and pressure drawdown is 40 MPa, the limit adsorption of natural gas in formation water is about 3.7% of reservoir reserves. However, when reservoir pressure declines to 40 MPa in the development of gas reservoirs, the overall pressure drop of formation water is much less than 40 MPa. Therefore, the desorption of water soluble gas can be negligible in contrast to reservoir reserves. As it is very difficult to dissolve gas in water completely, the experiment only considers the elastic expansion energy of water soluble gas, which ignores the desorption effect due to pressure drop. The experimental system of water invasion in gas reservoirs with edge and bottom aquifer was shown in Figure 1, which was composed of nitrogen gas source, ISCO pump, confining pressure pump, core holder, pressure sensor, gas flowmeter and data acquisition system, etc.

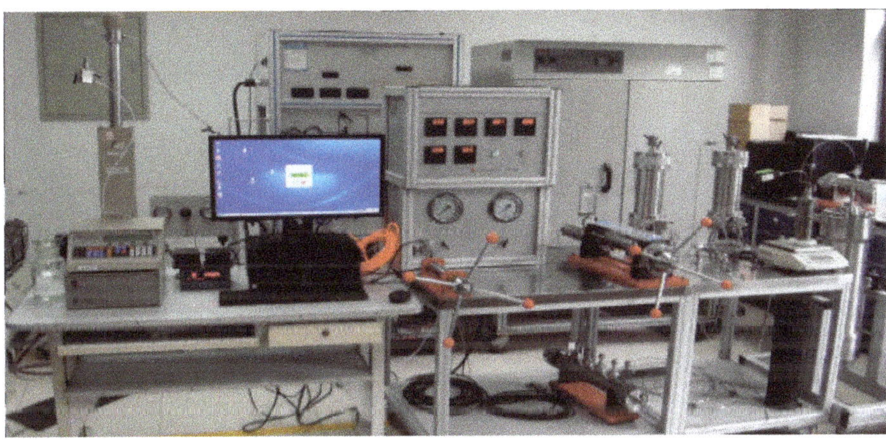

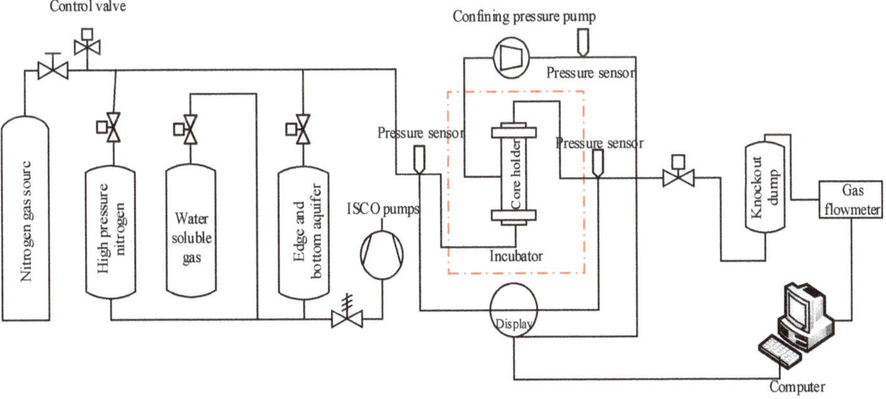

Figure 1. Experimental system of water invasion in gas reservoirs with edge and bottom aquifer.

2.2. Experimental Methods

Experimental core samples were obtained from carbonate gas reservoirs of Anyue gas field, which is located in Sichuan Province, China [25,26]. The reservoirs are characterized by strong heterogeneity, including dissolved vugs, pores and fractures, which are shown in Figure 2a,b, respectively. The core samples from NO. 1 to NO. 5 were dissolved pores while the sample NO. 6 and NO. 7 cores contained micro and macro fractures, respectively. The properties of carbonate core samples used in the study were summarized in Table 1. During the experiment, the edge-and-bottom aquifer (formation water) was 8.0×10^4 mg/L and the nitrogen gas source was 99.99% of purity. According to the experimental process, the experimental system was connected and the device tightness was checked. Then a confining pressure of 40 MPa was added to the experimental core, and a nitrogen of 30 MPa was saturated to the experimental core. Meanwhile, the experimental system of water invasion was synchronized to 30 MPa, which was the same as the reservoir pressure. Then the gas flowmeter was opened to simulate the production of gas reservoirs, and the pressure, flowrate and time was recorded until the abandonment pressure reached 3 MPa. Finally, the above experimental procedures were repeated after changing core samples, and the physical simulations with different water invasion energies, aquifers, production rates, permeabilities, fractures were conducted successively. The experiments of water invasion conducted on different conditions were summarized in Table 2.

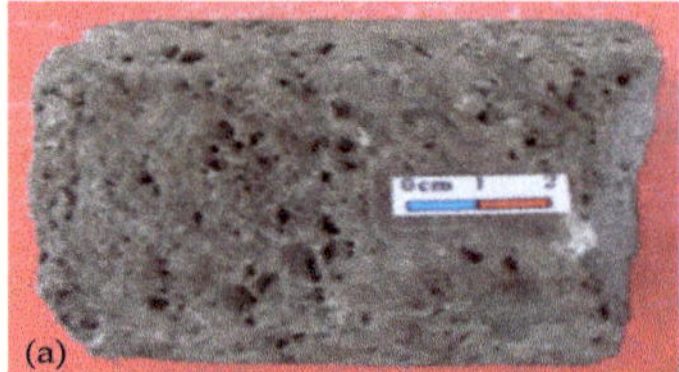

Figure 2. Core samples from the carbonate gas reservoir of Anyue gas field. (**a**) Core sample with vugs and pores; (**b**) Core sample with micro fractures.

Table 1. Some properties of carbonate core samples used in this study.

NO.	Length (cm)	Diameter (cm)	Permeability (mD)	Porosity (%)
1	10.105	10.438	0.001	4.10
2	9.807	10.021	0.021	4.30
3	10.156	10.440	0.101	5.10
4	10.350	10.159	1.010	8.80
5	10.153	10.357	15.310	10.60
6	10.255	10.257	0.110	5.30
7	10.250	10.257	10.120	5.40

Table 2. Different experiments of water invasion conducted in this study.

NO.	Fracture	Aquifer Size (Times)	Production Rate (mL/min)
1	no	0, 3, 7, infinite	500, 1000, 2000, 4000
2	no	0, 1, 3, 7, 12, infinite	500, 1000, 2000, 4000
3	no	0, 3, 7, infinite	500, 1000, 2000, 4000
4	no	0, 3, 7, infinite	500, 1000, 2000, 4000
5	no	0, 3, 7, infinite	500, 1000, 2000, 4000
6	yes	0, 3, 7, 12, infinite	500, 1000, 2000, 4000
7	yes	0, 3, 7, 12, infinite	500, 1000, 2000, 4000

3. Results and Discussion

3.1. Effect of Different Water Invasion Energies

Water supplies an extra mechanism to produce gas reservoirs, and it is important to understand the effects of water energies [22,23]. The core 4 is chosen to study the effects of different water invasion energies. The gas production rate and original saturation pressure are 1000 mL/min and 30 MPa, respectively. The experiments to understand water invasion are conducted when only considering the water elastic expansion effect, comprehensively water flexibility and elastic expansion of water soluble gas. The water invasion versus gas recovery in different water invasion energies is shown in Figure 3. From the result of Figure 3, we can see that water invasion with 3 times and 7 times aquifer are 0.021 PV and 0.055 PV when water invasion energy only considers the effect of water expansion. They are far less than water invasion considering water elastic expansion and elastic expansion of water soluble gas, which are 0.182 PV and 0.227 PV, respectively. The water invasion is very small with considering water elastic expansion whether it is 3 times or 7 times aquifer. The decline range of recovery factor is less than 5.0% compared with no aquifer. When the aquifer is 20 times, the dramatic phenomenon of water invasion occurs [21]. While comprehensively considering water invasion energies, the water invasion of 3 times aquifer is 0.182 PV and the recovery factor decreases by 17.6%, which has a strong influence on gas production. Therefore, during the simulation of water invasion we should not only consider the elastic expansion of water, but also consider the elastic expansion of water soluble gas.

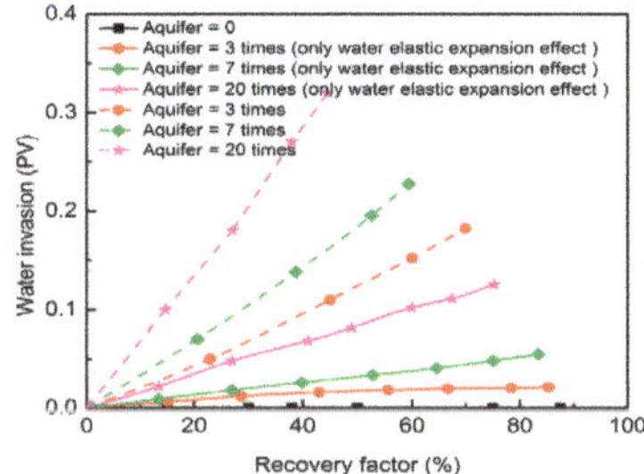

Figure 3. Water invasion versus recovery factor for different water invasion energies.

3.2. Effect of Different Aquifer Sizes

Nearly all gas reservoirs are surrounded by water-bearing rocks called aquifers [11]. The aquifer size has an important effect on water invasion and gas production in gas reservoirs [6,10]. The core 3 is used to study the effects of water invasion with different aquifer sizes. The gas production rate and original saturation pressure are 1000 mL/min and 30 MPa, respectively. The experiments of water invasion without aquifer and with 3 times, 7 times and infinite aquifer are performed, and the results are illustrated in Figure 4. From the result of the Figure 4a, it can be observed that the bottom-hole pressure decreases linearly with recovery factor when there is no aquifer. The pressure drop increases gradually with the increase of recovery factor, but the pressure drop is only 1.43 MPa until the production ends, shown in Figure 4b. This suggests that gas seepage resistance is very small without the edge-and-bottom aquifer and the recovery factor is high which can reach 83.26%.

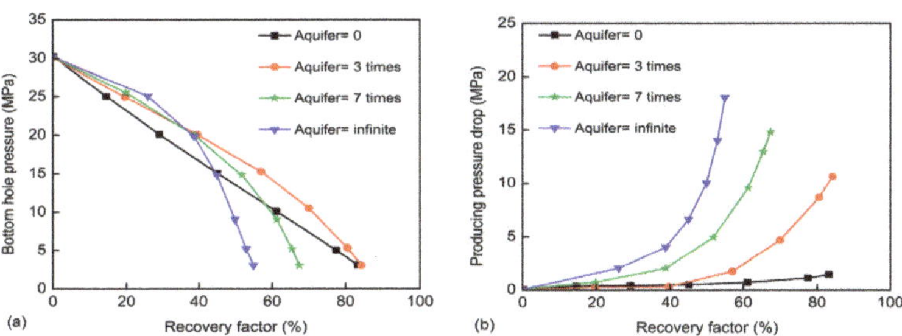

Figure 4. Bottom hole pressure and producing pressure drop versus recovery factor for different aquifers. (**a**) Bottom hole pressure; (**b**) producing pressure drop.

When there exists the edge-and-bottom aquifer, bottom-hole pressure changes with the increase of recovery factor, and there is a concave change in Figure 4a. This is because the early water invasion delays the decline of bottom-hole pressure. At the same recovery factor, the larger the aquifer is, the larger the bottom-hole pressure is. This shows that the early water invasion with the edge-and-bottom aquifer is the gas supply of energy, and the larger the aquifer is, the greater the energy supply is. In the late water invasion, the seepage resistance increases and bottom-hole pressure drops rapidly due to the existence of two phase flow. Thus, in the production and development of gas reservoirs, we can reversely infer the strength of edge-and-bottom aquifer according to the change of bottom-hole pressure so as to prevent the risk of water invasion. Moreover, there exists an obvious inflection point between pressure drop of gas reservoirs with edge-and-bottom aquifer. The bottom-hole pressure drops slowly before the point and the pressure drop increases slowly. However, the bottom-hole pressure drops sharply after the point and the pressure drop rises sharply. The larger the aquifer is, the lower the recovery factor is [20]. This is because it belongs to single-phase flow in the early water invasion, and the gas-water interface is relatively stable. With an increase of water invasion, the single-phase flow gradually shifts to the two-phase flow. Consequently, the seepage resistance and pressure dropdown increases and the recovery factor reduces.

From the result of Figure 5, water invasion gradually increases with the increase of aquifer size [4,11]. However, when the aquifer size increases to a certain value, the growth of water invasion changes slowly. With the increase of aquifer size, the recovery factor increases then decreases, and the recovery factor reaches the maximum with the 3 times aquifer. It shows that when the permeability is 0.1 mD, the small aquifer provides positive energy to gas reservoirs. Less water occupies the storage space of gas reservoirs, which displaces partial gas and improves the recovery factor. With the increase of aquifer size, water invasion energy increases. Water displaces gas and some gas will be trapped through the breaker, flow around and water lock. When water invasion is mainly governed by trapped gas, the recovery factor reduces.

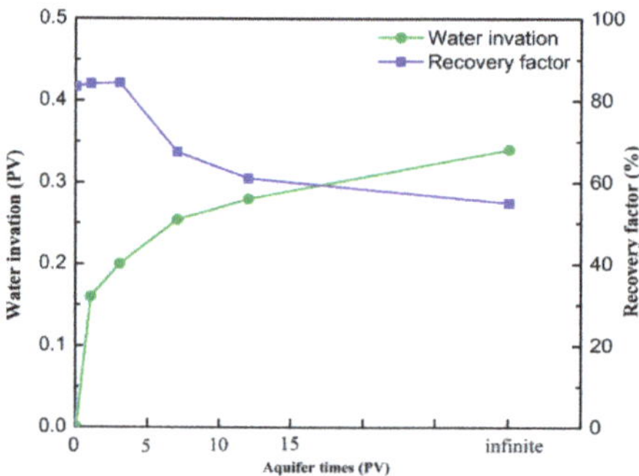

Figure 5. Water invasion and recovery factor versus aquifer times.

3.3. Effect of Different Production Rates

Gas production rate is a significant parameter in the development of gas reservoirs, which affects water coning and ultimate recovery [11,13]. In order to study the effects of different production rates, the core sample 2 is used to simulate water invasion in carbonate gas reservoirs. The aquifer size and original saturation pressure are 7 times and 30 MPa, respectively. The effects of gas production rates from 500 mL/min to 4000 mL/min are selected to simulate water invasion. Figure 6a,b shows bottom-hole pressure and producing pressure drop versus recovery factor for different production rates, respectively. It can be seen that when the recovery factor is less than 30%, the gas production rate has less impact on bottom-hole pressure and pressure drop. In the middle and later period, the gas production rate has a large effect on bottom-hole pressure and pressure drop. At the same recovery factor, the bigger the gas production rate is, the lower the bottom-hole pressure is and the larger the pressure drop is. This suggests that the bigger the gas production rate is, the more that the water invasion affects the recovery factor.

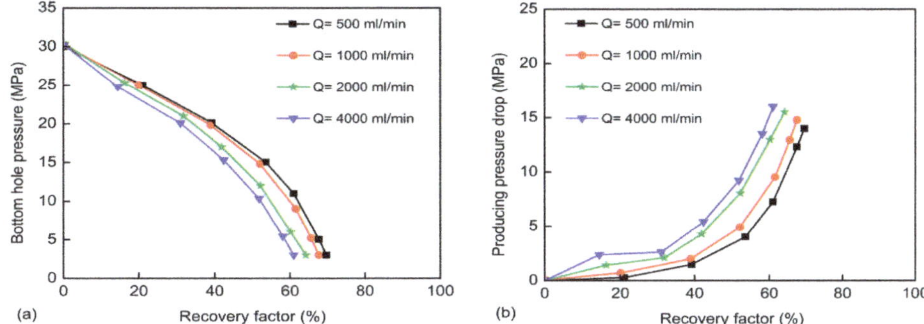

Figure 6. Bottom hole pressure and producing pressure drop versus recovery factor for different production rates. (**a**) Bottom hole pressure; (**b**) producing pressure drop.

Figure 7a,b shows water invasion versus time for different production rates and recovery factor versus gas production rate, respectively. From the result, it can be known that water invasion changes linearly with time in the same aquifer size, and the rate of water invasion approximates to a certain value. This suggests that the water invasion of gas reservoirs with dissolved pores belongs to a uniform advancing process. The greater the gas production rate is, the faster the water invasion is [4]. Gas production rate increases from 500 mL/min to 4000 mL/min, and recovery factor decreases from 69.66% to 60.82%. The excessively high production gas rate goes against the recovery factor.

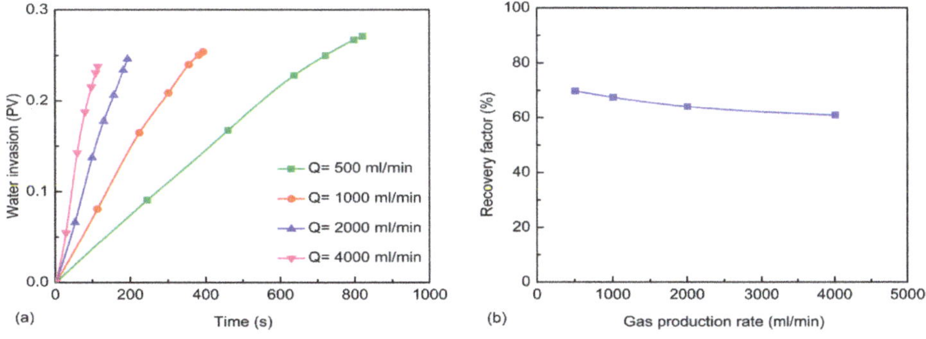

Figure 7. Water invasion versus time for different production rates and recovery factor versus gas production rate (**a**) Water invasion; (**b**) recovery factor.

3.4. Effect of Different Permeabilities

Permeability is a property of the reservoir rock that measures the capacity and ability of the formation to transmit fluid [11]. The core samples with different permeabilities are conducted to understand the effects of water invasion. The gas production rate and original saturation pressure are 1000 mL/min and 30 MPa, respectively. The experiment results without aquifer and with 3 times, 7 times and infinite aquifer are illustrated in Figure 8a. From the result of Figure 8a, it can be seen that the recovery factor increases gradually with the increase of permeability when there is no aquifer [21]. This is because the gas flow resistance in the low permeability core is far greater than its resistance in high permeability core, which is determined by the micro-pore structure characteristic of reservoirs. The recovery factor of the low permeability core with 0.001 mD is 54.09%, and it is considerably lower than that of 94.60% in high permeability with 10 mD. When the permeability is less than 0.1 mD, the 3 times aquifer promotes gas production and the recovery factor can increase by 3.7~8.5%. Once permeability exceeds 0.1 mD, water invasion reduces greatly, increasing the recovery factor, and the decrease is between 20.07% and 35.90%. This shows that the greater the permeability is, the more the aquifer size has an obvious effect on recovery factor at the same gas production rate. The gas reservoirs with different permeabilities correspond to an optimum aquifer, which can maximize the recovery factor. The lower the permeability is, the larger the corresponding aquifer is. The 0.1 mD core is 3 times aquifer while the 0.01 mD core is 7 times aquifer.

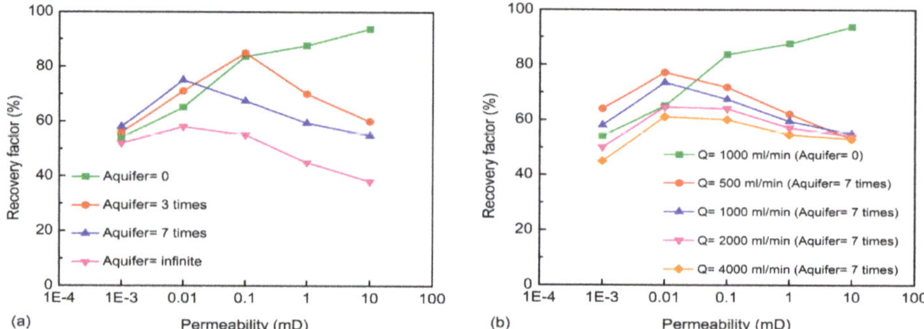

Figure 8. Recovery factor versus permeability for different aquifers and production rates. (**a**) Different aquifers; (**b**) different production rates.

Figure 8b shows water invasion with different production rates between 500 mL/min and 4000 mL/min in the 7 times aquifer. When the permeability is 0.001 mD, gas production rate increases from 500 mL/min to 4000 mL/min and the related recovery factor reduces from 64% to 45%. While the permeability is 10 mD, gas production rate increases from 500 mL/min to 4000 mL/min and recovery factor is almost kept at about 54%. This indicates that the increase of gas production rate reduces the recovery factor of gas reservoirs, but the impact on recovery factor decreases with the increase of permeability. Thus, in the low and extra-low permeability reservoirs, we should control gas production rate in the development of gas reservoirs and extend production time so as to achieve the high recovery factor while we should properly speed up gas production rate in the medium-high permeability reservoirs.

3.5. Effect of Different Fractures

In carbonate gas reservoirs, the pore structure of carbonate rock is more heterogeneous, including dissolved vugs, pores and fractures [26]. The existence of fractures has a strong influence on water and gas flow in gas reservoirs [14]. The core samples without different fractures are selected to study the effects of water invasion. The gas production rate is 1000 mL/min and original saturation pressure is 30 MPa, respectively. The experimental results without aquifer and with 3 times, 7 times and infinite aquifer are provided in Figure 9. From the result of Figure 9a, it is seen that the recovery factor of core 6 is higher than that of core 3 when there is no aquifer [21]. This suggests that the existence of fractures increases the discharge area of gas reservoirs when the permeability is close, which provides a prior seepage channel and increases the recovery factor of gas reservoirs [21]. When the aquifer increases from no aquifer to infinity, the recovery factor of core 6 decreases from 88.13% to 29.99%, which is far higher than that of core 3. Figure 9b shows water invasion versus aquifer times for different fractures. The water invasion of core 6 is less than that of core 3. When the core contains fractures, fractures are the main channels of water advance and small water influx can make recovery factor fall sharply. Hence, we should depict and know the location of fractures and aquifer so as to keep away from the connecting zone between fractures and aquifer in the well-spaced development of gas reservoirs.

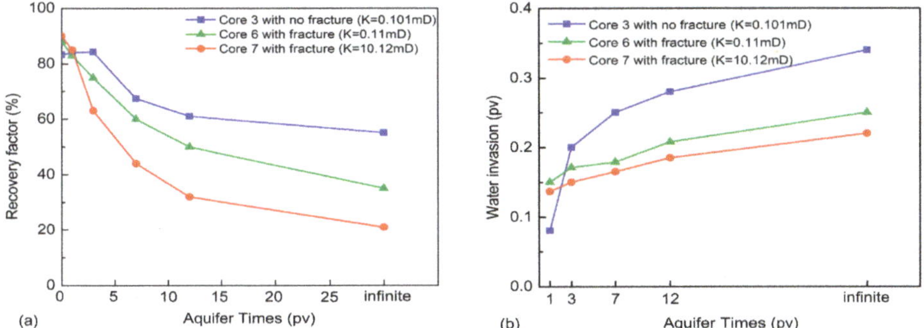

Figure 9. Recovery factor and water invasion versus aquifer times for different fractures. (**a**) Recovery factor; (**b**) water invasion.

In order to understand the effects of production rates for different fractures, the core samples are used to simulate water invasion from 500 mL/min to 4000 mL/min. The aquifer size is 7 times and the original saturation pressure is 30 MPa. Figure 10 illustrates recovery factor versus gas production rate for different fractures. From the result of Figure 10, it can be observed that the recovery factor of the cores with dissolved pores decreases with the increase of gas production rate. The recovery factor of the fracture cores increases and then decreases as gas production rate increases, and there exists an optimal gas production rate. That is to say, the production rate of gas reservoirs is equal to the rate which provides gas to fractures from base rocks. Thus, it is very significant to choose a reasonable production rate in the development of fractured gas reservoirs.

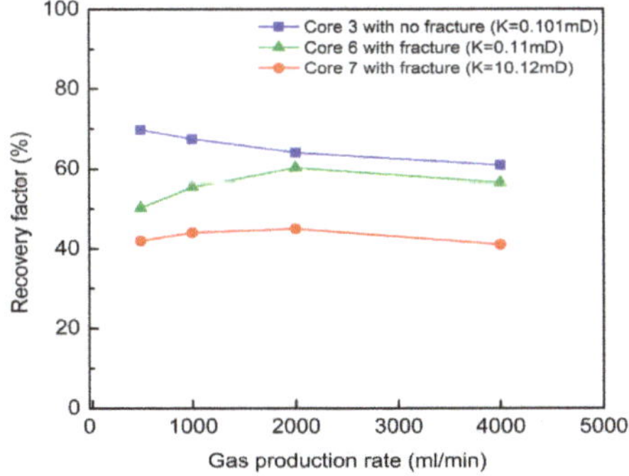

Figure 10. Recovery factor versus gas production rate for different fractures.

4. Conclusions

In this study, the experimental system of water invasion in gas reservoirs with edge and bottom aquifer was established and the elastic expansions of water and water soluble gas were considered. Seven full diameter core samples with dissolved pores and fractures were designed to simulate the process of water invasion in carbonate gas reservoirs. Then the effects of the related reservoir properties and production parameters were investigated. The following conclusions can be drawn from the study: During the physical simulation of water invasion energy source and performance, the elastic expansions of water and water soluble gas have a great influence. The aquifer size, gas production rate and permeability affects water invasion in gas reservoirs with dissolved pores. The bigger the aquifer is, the higher the water invasion is, and the greater the energy of gas reservoir provided by aquifer is. Moreover, the slower the bottom-hole pressure in early water invasion drops, the more quickly the bottom-hole pressure drops in the latter. Gas reservoirs with different permeabilities correspond to an optimum aquifer. The lower the permeability is, the larger the aquifer is. The higher the gas production rate is, the faster the water invasion is. However, gas production rate has less influence on recovery factor with the increase of permeability. The existence of fractures provides a prior seepage channel for water invasion in fractured gas reservoirs. The water will advance rapidly along the fractures, and the smaller water influx may result in a significant recovery factor.

Acknowledgments: This work was supported by National Science and Technology Major Project of the Ministry of Science and Technology of China Project (NO. 50150503-12 and NO. 2016ZX03005), and by the Project of PetroChina Research Institute of Petroleum Exploration & Development (NO. RIPED-LFFY-2017-JS-118). We thank the support from the Youth Foundation of Key Laboratory for Mechanics in Fluid Solid Coupling Systems, Chinese Academy of Sciences.

Author Contributions: Feifei Fang, Shusheng Gao, Huaxun Liu and Weijun Shen designed and discussed the experiment; Feifei Fang and Huaxun Liu conducted the experiment; Feifei Fang, Huaxun Liu, Qingfu Wang and Weijun Shen analyzed and discussed the results; Feifei Fang and Weijun Shen wrote the paper; Yang Li gave the help and suggestion in the revised manuscript.

Conflicts of Interest: The authors declare no conflict of interest.

References

1. Namani, M.; Asadollahi, M.; Haghighi, M. Investigation of water coning phenomenon in iranian carbonate fractured reservoirs. *Adv. Biomed. Res.* **2007**, *1*, 28–33.
2. Ould-Amer, Y.; Chikh, S.; Naji, H. Attenuation of water coning using dual completion technology. *J. Pet. Sci. Eng.* **2004**, *45*, 109–122. [CrossRef]
3. Shen, W.J.; Li, X.Z.; Liu, X.H.; Lu, J.L. Analytical comparisons of water coning in oil and gas reservoirs before and after water breakthrough. *Electron. J. Geotech. Eng.* **2014**, *19*, 6747–6756.
4. Shen, W.J.; Liu, X.H.; Li, X.Z.; Lu, J.L. Water coning mechanism in Tarim fractured sandstone gas reservoirs. *J. Cent. South Univ.* **2015**, *22*, 344–349. [CrossRef]
5. Jafari, I.; Jamshidi, S.; Masihi, M. Investigating the mechanism of water inflow in gas wells in fractured gas reservoirs and designing a controlling method. *Int. J. Chem. Environ. Biol. Sci.* **2012**, *3*, 482–487.
6. Zendehboudi, S.; Elkamel, A.; Chatzis, I.; Ahmadi, M.I.; Bahadori, A.; Lohi, A. Estimation of breakthrough time for water coning in fractured systems: Experimental study and connectionist modeling. *AIChE J.* **2014**, *60*, 1905–1919. [CrossRef]
7. Azim, R.A. Evaluation of water coning phenomenon in naturally fractured oil reservoirs. *J. Pet. Explor. Prod. Technol.* **2016**, *6*, 279–291. [CrossRef]
8. Amooie, M.A.; Soltanian, M.R.; Xiong, F.; Dai, Z.; Moortgat, J. Mixing and spreading of multiphase fluids in heterogeneous bimodal porous media. *Geomech. Geophys. Geo-Energy Geo-Resour.* **2017**, 1–20. [CrossRef]
9. Meyer, H.I.; Garder, A.O. Mechanics of two immiscible fluids in porous media. *J. Appl. Phys.* **1954**, *25*, 1400–1406. [CrossRef]
10. Blake, J.R.; Kueera, A. Coning in oils reservoirs. *Math. Sci.* **1988**, *13*, 36–47.
11. Ahmed, T. *Reservoir Engineering Handbook*, 4th ed.; Gulf Professional Publishing, Imprint Elsevier: Oxford, UK, 2010.
12. Muskat, M.; Wycokoff, R.D. An approximate theory of water-coning in oil production. *Trans. AIME* **1935**, *114*, 144–163. [CrossRef]
13. Høyland, L.A.; Papatzacos, P.; Skjaeveland, S.M. Critical rate for water coning: Correlation and analytical solution. *SPE Reserv. Eng.* **1989**, *4*, 495–502. [CrossRef]
14. Saad, E.D.S.; Darwich, T.; Asaad, Y. Water Coning in Fractured Basement Reservoirs. In Proceedings of the Middle East Oil Show, Manama, Bahrain, 11–14 March 1995.

15. Bahrami, H.; Shadizadeh, S.R.; Goodarzniya, I. Numerical Simulation of Coning Phenomena in Naturally Fractured Reservoirs. In Proceedings of the 9th Iranian Chemical of Engineering Congress, Iran University of Science and Technology (IUST), Tehran, Iran, 23–25 November 2004.

16. Cheng, K.H.; Jiang, T.W.; Wang, X.Y.; Mou, W.J.; Pan, Z.C. A study on water invasion mechanism of the bottom-water gas reservoir in the ordovician system of hetian-he gas field. *Nat. Gas Ind.* **2007**, *27*, 108–110.

17. Hu, S.; Lin, Q.; Tang, J. Analysis method of aquifer behavior and water invasion performance partly trapped in heterogeneous gas reservoir. *Nat. Gas Ind.* **2004**, *24*, 78–81.

18. Xiong, Y.; Yang, S.; Le, H.; Tang, J.; Yu, X. A new method of water influx performance analysis on fracture gas reservoirs with bottom water. *Nat. Gas Ind.* **2010**, *30*, 61–64.

19. Perez, E.; Garza, F.R.D.L.; Samaniego-Verduzco, F. Water Coning in Naturally Fractured Carbonate Heavy Oil Reservoir—A Simulation Study. In Proceedings of the SPE Latin America and Caribbean Petroleum Engineering Conference, Mexico City, Mexico, 16–18 April 2012.

20. Hu, Y.; Shao, Y.; Lu, Y.; Zhang, Y.F. Experimental study on occurrence models of water in pores and the influencing to the development of tight gas reservoir. *Nat. Gas Geosci.* **2011**, *22*, 176–181.

21. Shen, W.J.; Li, X.Z.; Liu, X.H.; Lu, J.L.; Jiao, C.Y. Physical simulation of water influx mechanism in fractured gas reservoirs. *J. Cent. South Univ. (Sci. Technol.)* **2014**, *45*, 3283–3287.

22. Jin, Z.H.; Johnson, S.E. Effects of elastic anisotropy on primary petroleum migration through buoyancy-driven crack propagation. *Geomech. Geophys. Geo-Energy Geo-Resour.* **2017**, 1–14. [CrossRef]

23. Fan, H.C.; Huang, Z.L.; Yuan, J.; Gao, G.; Tong, C.X. Solubility experiment of methane-rich gas and features of segregation and accumulation. *J. Jilin Univ. (Earth Sci. Ed.)* **2011**, *41*, 1033–1039.

24. Perera, M.S.A.; Ranjith, P.G.; Choi, S.K.; Airey, D. Numerical simulation of gas flow through porous sandstone and its experimental validation. *Fuel* **2011**, *90*, 547–554. [CrossRef]

25. Zou, C.N.; Du, J.H.; Xu, C.C.; Wang, Z.C.; Zhang, B.M.; Wei, G.Q.; Wang, T.S.; Yao, G.S.; Deng, S.H.; Liu, J.J.; et al. Formation, distribution, resource potential and discovery of the Sinian-Cambrian giant gas field, Sichuan Basin, SW China. *Pet. Explor. Dev.* **2014**, *41*, 278–293. [CrossRef]

26. Li, X.Z.; Gu, Z.H.; Wan, Y.J.; Liu, X.H.; Zhang, M.L.; Xie, W.R.; Su, Y.H.; Hu, Y.; Feng, J.W.; Yang, B.X.; et al. Geological characteristics and development strategies for Cambrian Longwanmiao Formation gas reservoir in Anyue gas field, Sichuan Basin, SW China. *Pet. Explor. Dev.* **2017**, *44*, 398–406. [CrossRef]

 *applied sciences*

Article

Power to Fuels: Dynamic Modeling of a Slurry Bubble Column Reactor in Lab-Scale for Fischer Tropsch Synthesis under Variable Load of Synthesis Gas

Siavash Seyednejadian [1,*], Reinhard Rauch [2], Samir Bensaid [1], Hermann Hofbauer [3], Gerald Weber [4] and Guido Saracco [1]

[1] Department of Applied Science and Technology, Politecnico di Torino, Corso Duca degli Abruzzi 24, 10129 Torino, Italy; samir.bensaid@polito.it (S.B.); guido.saracco@polito.it (G.S.)
[2] Engler-Bunte-Institut, Karlsruher Institute of Technology, Engler-Bunte- Ring 3, 76139 Karlsruhe, Germany; reinhard.rauch@kit.edu
[3] Institute of Chemical Engineering, Technical University of Vienna, Getreidemarket 9/166, 1060 Vienna, Austria; hermann.hofbauer@tuwien.ac.at
[4] Institute Bioenergy 2020+, Wiener Strasse 49, 7540 Güssing, Austria; gerald.weber@bioenergy2020.eu
* Correspondence: siavash.seyednejadian@polito.it; Tel.: +39-011-090-4662

Received: 23 February 2018; Accepted: 25 March 2018; Published: 28 March 2018

Abstract: This research developed a comprehensive computer model for a lab-scale Slurry Bubble Column Reactor (SBCR) (0.1 m D_t and 2.5 m height) for Fischer–Tropsch (FT) synthesis under flexible operation of synthesis gas load flow rates. The variable loads of synthesis gas are set at 3.5, 5, 7.5 m^3/h based on laboratory adjustments at three different operating temperatures (483, 493 and 503 K). A set of Partial Differential Equations (PDEs) in the form of mass transfer and chemical reaction are successfully coupled to predict the behavior of all the FT components in two phases (gas and liquid) over the reactor bed. In the gas phase, a single-bubble-class-diameter (SBCD) is adopted and the reduction of superficial gas velocity through the reactor length is incorporated into the model by the overall mass balance. Anderson Schulz Flory distribution is employed for reaction kinetics. The modeling results are in good agreement with experimental data. The results of dynamic modeling show that the steady state condition is attained within 10 min from start-up. Furthermore, they show that step-wise syngas flow rate does not have a detrimental influence on FT product selectivity and the dynamic modeling of the slurry reactor responds quite well to the load change conditions.

Keywords: Power to Liquid; Fischer–Tropsch; dynamic modeling; lab-scale

---

1. Introduction

In the last decade, Carbon Capture Utilization (CCU) with the aid of Renewable Energy Source (RES) power has led to significant progress in the field of Power to Gas (PtG) and Power to Liquid (PtL) technologies. In order to reduce Greenhouse Gas (GHG) emissions, European Energy policy proposes that the share of renewable energy is 40% by 2020 and 80% by 2050 [1]. This electricity-to-fuel process stabilizes the electrical power grid by converting the fluctuating characteristics of Renewable Energy Sources into storable energy carrier e.g., gaseous hydrocarbons (H_2 or CH_4) in PtG. This process may also provide liquid fuel for the use of chemicals and transport in PtL technology [2]. In this respect, Power to Liquid is a strong candidate for the transformation of power into chemical electricity as there is no loss during long-term storage; it has wide-ranging applications in the transport sector, due to its high energy density which is compatible with existing infrastructures [3]. The main pathway in this transformation is *Fischer–Tropsch Synthesis*, which offers many advantages in the process design and for optimizing product selectivity. The feedstock of the FT process is attained using generated synthesis gas and steam/co-electrolysis can be employed for generating syngas out of water and carbon dioxide. This enables RES power to be coupled with electrolysis providing a carbon free cycle [4]. Other alternatives for syngas generation include biomass gasification and natural gas reforming. Syngas derived from autothermal reforming or gasification technology with an air separation unit must be employed in large scale FT facilities and those methods have the further benefit of generating a high syngas ratio (H_2/CO) [5]. However, CO_2 gasification using biomass technology is not able to provide a sufficient syngas ratio for producing liquid fuel. Therefore, steam electrolysis or a water gas shift reactor could be used for adjusting the H_2/CO ratio so that it is equal to 2 (usage syngas ratio) [6].

In the field of power to fuel technology, flexible analysis of a Low Temperature Fischer–Tropsch (LTFT) reactor under variable operating conditions presents more challenges compared to other value-added processes such as methanation and the Dimethyl ether process. These challenges involve more complexity in FT product selectivity and lower feasibility in the dynamic analysis of an LTFT reactor compared to methanation and the DME synthesis [7]. In this respect, the most suitable FT reactors for analyzing under flexible operation are the Multi-Tubular Fixed Bed Reactor and the Slurry Bubble Column Reactor. In the Fixed Bed Reactor (FBR), dynamic analysis addresses the feasibility of reducing the size of H_2 storage and optimizing the temperature profile along the catalyst bed [8], whereas, in the FT Slurry type, the analysis is focused on improving mass transfer phenomena (gas-to-liquid contact and interfacial mass transfer area), leading to enhanced selectivity and catalyst performance with regard to complex hydrodynamic features and scale up issues [9]. During recent years, the Slurry Bubble Column Reactor has been identified as the best option for Fischer–Tropsch synthesis due to its many advantages compared to the other reactors. These advantages include (1) flexible temperature control and excellent heat transfer; (2) efficient inter-phase contacting which results in higher productivity; (3) low pressure drop leading to reduced compression costs; (4) better use of catalyst surface (fine particles less than 100 μm) allowing suitable liquid-solid mass transfer [5,10].

However, this reactor presents several technical challenges in the design of the pilot plant as well as large scale due to following reasons: (1) potential formation of slug regime flow; (2) very little information on mass transfer data; (3) difficulty in scale up due to complex hydrodynamic features; (4) Obstacles to the separation of fine catalysts from the slurry phase.

This work focuses on the dynamic modeling of a Slurry Bubble Column Reactor for Fischer Tropsch synthesis in pilot plant scale under variable loads of synthesis gas. This transient calculation is performed in once-through conditions under a cobalt-supported catalyst to identify *the effect of load change conditions* on FT selectivity, CO conversion, the alpha value and temperature distribution of the slurry reactor in the Winddiesel Technology. Winddiesel technology developed by the Technical University of Vienna has already been tested at an FT demonstration plant based on synthesis gas from biomass steam gasification and steam electrolysis in cases of availability of renewable energy [11,12]. In the modeling of the Fischer–Tropsch Slurry Bubble Column reactor (FT-SBCR) in lab-scale, a set of appropriate hydrodynamic parameters is incorporated into coupled FT kinetics and mass transfer through MATLAB code. This new approach enables us to analyze the behavior of *all species* through the length of the reactor from start-up to steady state condition. Moreover, in this dynamic modeling, change in the superficial gas velocity due to chemical reaction is coupled to the set of PDEs using *the overall species transport equations*. This approach estimates the reliable calculation of the gradient of gas flow rate for bubbles which proposes a realistic prediction of the reactor performance [13]. However, the majority of slurry reactors models for superficial gas velocity within the reactor linearize the gas velocity with syngas conversion [14–16].

In dynamic modeling SBCR for FT process, a number of methodologies based on several well-developed hydrodynamic concepts such as the Axial dispersion model (ADM), Single bubble class (SBC), two bubble class (TBC), etc. with different FT kinetics have been proposed. In 2002, J.W.A de Swart and R. Krishna [14] developed a model to predict the steady state and dynamic behavior of a bubble column slurry reactor for Fischer Tropsch synthesis. Their numerical procedure was based on four partial differential equations solved using Method Of Lines (MOL). The results indicated that steady-state is achieved within about seven minutes from start-up and no thermal runways were observed in a reactor of commercial scale. Furthermore, the influence of the back-mixing of the liquid phase on hydrogen conversion for two different reactor diameters (1, 7.5 m) was compared. It was concluded that at 1 m diameter the axial dispersion coefficient decreased, leading to a flatter velocity profile in the liquid phase, which in turn results in higher conversions of hydrogen [14].

In 2005, Rados et al. [13] simulated an FT-SBCR with two chemical reaction systems. They analyzed the hydrodynamic behavior of two bubble class models by assuming linear first-order reaction kinetics. They also considered the influence of the Axial Dispersion Model (ADM) on conversion and reactor diameter and compared this effect in ideal reactors i.e., plug flow (PF) and completely stirred tank (CST) [13]. In 2009, Hooshyar et al. [15] developed a dynamic slurry FT for both single and double bubble class at churn turbulent flow regime. They concluded that there is no discrepancy between single and double class models in terms of concentration, temperature and conversion profile. Thus, they considered the single bubble class as a less complex reliable model to analyze the slurry bubble column reactors [15]. In 2008, Laurent Sehabiague [17] et al. developed a computer model for a large-scale FT slurry reactor. The simulator was used to optimize superficial gas velocity and reactor geometry for producing 10,000 (barrels/day) of liquid fuels. Different operating conditions were also used to find the maximum space time yield (STY). However, the condition for maximum productivity was considered as the optimum operating condition because of lower operating and capital cost [17].

Accordingly, almost all dynamic modeling of FT slurry reactors has focused on the commercial scale of slurry bubble column reactors and there has been no detailed transient computer model at pilot scale (reactor diameter < 1 m) for SBCR so far. In fact, the modeling of laboratory scale FT slurry is a more difficult task because of the wealth of dynamic features, such as the prediction of rise velocity of small bubbles and wall effects, which make hydrodynamic parameters quite sensitive to system properties, and the presence of impurities [18]. Thus, the present work investigates, for the first time, the transient analysis of a slurry FT reactor in lab-scale under supported catalyst cobalt for all key components of Fischer–Tropsch synthesis developed by a MATLAB code. This comprehensive modeling enables us to predict the behavior of several slurry reactor parameters such as CO conversion, FT selectivity, α-value and temperature profile under variable loads of synthesis gas. Furthermore, the results of the modeling are in good agreement with the experimental data. The experimental data is adopted from a Master of Science thesis which was conducted using the Fischer–Tropsch research plant. The plant is located in town of Güssing in Austria [11,12]. Appendixs A and B reflect the experimental measurements for both base load (syngas flow rate of 5 m^3/h) and change load conditions (syngas flow rate of 3.5 m^3/h and 7.5 m^3/h) in one specific run [12].

2. Fischer–Tropsch Reaction Scheme

Fischer–Tropsch synthesis consists of a set of polymerization reactions leading to a blend of linear paraffins of different carbon numbers. In the present investigation, the rate expression of Fischer–Tropsch synthesis under catalyst cobalt is employed based on Yates and Satterfield 1991 [19] (see the relation of reaction kinetics in Table 1). In this reaction kinetics ($-R_{co}$), by modifying parameters a and b, the model predictions of CO conversion correspond to the experimental measurements. Anderson-Schulz Flory model gives an indication of the distribution for n-paraffins based on their mass fraction (W_n) which can be expressed with the relation of ASF distribution of products as mentioned in Table 1 (W_n relation). W_n is defined by parameter α (Chain growth probability factor). In fact, α-value reflects the distribution of the weight percentage of products with regard to their carbon number [20]. Factor α strongly depends on temperature, pressure and the catalyst used in the process. Generally, in our model, alpha correlation is employed in terms of temperature as mentioned in the relation of chain growth probability factor (Table 1), which is described by Song et al. (2004) [21]. In the section of model comparison, in order to calculate α-value a semi-logarithmic plot of mass fraction against carbon number is considered (logarithmic relation in Table 1) producing a straight line. The slope of the line is given the α-value. The calculation is performed for different operating conditions based on both model and experimental data. The products of Fischer–Tropsch synthesis under catalyst cobalt are predominantly paraffins through the generic reaction in Table 1 (paraffin reaction form). Therefore, the rate of paraffin formation based on Anderson-Sculz-Flory (ASF) equation can be calculated from r_i relation in Table 1 [22]. Also, Table 2 illustrates the operating conditions and several lab parameters. Details of the FT reactor and experimental setup are mentioned in references [11,12].

Table 1. Kinetic Characteristics of Fischer–Tropsch Slurry Bubble Column reactor (FT-SBCR): reaction parameters, product distribution.

Reaction Characteristics	Relations	Constants, Parameters and Paraffin Reaction Form	
Reaction Kinetics	$-R_{co} = \dfrac{a\,P_{co}P_{H_2}}{(1+bP_{co})^2}$	$a = 1.59064 \times 10^{-12}$	$b = 7.99389 \times 10^{-6}$
Chain grow probability factor (Song et al.)	$\alpha = \left(A\dfrac{y_{co}}{y_{H_2}+y_{co}} + B\right)[10.0039(T-533)]$	$A = 0.2332$ $y_{co} = 0.2$	$B = 0.633$ $y_{H_2} = 0.4$
Anderson Sculz-Flory distribution of products.	$W_n = n\,(1-\alpha)^2.\alpha^{n-1}$	$\log\dfrac{W_n}{n} = n\log(\alpha) + \log\dfrac{(1-\alpha)^2}{\alpha}$	
Rate of paraffin formation based on ASF distribution.	$r_i = R_{co}\alpha^{n-1}$	$n\,CO + (2n+1)H_2 \rightarrow C_nH_{2n+2} + nH_2O$	

Table 2. Operating conditions and liquid properties of lab-scale Fischer–Tropsch Slurry Bubble Column.

Operating Condition	
Reactor Temperature	503 K
Reactor Pressure	20 bar
H_2/CO	2
Reactor diameter	0.1 m
Reactor height	2.5 m
Volumetric Flow Rate (loads)	$3.5, 5, 7.5\ m^3/h$
Liquid Phase Properties	
Liquid Density	$715\ kg/m^3$
Surface tension	0.023 N/m
Liquid Viscosity	$3 \times 10^{-3}\ Pa.s$

3. Modeling Activity

3.1. Model Framework

A detailed computer model for a pilot plant FT-SBCR (D = 0.1 m, H = 2.5 m) for all the key components of FT synthesis reactor was developed. A set of Partial Differential Equations (PDEs) of transport species including mass transfer and kinetics was successfully coupled with hydrodynamic parameters. In general, 10 equations for the gas phase (single bubble class) and 10 equations for the liquid phase need to be solved simultaneously. The reactor model was established with these assumptions: (1) an Axial Dispersion Model (ADM) in the form of a convection-diffusion phenomenon with a single class of gas bubble diameter is considered (ADM-SBCD); (2) the gas-liquid mass transfer resistance is positioned on both the gas and liquid side; (3) based on the non-dimensional form of Peng-Robinson equation-of-state (PR-EOS), the compressibility factor of the gas phase corresponds to near unity ($Z \approx 1$). The PR-EOS is a suitable fluid model for FT systems to predict vapor-liquid compositions and flow rates inside the slurry reactor [23,24]; (4) considering a low pressure drop in the slurry reactor, the operating pressure is assumed to be constant along the reactor height; (5) the reactor operates under isothermal condition; (6) by assuming constant pressure and temperature, the overall continuity balance occurs at the inlet and outlet of the reactor (Equation (3)); (7) liquid-solid mass transfer resistance may be ignored and consequently solid suspension in liquid is simulated as a single pseudo-homogeneous slurry phase in this work; (8) chemical reaction through Langmuir Hinshelwood Hougen Watson (LHHW) kinetics in liquid phase is considered.

Based on these model assumptions, the mass balances of each component in gas and liquid phases can be derived as follows:

Gas phase equations:

$$\underbrace{\frac{\partial\left(\varepsilon_g C_{g,i}\right)}{\partial t}}_{\text{Accumulation}} = \underbrace{\frac{\partial}{\partial z}\left(D_g\,\varepsilon_g\frac{\partial C_{g,i}}{\partial z}\right)}_{\text{Axial Dispersion}} - \underbrace{\frac{\partial}{\partial z}\left(\varepsilon_g U_g C_{g,i}\right)}_{\text{Convection}} - \underbrace{(K_l a)_{g,i}\varepsilon_l\left(\frac{C_{g,i}}{A_i} - C_{l,i}\right)}_{\text{Mass Transfer}} \tag{1}$$

Liquid phase equations:

$$\underbrace{\frac{\partial\left(\varepsilon_l C_{l,i}\right)}{\partial t}}_{\text{Accumulation}} = \underbrace{\frac{\partial}{\partial z}\left(\varepsilon_l D_{ax,l}\frac{\partial C_{l,i}}{\partial z}\right)}_{\text{Axial Dispersion}} - \underbrace{\frac{\partial}{\partial z}\left(\varepsilon_l U_l C_{l,i}\right)}_{\text{Convection}} + \underbrace{(K_l a)_{g,i}\varepsilon_l\left(\frac{C_{g,i}}{Ai} - C_{l,i}\right)}_{\text{Mass Transfer}} - \underbrace{\varepsilon_l \rho_{cat}\varepsilon_{cat} r_i}_{\text{Reaction}} \tag{2}$$

Total concentration:

$$\sum_{i}^{n} C_{g,i} = C_{tot,in} = \frac{P}{ZRT} = C_{tot,out} \tag{3}$$

In the above equations, A_i denotes the Henry's constant (H_i) for syngas and light hydrocarbons (CO, H_2, H_2O, C_1 and C_2). By employing Henry's law solubility factor ($P_i = x_i \times H_i^\infty$), the Henry's constant for each reactant gas was calculated based on the heat of solution for each component and operating temperature [25].

$$A_i = H_i/RT \tag{4}$$

$$H_i = H_i^* \exp\left(-\frac{\Delta H_{s,i}}{RT}\right) \tag{5}$$

where, the value of parameters H_i^* and $\Delta H_{s,i}$ are listed in reference [25].

For heavier components (C_4^+), Ai was calculated using Raoult's law for gas-liquid phase at the equilibrium as follows:

$$A_i = H_i^\infty/RT.1/C_{tot}^L \tag{6}$$

$$C_{tot}^L = \rho_L/MW_{L,avg.} \tag{7}$$

where, C_{tot}^L is the total concentration of liquid components, ρ_L is the liquid density, $MW_{L,avg.}$ is the average molecular weight of liquid components and H_i^∞ is the Henry's constant at infinite dilution which is defined as follow:

$$H_i^\infty = \gamma_i^\infty \, P_{i,sat} \tag{8}$$

where γ_i^∞ is the activity coefficient for heavier components and $P_{i,sat}$ is the vapor pressure of component i which is calculated from asymptotic behavior correlations (extension of the Antoine equations). By supposing ideal behavior of FT mixture due to long-chain n-paraffins, the Henry's constant for the mixture can be expressed by:

$$LnH_{i,mix}^\infty = \sum_{j} x_j Ln\, H_j^\infty \tag{9}$$

where j is the components in the solvent. Since LnH_i^∞ and $Ln\gamma_i^\infty$ is asymptotically linear with solute carbon number (m), the formula for calculating the infinite-dilution activity coefficient becomes as follow:

$$Ln\gamma_m^\infty = Ln\gamma_r^\infty \frac{(n-m)}{(n-r)} \tag{10}$$

where n is the carbon number of solvent (n-paraffin) and r denotes the carbon number of reference solute (n-C_6H_{14}) in the same solvent. They are described in more detail in references [24,26].

As mentioned in the model assumption, since the reactor is supposed to be operated under constant pressure and temperature, the total concentration is constant. As a consequence, the variation of the gas flow rate due to chemical reaction and mass transfer is determined by assuming the constant total concentration of components in single bubble gas diameter (Equation (3)). Therefore, the Equation (11) as a sub-model in the form of a gas state equation is incorporated into the SBCD model. This sub-model used for behavior of superficial gas velocity inside the reactor is a reliable approach since it considers the concentrations of all gaseous components [13].

$$\frac{\partial U_g}{\partial z} = -\frac{1}{C_{tot}} \sum_{i=1}^{n} K_l \times a \times \varepsilon_l \times \left(\frac{C_{g,i}}{A_i} - C_{l,i}\right) \tag{11}$$

Table 3 depicts the initial and boundary conditions of Equations (1), (2) and (11). The initial values in the model are set based on pressure, temperature and concentration of species. The boundary conditions adopted from Danckwerts' type are defined for the gas and liquid at the inlet and outlet of the reactor. For the gas phase at the reactor inlet, the concentration is taken from a syngas composition under the operating conditions of the laboratory, calculated based on Table 3. The inlet superficial gas velocity for single bubble gas diameter is calculated from Equation (14).

The effective gas-liquid interfacial area for mass transfer of small bubbles between two phases can be expressed as follows [27]:

$$a = 6 \times \varepsilon_g / d_B \tag{12}$$

The mass transfer coefficient (K_l) is calculated from the following empirical correlation. It is applicable in a wide range of operating conditions which leads to a good prediction of gas-liquid mass transfer [27]:

$$K_l a = 1.77 \times \sigma^{-0.22} \times \exp(1.65 U_l - 0.65 \mu_l) \times \varepsilon_g^{1.2} \tag{13}$$

Table 3. Initial and boundary conditions for the slurry reactor model.

Initial Condition (t = 0)	Reactor Inlet (z = 0)	Reactor Outlet (z = H)
$C_{gi} = C_{gi,in}$	$C_{gi} = C_{gi,in}$ $C_{g,i0} = \frac{P_i}{ZRT}$	$\frac{\partial C_{g,i}}{\partial z} = 0$
$C_{li} = C_{gi,in}/A_i$	$\varepsilon_l D_l \left(\frac{\partial C_{l,i}}{\partial z}\right)_{z=0} = U_l \left(C_{l,i} - C_{g,i0}/A_i\right)$	$\frac{\partial C_{g,i}}{\partial z} = 0$
$U_i = U_{g,in}$	$U_{gi} = U_{g,in}$	$\frac{\partial U_g}{\partial z} = 0$

The superficial gas velocity is defined as the volumetric gas flow rate divided by the cross-sectional area of the reactor above the gas distributor [28]:

$$U_{g,in} = \frac{V_f}{A_r} \tag{14}$$

The initial bubble size depending on buoyancy forces and surface tension is derived by the theoretical Davidson and Schuler expression [29]:

$$d_B = \left[\frac{6\sigma \, d_0}{g \left(\rho_{SL} - \rho_G\right)}\right]^{1/3} \tag{15}$$

The gas hold up can be expressed as the volume of the gas phase divided by the reactor volume consisting of gas volume, liquid volume and volume of catalyst used [28]:

$$\varepsilon_g = \frac{V_{gas}}{V_{gas} + V_{liquid} + V_{cat}} \tag{16}$$

This is calculated to be 0.161 ($m_G^3\ m_R^{-3}$) and the catalyst volume fraction (ε_{cat}) is calculated to be 0.34 ($m_{cat}^3\ m_L^{-3}$). It should be noted that in the Equations (1) and (2), ε_L is the liquid holdup ($m_L^3\ m_R^{-3}$) and r_i denotes reaction rate of FT species (mol $kg_{cat}^{-1}\ s^{-1}$). Moreover, the velocity of liquid inside the reactor is assumed to be 0.00089 m/s.

In the slurry bubble column reactors, the gas bubble coalescence occurs in a short time with an increase of the column diameter, and the large bubbles collect around the center of the column in the operation of a churn turbulent regime. This is called heterogeneous flow, which usually occurs when the superficial gas velocity is greater than 0.05 m/s. When the gas-liquid mixture reaches the surface, the bubbles disengage, allowing the degassed liquid to recirculate. Therefore, the main cause of back-mixing and liquid dispersion in the lab-scale slurry FT reactor is attributed to downward velocity of the liquid in the wall region and upward direction ($V_L(r)$) in the central axis [30]. Based on Riquarts correlation [31] the magnitude of $V_L(0)$ depends on the column diameter, superficial gas velocity and kinematic viscosity of the liquid phase as described in Equation (17). The experimental data [30,31] show that the liquid phase axial dispersion coefficient ($D_{ax,L}$) has a direct proportionality to column reactor diameter and centre-line liquid velocity as mentioned in Equation (18).

$$V_L(0) = 0.2 \times (g\ D_t)^{1/2} \times (U^3_{g,in}/g\ \nu_L)^{1/8} \tag{17}$$

$$D_{ax,L} = 0.31 \times V_L(0) \times D_t \tag{18}$$

According to reference [30,31], the two correlations given above are the most appropriate ones for the estimation of axial dispersion coefficient which was recommended in all systems (including slurry). The axial dispersion coefficient of the gas phase for the small bubbles is equal to that of the liquid phase according to the relationship which was proposed by Sehabiague et al. [10].

3.2. Computer Solution Procedure

In this transient calculation, a MATLAB pdepe (Partial Differential Equations Parabolic Elliptic) solver is implemented. The solver converts a set of PDEs to ODEs (Ordinary Differential equations) using an accurate spatial discretization based on a specified grid size. The solution domain is equal to the length scale of the reactor and the optimized discretization is chosen based on a balance between the desired level of accuracy and the affordable CPU time [8].

For solving equations, the mass balance of 10 components of H_2, CO, H_2O, CO_2, CH_4, C_2H_6, C_4H_{10}, $C_{10}H_{22}$, $C_{18}H_{38}$, and $C_{30}H_{62}$ in two phases, gas and liquid, are considered.

4. Results and Discussions

The simulation of the FT slurry reactor begins with the start-up after all heating devices are switched on and the alarm values are set. Once an operating parameter such as temperature or pressure, attains its alarm value, the research plant is automatically switched off and this is called an alarm shut down (ASD). The alarm values are vitally important since they ensure a safe operation of the plant. Then, the FT plant is started under N_2 flow as an inert gas in the manual mode until the FT reactor reaches a temperature of 453 K, at which the wax in the FT reactor is liquid. Afterwards, the plant is switched to the automatic mode and the FT reactor reaches the required operating temperature of 503 K and the plant starts to operate under syngas. The time required for the start-up stage before switching the plant to automatic mode is about 2–3 h.

4.1. Species Distribution

4.1.1. The Behavior of Components H_2, CO

Firstly, the aforementioned hydrodynamic parameters are implemented in the computer model then the equations related to small bubbles of gas species and the liquid phase are coupled to obtain the concentration behavior of all FT key components from the beginning to the steady state conditions through the length of the reactor. As Figure 1 shows, at $\tau = 0$ there is no carbon monoxide and hydrogen in either of the two phases. At normal operation ($\tau > 0$), these species enter the reactor with their own inlet values (which are already calculated based on the boundary conditions in Table 3; $C_{g,in}$), initiating the reaction. At the reactor inlet a maximum value is seen and liquid is supposed to be saturated with gas phase. The CO and H_2 are consumed due to the chemical reaction and their concentration decreases across the reactor length to the equilibrium values in two phases. In this respect, "wall effects" cause small bubbles to dissolve faster into the liquid phase before reaching steady state values. The concentration behavior of species (mol/m^3_{gas} for gaseous components and mol/m^3_{liquid} for liquid components) shows that the steady state values are reached at 10 min, as expected from the experimental data [12]. The CO conversion in this base load condition equals 58%, which is slightly overestimated compared to the experimental data. The run is performed for T = 503 K, P = 20 bar, U_g = 0.17 m/s, ε_{cat}= 0.34.

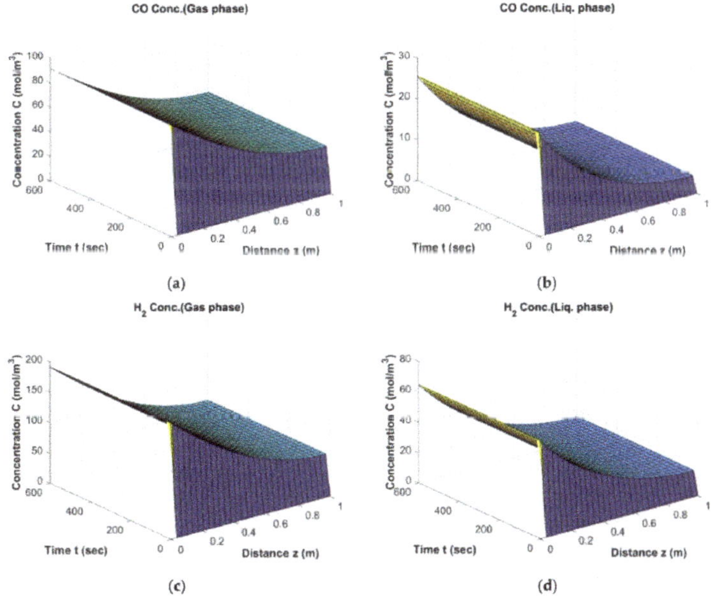

Figure 1. Results of Fischer–Tropsch (FT) Simulation; The behavior of CO, H_2 in gas phase (**a,c**) and liquid phase (**b,d**) from start-up to steady state conditions (D_t = 0.1 m, H = 2.5 m, ε_{cat} = 0.34, $U_{g,in}$ = 0.17 m/s).

4.1.2. The Behavior of Components H_2O, CO_2

As Figure 2 illustrates, water vapor, which is one of the products of FT synthesis, increases over the reactor height and over the time. Since the FT reaction occurs under a co-supported catalyst, no water gas shift reaction is promoted and the inlet CO_2 acts as an inert through the process. The CO_2 concentration increases due to the volume reduction of off-gas and syngas consumption. It is worth noting that the presence of diluents such as CO_2, CH_4 or N_2 has a beneficial influence in slurry bubble column reactors. The inert (here as carbon dioxide) enables supplementary mixing energy to the slurry system to maintain catalyst suspension. On the other hand, in FBRs we need to avoid diluents since they elevate the pressure drop across the reactor bed [32].

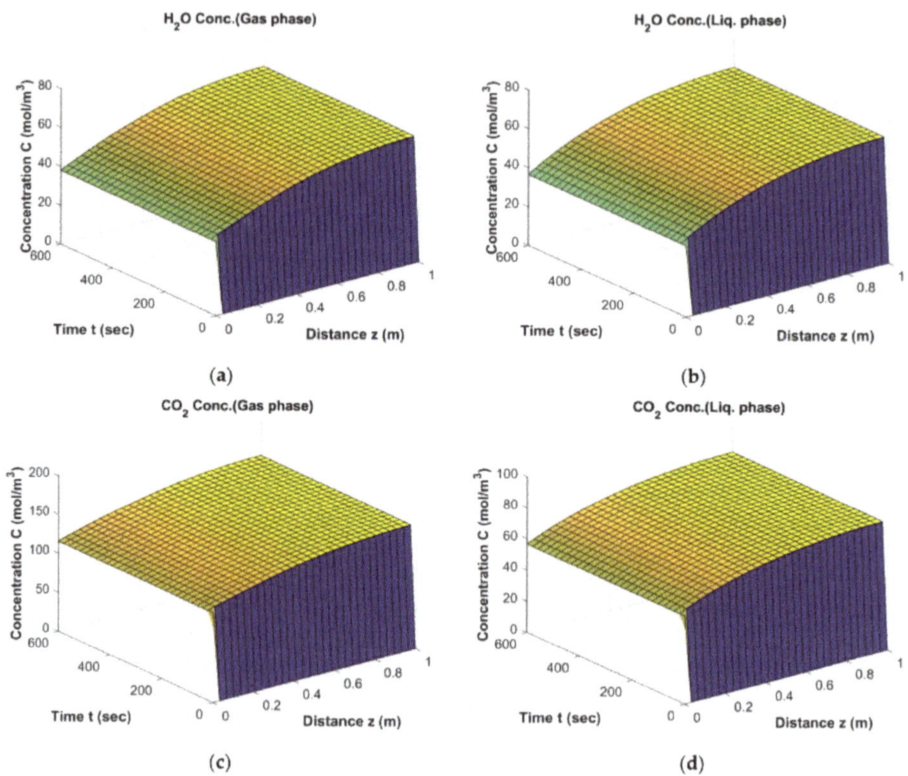

Figure 2. Results of FT Simulation; The behavior of H$_2$O and CO$_2$ in gas phase (**a,c**) and liquid phase (**b,d**) from start-up to steady state conditions (D_t = 0.1 m, H = 2.5 m, ε_{cat} = 0.34, $U_{g,in}$ = 0.17 m/s).

4.1.3. The Behavior of Components CH$_4$, C$_2$H$_6$, C$_4$H$_{10}$

Methane, ethane and butane increase over time and achieve their highest concentration at the final time in the reactor outlet. These highest values in the gas phase reach around 85, 35 and 50 mol/m^3 for CH$_4$, C$_2$H$_6$ and C$_4$H$_{10}$, respectively as shown in Figure 3.

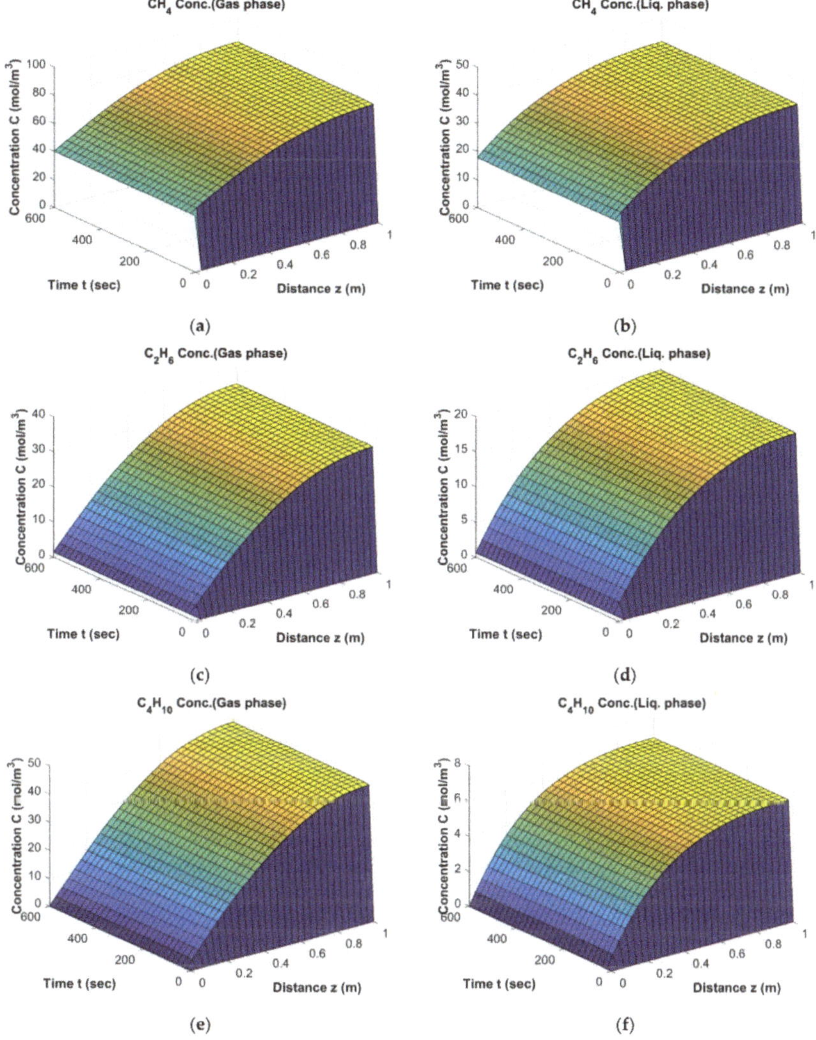

Figure 3. Results of FT Simulation; The behavior of CH$_4$, C$_2$H$_6$ and C$_4$H$_{10}$ in gas phase (**a,c,e**) and liquid phase (**b,d,f**) from start-up to steady state conditions (D$_t$ = 0.1 m, H = 2.5 m, ε_{cat} = 0.34, U$_{g,in}$ = 0.17 m/s).

4.1.4. The Behavior of Liquid Products C$_{10}$H$_{22}$, C$_{18}$H$_{38}$, C$_{30}$H$_{62}$

In this mathematical modeling, three components are considered as representatives of each specific carbon cut. C$_{10}$H$_{22}$, C$_{18}$H$_{38}$ and C$_{30}$H$_{62}$ were introduced as naphtha, diesel and wax respectively, which are derived from three condensers with reaction water in each condenser and collected in each related drum [12]. The results of modeling show that all three sets of products have a similar profile through the slurry reactor. As illustrated in Figure 4, the liquid products have an upward trend over time through the reactor height and the highest magnitude belongs to the middle distillate whereas diesel and wax stand at lower values respectively.

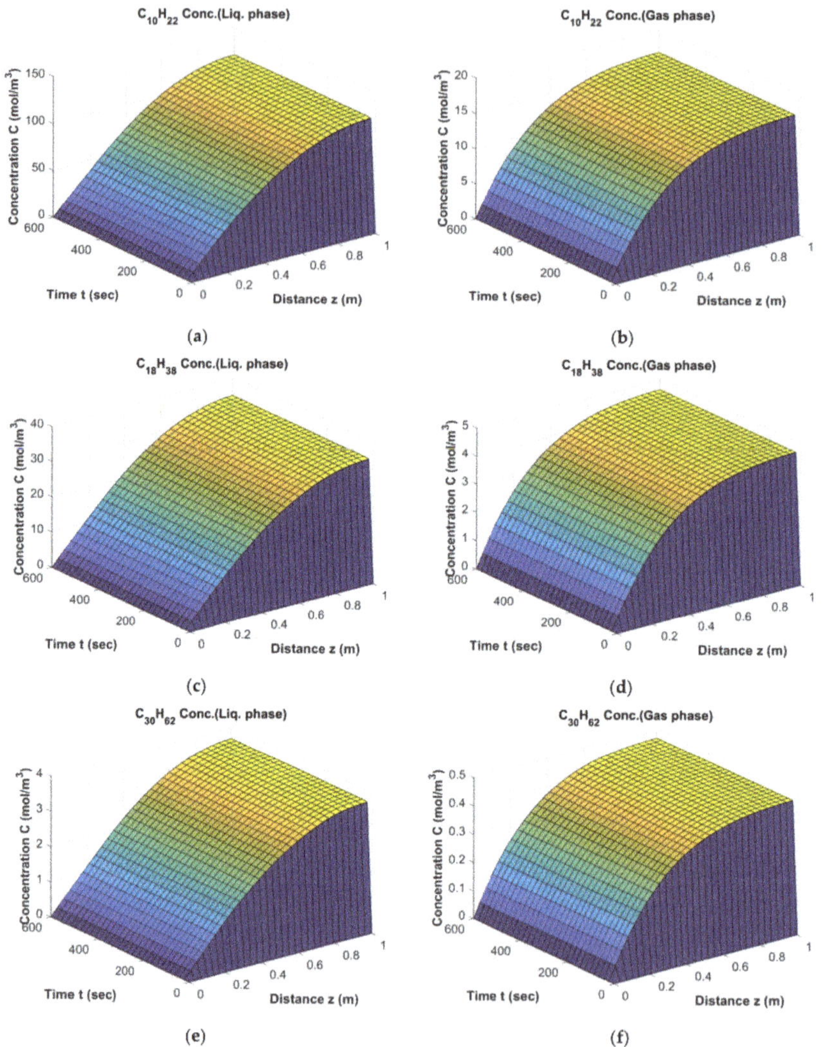

Figure 4. Results of FT simulation; The behavior of $C_{10}H_{22}$, $C_{18}H_{38}$ and $C_{30}H_{62}$ in liquid phase (**a,c,e**) and gas phase (**b,d,f**) from start-up to steady state conditions (D_t = 0.1 m, H = 2.5 m, ε_{cat} = 0.34, $U_{g,in}$ = 0.17 m/s).

These simulations results relate to base load conditions (volumetric gas flow rate 5 m^3/h). The computer model is also able to work quite well under variable loads of synthesis gas. Therefore, the flow rate in the model was changed in the range of 3.5–7.5 m^3/h to find maximum CO conversion and FT product selectivity.

4.2. Model Comparison with Experimental Data

The comparison of the model results with experimental data was conducted based on the composition (volume %) of syngas and off-gas which were measured by a GC device under base load and change load conditions. Figures 5–7 show a comparison of the predicted values by the computer model with the measured data from the conducted experiments. In general, the results show that the predicted model is in good agreement with the experimental data. When compared with FT products only in the case of naphtha (C_8–C_{10}) is there a noticeable difference between model and laboratory data in all three operating conditions. This is due to the volatility of this carbon cut and difficulty in collecting them together. In product distribution (Figures 5c, 6c and 7c) only, two groups (C_8–C_{15}) and (C_{30}–C_{40}) show a deviation as well as discontinuities in the model prediction. The deviation is probably due to the neglect of olefin formation in the reactor modeling and the discontinuities can be attributed to the model equations which were solved for specific classes of FT species (C_1, C_2, C_4, C_{10}, C_{18} and C_{30}). As mentioned earlier, the α-value can be derived from the slope of the drawn line in all three load conditions. In the model, the α-value for volumetric flow rate of 3.5, 5 and 7.5 m^3/h is calculated to be 0.89, 0.9 and 0.88 respectively. It shows that change load conditions have almost no influence on the α-value or catalyst selectivity as predicted from the experiment.

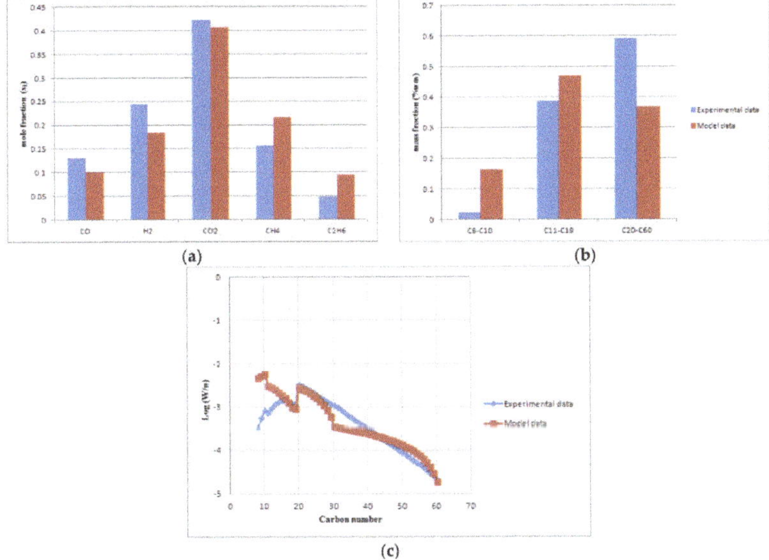

Figure 5. The comparison of the predicted model with experimental data at base load conditions (volumetric gas flow rate 5 m^3/h, T = 503 K, ε_{cat} = 0.34, P = 20 bar): (**a**) off-gas molar fractions; (**b**) FT products (naphta, diesel, wax); (**c**) ASF distribution.

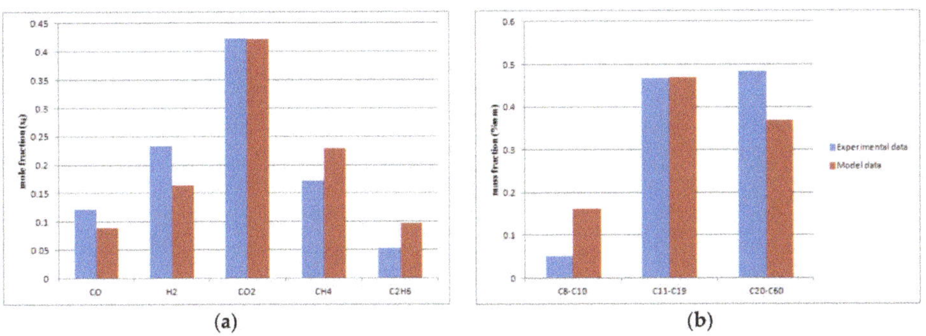

Figure 6. *Cont.*

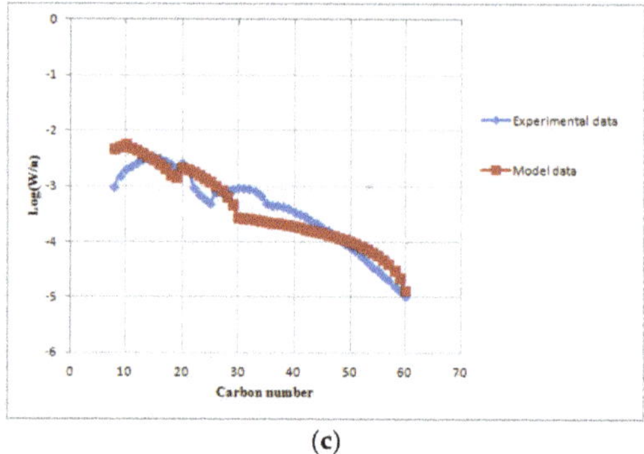

(c)

Figure 6. The comparison of the predicted model with experimental data at change load conditions (volumetric gas flow rate 3.5 m^3/h, T = 503 K, ε_{cat} = 0.34, P = 20 bar): (**a**) off-gas molar fractions; (**b**) FT products (naphta, diesel, wax); (**c**) ASF distribution.

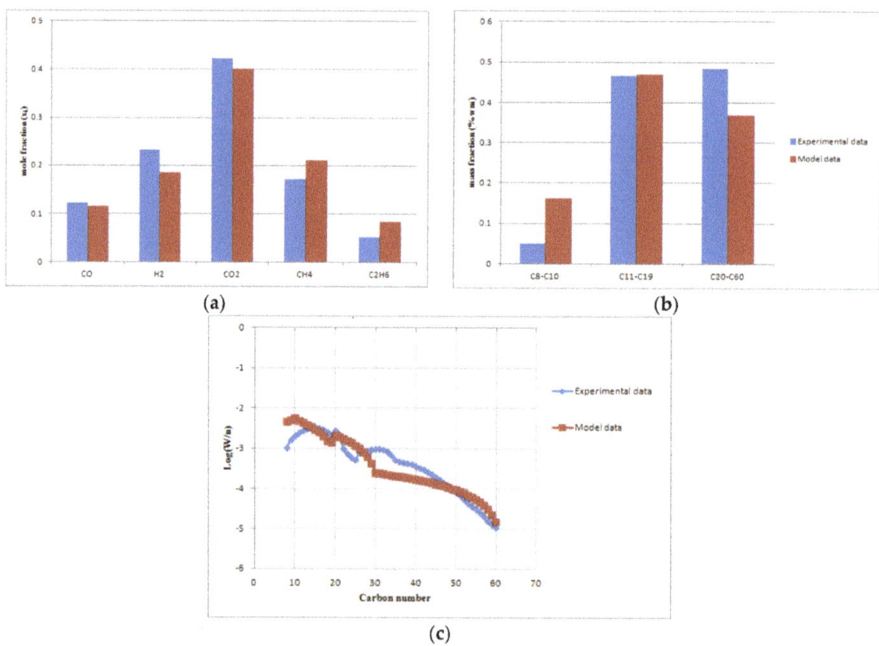

(a)

(b)

(c)

Figure 7. The comparison of the predicted model with experimental data at change load conditions (volumetric gas flow rate 7.5 m^3/h, T = 503 K, ε_{cat} = 0.34, P = 20 bar): (**a**) off-gas molar fractions; (**b**) FT products (naphtha, diesel, wax); (**c**) ASF distribution.

The results show that the CO conversion both in the model and experiments has lower values at a higher superficial velocity of syngas. This is due to the decreasing residence time of reactants at higher values of velocity. Thus, the maximum CO conversion occurs at 3.5 m^3/h which is equal to 60%. As Figure 8 shows, the CO conversion also is in good agreement with the experimental data; however, the model values are slightly overestimated. This is due to larger area for gas-liquid mass transfer in the result of the assumption of the single bubble class diameter.

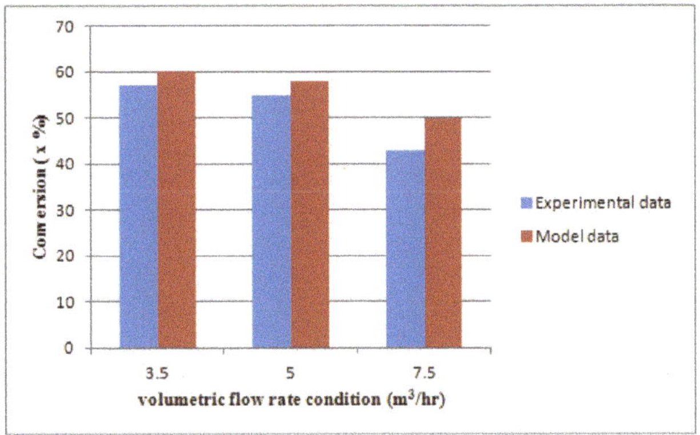

Figure 8. CO Conversion at base and change load condition based on experimental and model data.

4.3. The Effect of Temperature on Product Selectivity

One strength of FT reactor modeling is that analysis of the product selectivity with reaction temperatures can be carried out to find the desired operating conditions. Figures 9–11 illustrate the variation of product selectivity based on mass fraction with operating temperature at three loads of volumetric flow rates (3.5, 5, 7.5 m^3/h), respectively. As shown in these figures, for all three loads with increasing temperature the light gaseous (C_1–C_4), naphtha (C_5–C_{10}) and diesel increase whereas, heavier liquid fuels such as wax (C_{20}–C_{60}) tend to decrease. This trend can be expected since higher temperatures tend to shift the α-parameter to lower values thus producing more light hydrocarbons (C_1–C_4). In the design of the lab-scale SBCR for the FT process, it is advisable to operate the reactor under a narrow temperature range (483–503 K). This prevents catalyst deactivation, avoiding higher increases in methane as well as obtaining selectivity of diesel products [12].

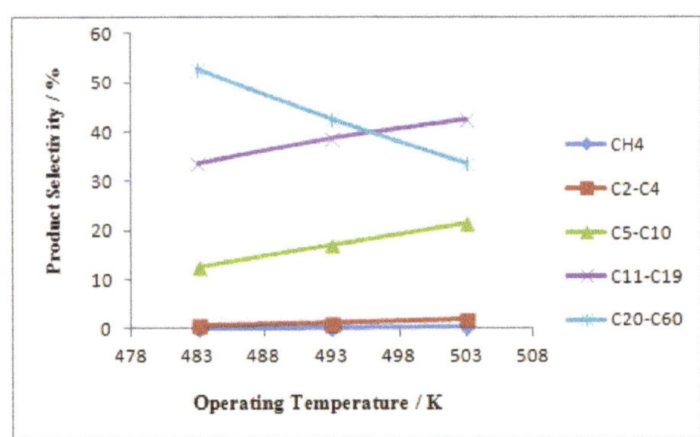

Figure 9. Influence of operating temperature on product selectivity; D_t = 0.1 m, H = 2.5 m, V_f = 3.5 m^3/h, P = 20 bar, ε_{cat} = 0.34.

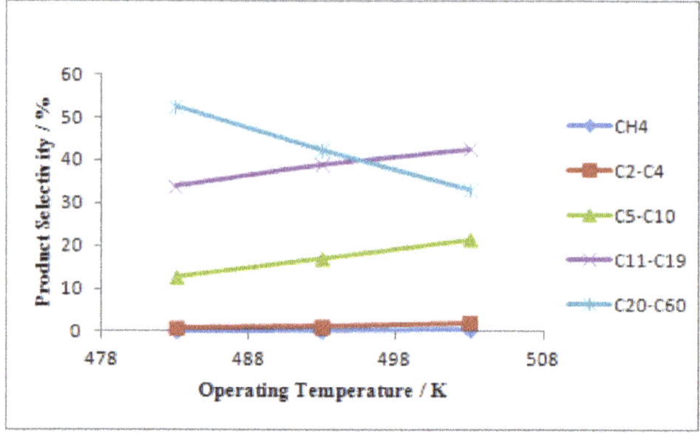

Figure 10. Influence of operating temperature on product selectivity; D_t = 0.1 m, H = 2.5 m, V_f = 5 m^3/h, P = 20 bar, ε_{cat} = 0.34.

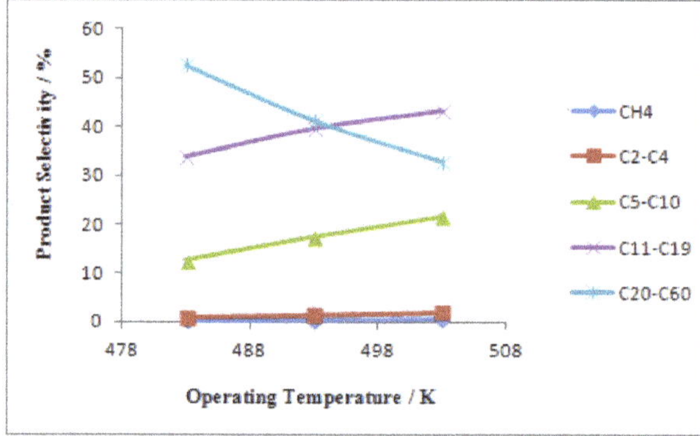

Figure 11. Influence of operating temperature on product selectivity; D_t =0.1 m, H = 2.5 m, V_f = 7.5 m^3/h, P = 20 bar, ε_{cat} = 0.34.

It was concluded that by increasing the load of syngas flow rate, the selectivity of wax and diesel remain constant in each corresponding temperature. It was also concluded that there is a homogenous temperature profile within the reactor.

4.4. The Behavior of the Species Inside the Reactor

The experimental data of the research FT reactor in laboratory scale is only able to reflect information about the reactor outlet. However, this mathematical modeling attempts to predict the dynamic behavior of the system through the height of the reactor which includes the concentration change of all species along the reactor length. Figure 12a,b show the molar concentration and conversion of syngas (CO and H_2) along the reactor bed at base load condition, respectively. The slightly higher values of H_2 conversion compared to CO conversion is attributed to the stochiometric ratio of H_2/CO taking the value 3 for producing methane, which then shifts to 2 in paraffin formation.

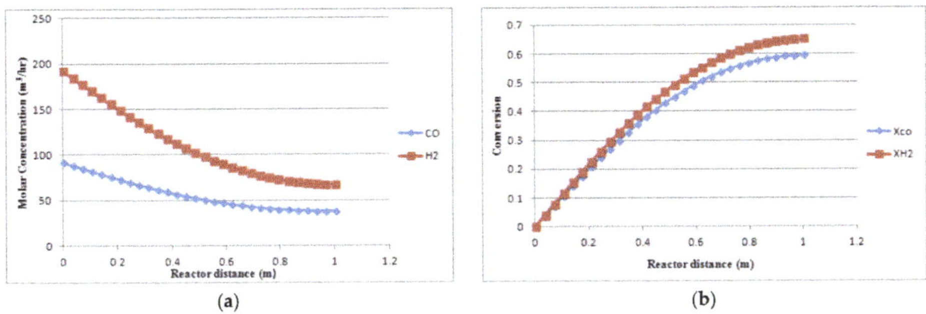

Figure 12. The syngas variation on the reactor inside, (a) Molar concentration of CO and H_2; (b) CO and H_2 conversion.

The molar concentration of other FT species along the reactor height at three load conditions is also investigated as shown in Figure 13. It shows that all products increase with more or less the same intensity apart from naphtha (C_5–C_{10}) and diesel (C_{11}–C_{19}) which shift to lower values at higher syngas flow rate loads.

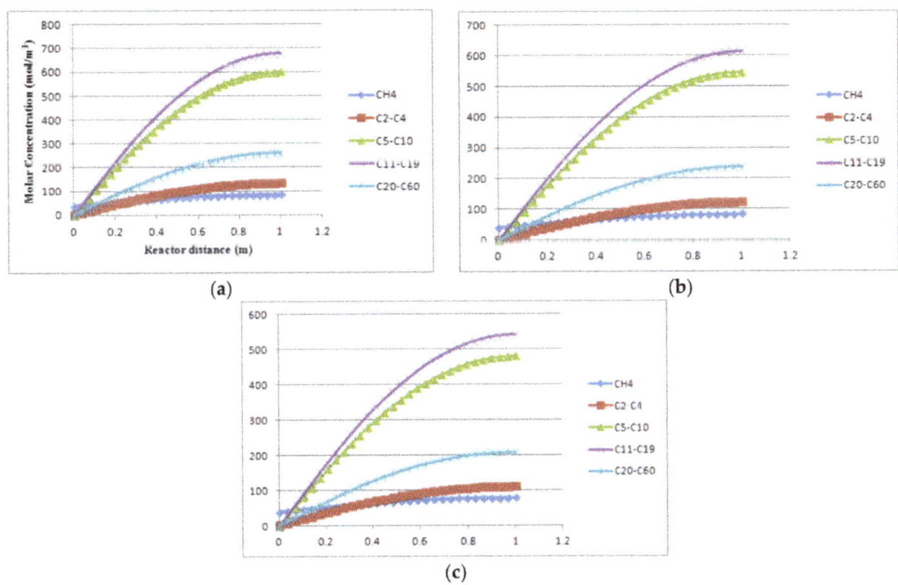

Figure 13. The molar concentration of the FT products along the reactor bed at three load conditions: (a) T = 503 K, P = 20 bar, V_f = 3.5 m³/h; (b) T = 503 K, P = 20 bar, V_f =5 m³/h; (c) T = 503 K, P = 20 bar, V_f = 7.5 m³/h.

Furthermore, Figure 14 illustrates the reduction of superficial gas velocity at three different loads of conditions based on the overall mass balance (Equation (11)).

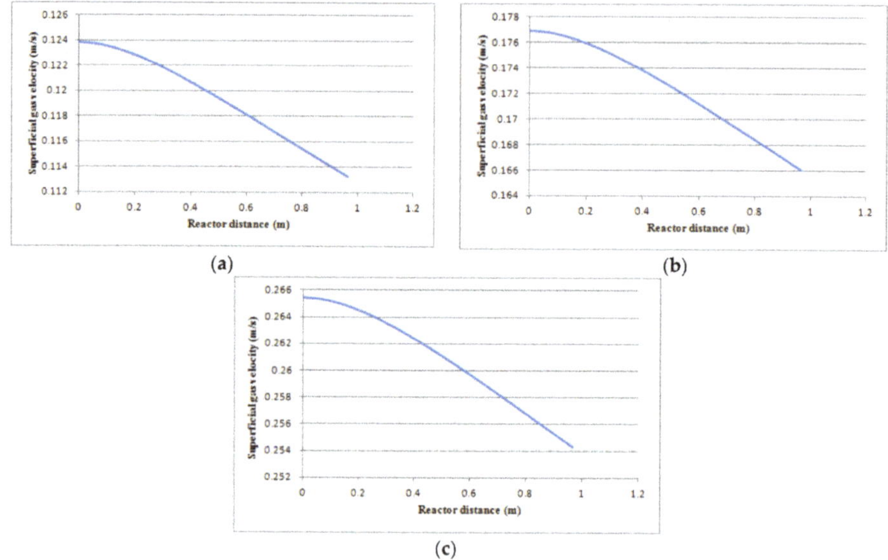

Figure 14. Reduction of superficial gas velocity as a function of reactor height at three different operating load conditions: (**a**) T = 503 K, P = 20 bar, V_f = 3.5 m^3/h; (**b**) T = 503 K, P = 20 bar, V_f = 5 m^3/h; (**c**) T = 503 K, P = 20 bar, V_f = 7.5 m^3/h.

5. Conclusions

A rigorous computer model for a lab-scale FT slurry reactor was developed to investigate flexible reactor operation. This flexibility was performed by a step-change of syngas flow rate load (3.5, 5, 7.5 m^3/h) in a low-temperature Fischer–Tropsch synthesis. It was found that the dynamic simulation is not only able to predict all Fischer–Tropsch components over the reactor bed but can also describe the behavior of superficial gas velocity as a sub-model using the overall gas mass balance.

The effect of a step-change volumetric syngas flow on the performance of FT slurry reactor was investigated. The results show that the temperature distribution of the slurry reactor remains constant under base load and change load conditions. It can be concluded that load change conditions do not have a negative influence on the temperature distribution inside the reactor and the presented dynamic model of the slurry reactor responds quite well to the load change conditions.

Acknowledgments: The authors acknowledge financial support by the Austrian government through the "Klima und Energiefonds" financed Project Winddiesel_klienIF within the "e!Mission.at" funding scheme. The project Winddiesel_klienIF is executed in cooperation with Energie Burgenland AG, Bilfinger SEBUR, Güssing Energy Technologies GmbH, Renewable Power Technologies Umwelttechnik GmbH, Energy & Chemical Engineering GmbH and the Institute of Chemical Engineering from TU WIEN.

Author Contributions: R.R., H.H. and G.W. prepared the materials and designed the experiments and S.S., R.R., S.B. and G.S. contributed to the paper writing.

Conflicts of Interest: The authors declare no conflict of interests.

Nomenclature

A_r (m^3)	Reactor cross sectional area
a (m^2 m^{-3})	Effective gas-liquid interfacial area per unit bed volume
C_i (mol/m^3)	Concentration of component i
C_{tot} (mol/m^3)	Total concentration of gaseous components
C_{tot}^L (mol/m^3)	Total concentration of liquid components
$D_{ax,L}$ (m^2/s)	Liquid phase axial dispersion coefficient
D_g (m^2/s)	Gas phase axial dispersion coefficient
D_t (m)	Reactor diameter
d_0 (m)	Gas distributor diameter
d_B (m)	Bubble diameter
g (m/s^2)	Acceleration due to gravity
H (m)	Reactor height
H_i (m$_G^3$ m$_L^{-3}$)	Henry's solubility constant of gaseous component i
H_i^* (MPa. m^3/kmol)	Coefficient in Henry's law for gaseous component i
$-\Delta H_{s,i}$ (kJ/kmol)	Heat of solution for gaseous component i
K_l a (s^{-1})	Volumetric Mass transfer coefficient
$MW_{ave,L}$ (kg mol^{-1})	Average molecular weight of liquid components
P (Pa or bar)	Pressure
Pi,sat (bar)	Vapor pressure of component i
R (J mol^{-1} K^{-1})	Gas constant
R_{co} (mol kg$_{cat}^{-1}$ s^{-1})	Carbon monoxide consumption rate
r_i (mol kg$_{cat}^{-1}$ s^{-1})	Chemical reaction rate of component i
T (K)	Temperature
t (s)	Time
U_g (m/s)	Superficial gas velocity
U_l (m/s)	Liquid velocity
V_{cat} (m^3)	Volume of catalyst in column
V_f (m^3/h)	Volumetric gas flow rate
V_{gas} (m^3)	Volume of gas in column
V_{liquid} (m^3)	Volume of liquid in column
$V_L(0)$ (m/s)	Center-line liquid velocity
z (m)	Axial coordinate
Z	Compressibility factor, dimensionless

Greek Letters

α	Chain growth probability factor, dimensionless
ε_g (m$_G^3$ m$_R^{-3}$)	Gas holdup, dimensionless
ε_l (m$_L^3$ m$_R^{-3}$)	Liquid holdup, dimensionless
ε_{cat} (m$_{cat}^3$ m$_L^{-3}$)	Catalyst volume fraction
γ_i	Activity coefficient for component i, dimensionless
ρ_{cat} (kg/m^3)	Catalyst density
ρ_{sL} (kg/m^3)	Slurry density
μ_l (Pa.s)	Liquid viscosity
σ (N/m)	Surface tension
ρ_G (kg/m^3)	Gas density
ν (m^2/s)	Kinematic viscosity of phase
τ	Dimensionless time coordinate

Abbreviations

ADM-SBCD	Axial Dispersion Model- Single Bubble Class Diameter
ASF	Anderson-Sculz-Flory
CCU	Carbon Capture Unit
CPU	Central Processing Unit
FBR	Fixed Bed Reactor
FT	Fischer–Tropsch
GHG	Greenhouse Gas
MOL	Method Of Lines
ODE	Ordinary Differential Equation
PDE	Partial Differential Equation
PDEPE	Partial Differential Equations Parabolic Elliptic
RES	Renewable Energy Sources
SBCR	Slurry Bubble Column Reactor

Appendix A

Table A1. Average composition of Inlet/Outlet Gases under BL condition [12].

Inlet/Outlet Gas	N_2	CO_2	CH_4	C_2H_4	C_2H_6	CO	C_3H_6	C_3H_8	H_2
Syngas (vol. %)	4.6	24.2	8.4	2.5	0.3	19.1	~0	~0	40.1
Off-gas (vol. %)	7.0	39.0	14.4	0.1	4.5	12.1	~0	0.1	22.6

Table A2. Average composition of Inlet/Outlet Gases under CL condition [12].

Inlet/Outlet Gas	N_2	CO_2	CH_4	C_2H_4	C_2H_6	CO	C_3H_6	C_3H_8	H_2
Syngas (vol. %)	2.5	25.1	9.5	2.8	0.2	19.2	~0	~0	40.1
Off-gas (vol. %)	3.6	40.6	16.5	~0	5.0	11.7	~0	0.1	22.4

Appendix B

Table A3. Mass fraction of the FT products measured by off-line device [12].

Products	C_8	C_9	C_{10}	C_{11}	C_{12}	C_{13}	C_{14}	C_{15}	C_{16}
BL	0.0039	0.0072	0.012	0.0186	0.0271	0.0361	0.0442	0.0506	0.0536
CL	0.0095	0.0167	0.0238	0.0323	0.0405	0.0487	0.0555	0.0606	0.0624

Products	C_{17}	C_{18}	C_{19}	C_{20}	C_{21}	C_{22}	C_{23}	C_{24}	C_{25}
BL	0.0538	0.0524	0.0505	0.048	0.0454	0.0427	0.0395	0.0364	0.0334
CL	0.0604	0.0564	0.0498	0.0415	0.0333	0.0262	0.0206	0.0177	0.016

Products	C_{26}	C_{27}	C_{28}	C_{29}	C_{30}	C_{31}	C_{32}	C_{33}	C_{34}
BL	0.0307	0.0285	0.0265	0.0244	0.0229	0.0214	0.0193	0.0178	0.0158
CL	0.016	0.0161	0.0181	0.0202	0.0218	0.0227	0.0222	0.0213	0.0177

Products	C_{35}	C_{36}	C_{37}	C_{38}	C_{39}	C_{40}	C_{41}	C_{42}	C_{43}
BL	0.0143	0.0127	0.0115	0.0107	0.0098	0.0088	0.0081	0.0072	0.0063
CL	0.0135	0.0128	0.0129	0.0126	0.0122	0.011	0.0103	0.0094	0.0083

Products	C_{44}	C_{45}	C_{46}	C_{47}	C_{48}	C_{49}	C_{50}	C_{51}	C_{52}
BL	0.0059	0.0053	0.0046	0.0043	0.0039	0.0034	0.0031	0.0028	0.0025
CL	0.0075	0.0065	0.0058	0.0049	0.0043	0.0036	0.0032	0.0027	0.0022

Products	C_{53}	C_{54}	C_{55}	C_{56}	C_{57}	C_{58}	C_{59}	C_{60}	
BL	0.0022	0.002	0.0018	0.0016	0.0014	0.0012	0.0011	0.0009	
CL	0.0018	0.0015	0.0013	0.0011	0.0009	0.0007	0.0006	0.0005	

BL: Base Load condition (V_f =5 m^3/h). CL: Change Load condition (V_f =3.5, 7.5 m^3/h).

References

1. Varone, A.; Ferrari, M. Power to liquid and Power to gas: An option for the German *Energiewende*. *J. Renew. Sustain. Energy Rev.* **2015**, *45*, 207–218. [CrossRef]
2. Eilers, H.; González, M.I.; Schaub, G. Lab-scale experimental studies of Fischer–Tropsch kinetics in a three-phase slurry reactor under transient reaction conditions. *Catal. Today* **2015**, *275*, 164–171. [CrossRef]
3. König, D.H.; Freiberg, M.; Dietrich, R.-U.; Wörner, A. Techno-economic study of the storage of fluctuating renewable. *Fuel* **2015**, *159*, 289–297. [CrossRef]
4. Cinti, G.; Baldinelli, A.; di Michele, A.; Desideri, U. Integration of Solid Oxide Electrolyzer and Fischer-Tropsch: A sustainable pathway for synthetic fuel. *Appl. Energy* **2016**, *162*, 308–320. [CrossRef]
5. Maitlis, P.M.; de Klerk, A. *Greener Fischer Tropsch Process for Fuels and Feedstocks*; John Wiley & Sons: Weinheim, Germany, 2013; pp. 40–58. ISBN 9783527326051.
6. Gross, P.; Rauch, R.; Hofbauer, H.; Aichering, C.; Zweiler, R. Winddiesel Technology—An alternative to power to gas systems. In Proceedings of the 23rd European Biomass Conference and Exhibition, Vienna, Austria, 1–4 June 2015.
7. Gonzalez, M.I. Gaseous Hydrocarbon Synfuels from H_2/CO_2 based on Renewable Electricity-Kinetics, Selectivity and Fundamentals of Fixed-Bed Reactor Design for Flexible Operation. Ph.D. Thesis, Karlsruher Institute für Technologie (KIT), Karlsruher, Germany, 2016.
8. Gonzalez, M.I.; Eilers, H.; Schaub, G. Flexible Operation of Fixed bed Reactors for a Catalytic Fuel Synthesis-CO_2 Hydrogenation as Example Reaction. *Energy Technol.* **2016**, *4*, 90–103. [CrossRef]
9. Lucero, A. *Improved Fischer Tropsch Slurry Reactors*; Worked Performed under Cooperative Agreement JSR Task 26 Under DE-FC26-98FT40323; PowerEnerCat, Inc.: Lakewood, CO, USA, 2009.
10. Sehabiague, L.; Morsi, B.I. Modeling and Simulation of Fischer-Tropsch Slurry Bubble Column Reactor Using different kinetic rate expressions for Iron and Cobalt Catalysts. *Int. J. Chem. React. Eng.* **2013**, *11*, 309–330. [CrossRef]
11. Sauciuc, A.; Abosteif, Z.; Weber, G.; Potetz, A.; Rauch, R.; Hofbauer, H.; Schaub, G.; Dumitrescu, L. Influence of operating conditions on the performance of biomass-based Fischer-Tropsch synthesis. *Biomass Convers.* **2012**, *2*, 253–263. [CrossRef]
12. Abulmfalfel, R. Winddiesel: Power to Liquids Assisted by Biomass Steam Gasification. Master's Thesis, Karlsruhe Institute of Technology, Karlsruher, Germany, March 2017.
13. Novica, R.; Al-Dahhan, M.H.; Duduković, M.P. Dynamic modeling of slurry bubble column reactors. *Ind. Eng. Chem. Res.* **2005**, *16*, 6086–6094. [CrossRef]
14. De Swart, J.W.A.; Krishna, R. Simulation of transient and steady state behavior of a bubble column slurry reactor for Fischer-Tropsch synthesis. *Chem. Eng. Process.* **2002**, *41*, 35–47. [CrossRef]
15. Hooshyar, N.; Fatemi, S.; Rahmani, M. Mathematical Modeling of Fischer-Tropsch Synthesis in an industrial Slurry Bubble Column. *Chem. React. Eng.* **2009**, *7*. [CrossRef]
16. Maretto, C.; Krishna, R. Modelling of a bubble column slurry reactor for Fischer-Tropsch synthesis. *Catal. Today* **1999**, *52*, 279–289. [CrossRef]
17. Sehabiague, L.; Lemoine, R.; Behkish, A.; Heintz, Y.J.; Sanoja, M.; Oukaci, R.; Morsi, B.I. Modeling and optimization of a large-scale slurry bubble column reactor for producing 10,000 bbl/day of Fischer–Tropsch liquid hydrocarbons. *J. Chin. Inst. Chem. Eng.* **2008**, *39*, 169–179. [CrossRef]
18. Krishna, R.; Urseanu, M.I.; van Baten, J.M.; Ellenberger, J. Liquid phase dispersion in bubble columns operating in the churn-turbulent flow regime. *Chem. Eng. J.* **2000**, *78*, 43–51. [CrossRef]
19. Ian, C.Y.; Satterfield, C.N. Intrinsic kinetics of the Fischer-Tropsch synthesis on a cobalt catalyst. *Energy Fuels* **1991**, *5*, 168–173. [CrossRef]
20. Baliban, R.C.; Elia, J.A.; Floudas, C.A. Toward Novel Hybrid Biomass, Coal, and Natural Gas Processes for Satisfying Current Transportation Fuel Demands, 1: Process Alternatives, Gasification Modeling, Process Simulation, and Economic Analysis. *Ind. Eng. Chem. Res.* **2010**, *49*, 7343–7370. [CrossRef]
21. Song, H.-S.; Ramkrishna, D.; Trinh, S.; Wright, H. Operating Strategies for Fischer-Tropsch Reactors: A Model-Directed Study. *Korean J. Chem. Eng.* **2004**, *21*, 308–317. [CrossRef]
22. Joseph William, P. *A Fischer-Tropsch Synthesis Reactor Model Framework for Liquid Biofuels Production*; Sandia National Laboratories: Livermore, CA, USA, 2012.

23. Watson, L.A. Solar-Boosted Biomass-to-Liquid (BtL) Process Modelling. Bachelor Thesis, Australian National University, Canberra, Australia, 2011.
24. Marano, J.J.; Holder, G.D. Characterization of Fischer-Tropsch liquids for vapor-liquid equilibria calculations. *Fluid Phase Equilib.* **1997**, *138*, 1–21. [CrossRef]
25. Shah, Y.T.; Dassori, C.G.; Tierney, J.W. Multiple steady states in non-isothermal FT slurry reactor. *Chem. Eng. Commun.* **1990**, *88*, 49–61. [CrossRef]
26. Marano, J.J.; Holder, G.D. General equation for correlating the thermophysical properties of *n*-paraffins, *n*-olefins, and other homologous series. 2. Asymptotic behavior correlations for PVT properties. *Ind. Eng. Chem. Res.* **1997**, *36*, 1895–1907. [CrossRef]
27. Lau, R.; Peng, W.; Velazquez-Vargas, L.G.; Yang, G.Q.; Fan, L.S. Gas-liquid mass transfer in high-pressure bubble columns. *Ind. Eng. Chem. Res.* **2004**, *43*, 1302–1311. [CrossRef]
28. Loipersböck, J.; Weber, G.; Gruber, H.; Kratky, J.; Capistrano, R.; Hofbauer, H.; Rauch, R. Upscaling and operation of a biomass-derived Fischer-Tropsch pilot plant producing one barrel per day. In Proceedings of the 25th European Biomass Conference and Exhibition, Stockholm, Sweden, 12–15 June 2017; pp. 1088–1093. [CrossRef]
29. Behkish, A. Hydrodynamic and Mass Transfer Parameters in Large-Scale Slurry Bubble Column Reactors. Ph.D. Thesis, University of Pittsburgh, Pittsburgh, PA, USA, 2005.
30. Krishna, R.; Sie, S.T. Design and Scale up of the Fischer Tropsch Bubble Column Slurry Reactor. *Fuel Process. Technol.* **2000**, *64*, 73–105. [CrossRef]
31. Krishna, R. A Scale-up Strategy for a Commercial Scale Bubble Column Slurry Reactor for Fischer-Tropsch Synthesis. *Oil Gas Sci.* **2000**, *4*, 359–393. [CrossRef]
32. Eric, H.; Iglesia, E. Slurry Bubble Column (C-2391). U.S. Patent No. 5,348,982, 20 September 1994.

Article

Dynamics of a Partially Confined, Vertical Upward-Fluid-Conveying, Slender Cantilever Pipe with Reverse External Flow

Xinbo Ge [1,2], Yinping Li [1,2,*], Xiangsheng Chen [1,2], Xilin Shi [1,2,*], Hongling Ma [1,2], Hongwu Yin [1,2], Nan Zhang [3] and Chunhe Yang [1,2]

[1] State Key Laboratory of Geomechanics and Geotechnical Engineering, Institute of Rock and Soil Mechanics, Chinese Academy of Sciences, Wuhan 430071, China; gexinbo@163.com (X.G.); chenxs15@163.com (X.C.); HonglingMa@outlook.com (H.M.); hong5yin@163.com (H.Y.); chyang@whrsm.ac.cn (C.Y.)

[2] University of Chinese Academy of Sciences, Beijing 100049, China

[3] State Key Laboratory of Coal Mine Disaster and Control, Chongqing University, Chongqing 400044, China; zhangnan0032@foxmail.com

[*] Correspondence: ypli@whrsm.ac.cn (Y.L.); xlshi@whrsm.ac.cn (X.S.); Tel.: +86-027-8719-7469 (Y.L. & X.S.)

Received: 18 February 2019; Accepted: 2 April 2019; Published: 4 April 2019

Abstract: A linear theoretical model is established for the dynamics of a hanging vertical cantilevered pipe which is subjected concurrently to internal and reverse external axial flows. Such pipe systems may have instability by flutter (amplified oscillations) or static divergence (buckling). The pipe system under consideration is a slender flexible cantilevered pipe hanging concentrically within an inflexible external pipe of larger diameter. From the clamped end to the free end, fluid is injected through the annular passage between the external pipe and the cantilevered pipe. When exiting the annular passage, the fluid discharges in the counter direction along the cantilevered pipe. The inflexible external pipe has a variable length and it can cover a portion of the length of the cantilevered pipe. This pipe system has been applied in the solution mining and in the salt cavern underground energy storage industry. The planar motion equation of the system is solved by means of a Galerkin method, and Euler–Bernoulli beam eigenfunctions are used as comparison functions. Calculations are conducted to quantify the effects of different confinement conditions (i.e., the radial confinement degree of the annular passage and the confined-flow length) on the cantilevered pipe stability, for a long leaching-tubing-like system. For a long system, an increase in the radial confinement degree of the annular passage and the confined-flow length gives rise to a series of flutter and divergence. Additionally, the effect of the cantilevered pipe length is studied. Increasing the cantilevered pipe length results in an increase of the critical flow velocity while a decrease of the associated critical frequency. For a long enough system, the critical frequency almost disappears.

Keywords: salt cavern; leaching tubing; flutter instability; flow-induced vibration; internal and reverse external axial flows

---

1. Introduction

This paper focuses on flow-induced vibration of a slender flexible cantilevered pipe which is hanging concentrically inside a shorter inflexible external pipe, thus constituting an annular passage between the cantilevered pipe and the shorter external pipe, as shown in Figure 1. The entire pipe system is submerged in incompressible fluid and is placed within a closed cavity, with the upper portion outside the cavity. As shown in Figure 1a, from the fixed end to the free end of the shorter inflexible external pipe, fluid is injected as external flow into the closed cavity through the annular passage, and flows out as a free jet into stagnant fluid. In consideration of conservation of mass, initially stagnant fluid in the closed cavity is then forced through the cantilevered pipe, as a bounded reverse internal flow, discharging from the closed cavity upwards along the cantilevered pipe. Comparing the pipe configuration shown in Figure 1a with that given in Figure 1b, it is clear that the fluid flowing directions are opposite from each other. Contrary to the direction of fluid circulation illustrated in Figure 1a, fluid [1] is injected as internal flow into the closed cavity through the cantilever pipe and discharges through the annular passage, as shown in Figure 1b.

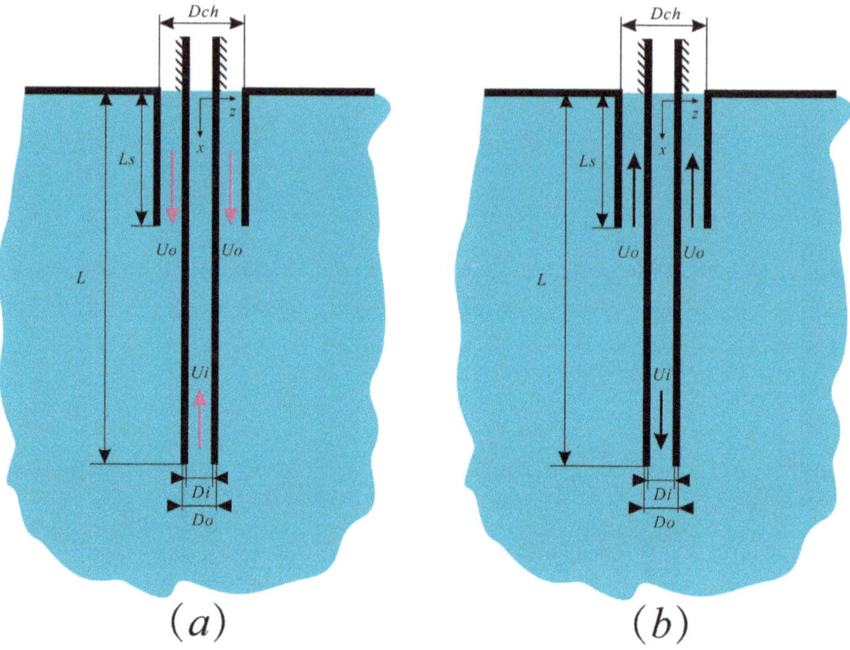

Figure 1. Schematic view of the pipe configuration. A hanging flexible cantilevered pipe conveying fluid is partially confined inside a shorter inflexible external pipe. (**a**) the system considered in this paper; (**b**) the system considered in [1].

The pipe configuration shown in Figure 1 has been applied in solution mining and in the salt cavern underground energy storage industry. Figure 2 shows the schematic view of pipe configurations applied in the salt cavern underground energy storage. Figure 2a,b correspond to Figure 1a,b, respectively. In fact, the pipe failure or excessive bending caused by flow-induced vibrations is a well-known issue in the salt cavern storage industry [2,3]. For example, excessive bending of the inner tubing in a Chinese salt cavern storage is shown in Figure 3. This paper focuses on the pipe model shown in Figure 1a, and the motivation is desired to potentially improve the design of slender cantilever pipes in order to avoid issues cause by flow-induced vibration.

For several decades, a few researchers have been done the work on conveying-fluid pipes subjected simultaneously to both internal and external axial flows. Most of the work done involves concurrent flows, i.e., internal and external axial flows in the same direction. In [4], Cesari and Curioni investigated the buckling instability of pipes subjected to internal and external axial fluid flow. In [5], Hannoyer et al. investigated the dynamics and stability of both clamped–clamped cylindrical tubular beams conveying fluid, which simultaneously are subjected to independent axial external flows. They also studied the dynamics of cantilevered pipes fitted with a tapered nozzle at the free end. They provided theoretical and experimental results to support their model. In [6], Païdoussis and Besancon studied various aspects of the dynamic characteristic of clusters of tubular pipes which were subjected to internal flows and concurrently surrounded by bounded outer axial flows. To establish general characteristics of free motions, they obtained the eigenfrequencies of the system and studied their evolution by increasing either external or internal flows. In [7], Wang and Bloom established a mathematical model to investigate the vibration and stability of a submerged and inclined concentric pipe system which subjected to internal and external flow, and obtained the resonant frequencies of the system. In [8], Païdoussis et al. developed a theoretical model to study the vibration of a hanging tubular cantilever which was centrally inside a cylindrical container, with fluid flowing within the cantilever, and discharging from the free end. The configuration thus resembles that of a drill-string with a floating fluid-powered drill-bit. Of particular interest is the study in [1]. Moditis and Païdoussis preliminarily discussed the dynamics of the pipe configuration shown in Figure 1b. Furthermore, they developed a theoretical model and carried out corresponding experiments.

Because the length of the cantilevered pipe system in practical applications is on the order of one kilometer, how the system behavior evolves is of interest as the length increases. Theoretical and experimental studies [9,10] regarding long hanging vertical pipes conveying fluid have shown that both the critical flow velocity and associated frequency for instability tend to be asymptotic toward different limiting values with increasing the length of the pipes. The literature related to aspirating cantilevered pipes was believed to be useful for analyzing the dynamics of a vertical hanging pipe shown in Figure 1a. For example, in [11,12], the authors theoretically and experimentally considered the dynamic stability of a submerged cantilever pipe aspirating fluid, which could be applied in deep ocean mining.

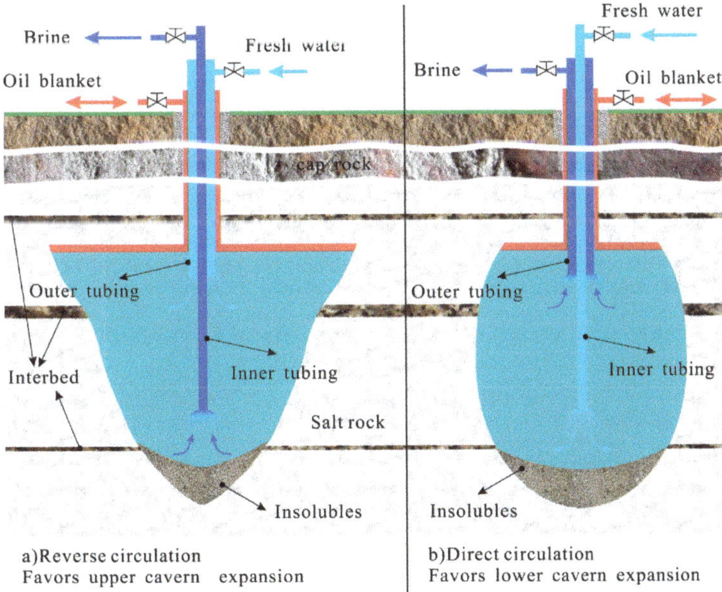

Figure 2. Schematic view of the salt cavern underground energy storage application. Fresh water or light brine is injected into the inner tubing, and high-concentration brine is expelled through the annular space in between the inner and outer tubings. This process is known as direct circulation, as distinguished from reverse circulation. Direct and reverse circulations are alternately employed until the salt cavern achieves the design shape and dimensions.

Figure 3. Excessive bending of the inner tubing at a Chinese salt cavern storage [3].

Several key dimensions of the system under consideration are defined in Figure 1a. The x-axis lies in the undeflected centerline of the internal cantilever pipe, while the z-axis lies along the lateral direction and is perpendicular to the x-axis. L is the length of the cantilever pipe, while L_s is the length of the shorter inflexible external pipe (i.e. the length of the annular passage). U_i is the internal flow velocity (area average flow velocity) inside the cantilevered pipe, and U_o is the external flow velocity (area average flow velocity) in the annular passage with reference to the cantilever pipe.

Firstly, the present paper studies the linear idealized system shown in Figure 1a. We present a derivation of a theoretical model as well as a method of solving it are discussed. Secondly, we give theoretical results for slender leaching-tubing-like systems which are associated with the industrial application of salt cavern storage. We also obtained some unexpected and interesting results by numerical simulation. Finally, we present general conclusions.

2. Problem Formulation

2.1. Derivation of the Motion Equation of Theoretical Model

The derivation of the linear theoretical model is carried out as follows. Take a small element of length δx of the cantilevered pipe into consideration, as shown in Figure 4a, under the action of fluid-related and structural forces and moments. Force balances in the x- and z-directions renders, respectively

$$\frac{\partial F_T}{\partial x} - \frac{\partial}{\partial x}\left(Q\frac{\partial w}{\partial x}\right) + M_p g - F_{it} + F_{et} - (F_{in} + F_{en})\frac{\partial w}{\partial x} = 0 \tag{1}$$

$$\frac{\partial Q}{\partial x} + \frac{\partial}{\partial x}\left(F_T\frac{\partial w}{\partial x}\right) + F_{in} + F_{en} + (F_{et} - F_{it})\frac{\partial w}{\partial x} - M_p\frac{\partial^2 w}{\partial t^2} = 0 \tag{2}$$

where w is the lateral deflection; F_T is the axial tension; M_p is the mass per unit length of the cantilever pipe; Q is the transverse shear force in the cantilevered pipe; g is the gravitational acceleration; F_{in} and F_{it} are the normal and tangential hydrodynamic forces because of the internal flow U_i, respectively; F_{en} and F_{et} are the normal and tangential hydrodynamic forces because of the external flow, respectively.

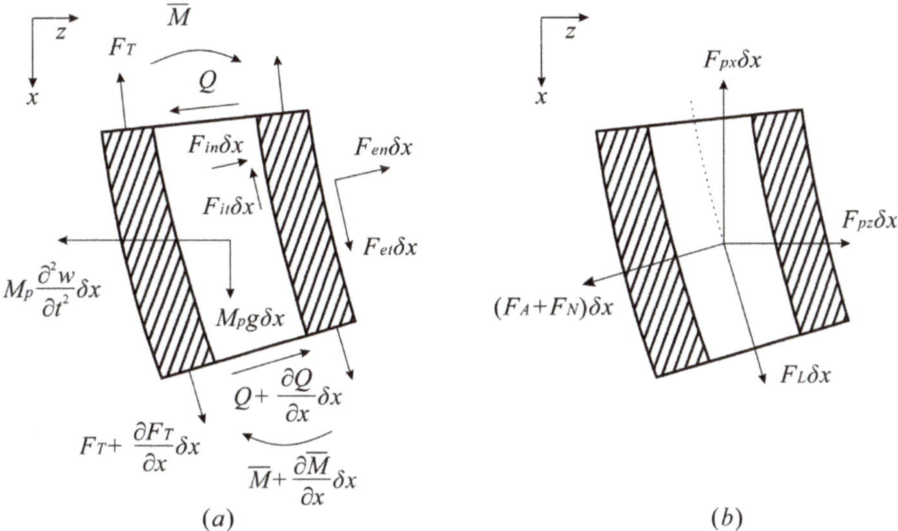

Figure 4. (**a**) Forces acting on an element of length δx of the cantilevered pipe. (**b**) Forces due to the outside fluid acting on an element δx of the cantilevered pipe.

As approximated by the Euler–Bernoulli (E–B) beam theory, one obtains the following relationship

$$Q = -\frac{\partial}{\partial x}\left(EI\frac{\partial^2 w}{\partial x^2}\right) \tag{3}$$

where EI is the flexural rigidity. Internal dissipation in the cantilevered pipe material is relatively small compared to dissipation by the surrounding fluid so it can be neglected [13]. Substituting Equation (3) into Equations (1) and (2), and neglecting second-order terms, one finds the following expressions in the x- and z-directions, respectively

$$\frac{\partial F_T}{\partial x} - \left(F_{it} + F_{in}\frac{\partial w}{\partial x}\right) + \left(F_{et} - F_{en}\frac{\partial w}{\partial x}\right) + M_{pg} = 0 \tag{4}$$

$$EI\frac{\partial^4 w}{\partial x^4} - \frac{\partial}{\partial x}\left(F_T\frac{\partial w}{\partial x}\right) - \left(F_{in} - F_{it}\frac{\partial w}{\partial x}\right) - \left(F_{en} + F_{et}\frac{\partial w}{\partial x}\right) + M_p\frac{\partial^2 w}{\partial t^2} = 0 \tag{5}$$

The F_{in} and F_{it} (hydrodynamic forces because of internal flow) are obtained by a force balance on an element δx of internal fluid (Figure 5b) in the manner of Païdoussis [14,15]. The F_{in}' and F_{it}' in Figure 5b are a pair of interaction forces with the F_{in} and F_{it} in Figure 4a, $F_{in}' = -F_{in}$ and $F_{it}' = -F_{it}$. The resulting expressions are written as below in the x- and z-directions, respectively,

$$F_{it} + F_{in}\frac{\partial w}{\partial x} = -M_{fg} + A_f\frac{\partial p_i}{\partial x} \tag{6}$$

and

$$F_{in} - F_{it}\frac{\partial w}{\partial x} = -M_f\left(\frac{\partial^2 w}{\partial t^2} - 2U_i\frac{\partial^2 w}{\partial x\partial t} + U_i^2\frac{\partial^2 w}{\partial x^2}\right) - A_f\frac{\partial}{\partial x}\left(p_i\frac{\partial w}{\partial x}\right) \tag{7}$$

where M_f is the mass of internal fluid per unit length; p_i is the internal pressure in the cantilevered pipe; A_f is the cross-sectional area of the internal flow, and $M_f = \rho_f A_f$, in which ρ_f is the fluid density.

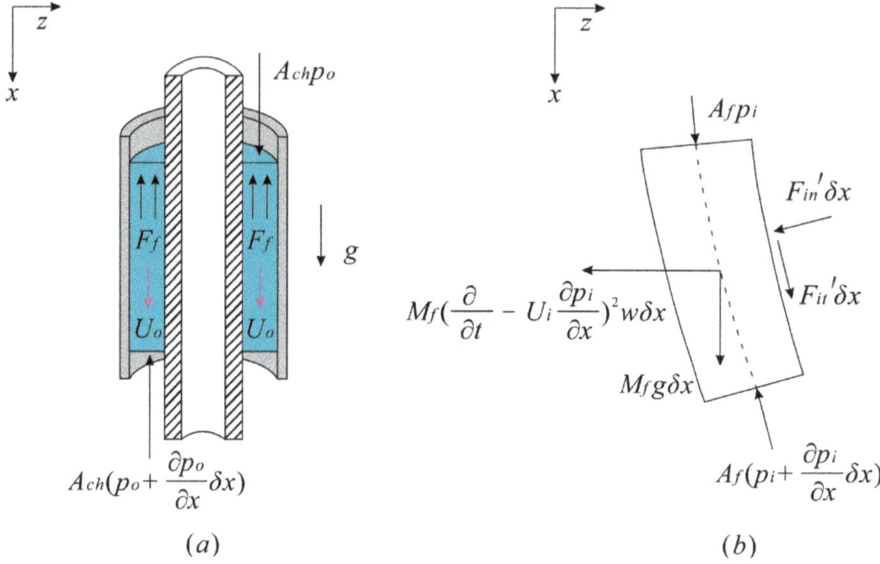

Figure 5. (a) Forces acting on an element of length δx of the annular flow. (b) Forces acting on an element δx of the internal flow.

Substituting Equations (6) and (7) in (4) and (5) renders the following equations in the x- and z-directions, respectively

$$\frac{\partial F_T}{\partial x} + \left(M_f g - A_f\frac{\partial p_i}{\partial x}\right) + \left(F_{et} - F_{en}\frac{\partial w}{\partial x}\right) + M_p g = 0 \tag{8}$$

$$EI\frac{\partial^4 w}{\partial x^4} - \frac{\partial}{\partial x}\left(F_T\frac{\partial w}{\partial x}\right) + \left[M_f\left(\frac{\partial^2 w}{\partial t^2} - 2U_i\frac{\partial^2 w}{\partial x\partial t} + U_i^2\frac{\partial^2 w}{\partial x^2}\right) + A_f\frac{\partial}{\partial x}\left(p_i\frac{\partial w}{\partial x}\right)\right] - \left(F_{en} + F_{et}\frac{\partial w}{\partial x}\right) + M_p\frac{\partial^2 w}{\partial t^2} = 0 \tag{9}$$

The field of flow outside the cantilevered pipe can be simplified because the cantilever pipe displacements w are considered to be small according to E–B beam theory. More specifically, the external flow is modelled as the superposition of a perturbation potential because of the cantilever pipe vibrations in the mean axial flow [15,16]. Hence, the effects of fluid viscosity are considered to be different from this potential flow, and the viscosity-related forces are added separately to the system. When linear investigations of fluid-structure interaction issues are carried out, this is a common approach. There are more details in [5,15].

The resultant force of all external hydrodynamic forces acting on the cantilevered pipe (Figure 4a) can be broken down into F_{en} and F_{et} in the directions perpendicular and tangential to the cantilever pipe centerline, respectively. Projection of the external hydrodynamic force in the x- and z-directions yields the following force-balance equations

$$F_{et} - F_{en}\frac{\partial w}{\partial x} = F_L - F_{px}, \tag{10}$$

and

$$F_{en} + F_{et}\frac{\partial w}{\partial x} = -(F_A + F_N) + F_{pz} + F_L\frac{\partial w}{\partial x}. \tag{11}$$

In Equations (10) and (11), F_A is the lateral inviscid hydrodynamic force; F_{px} and F_{pz} are the resultant forces due to external pressure in the x and z directions, respectively; F_N and F_L are the fluid frictional viscous forces perpendicular and tangential to the cantilever pipe centerline, respectively; and high-order terms have been omitted.

The added (hydrodynamic) mass M_a is the mass per unit length for the annular (external) fluid associated with motions of the cantilevered pipe. Based on the perturbation potential, for a thin boundary layer, the added (hydrodynamic) mass $M_a = \chi \rho_f A_o$ [17,18], where $A_o = \pi D_o^2/4$ is the external cross-sectional area of the cantilevered pipe. The parameter χ quantifies the effect of radial confinement degree (or narrowness) of the annular flow passage as

$$\chi = \frac{(D_{ch}/D_o)^2 + 1}{(D_{ch}/D_o)^2 - 1}, \quad \lim_{D_{ch}\to\infty}(\chi) = \lim_{D_{ch}\to\infty}\frac{(D_{ch}/D_o)^2 + 1}{(D_{ch}/D_o)^2 - 1} = 1. \tag{12}$$

In Equation (12), D_{ch} is the inner diameter of the passage (the inflexible external pipe) and D_o is the outer diameter of the cantilever pipe as shown in Figure 1. Regarding an unconfined pipe, $D_{ch} \to \infty$ and as a result M_a is equal to $\rho_f A_o$, i.e., the displaced mass of the fluid per unit length. Hence, considering the variation of confinement along the cantilevered pipe, the added (hydrodynamic) mass may be written as

$$M_a = [\chi + (1 - \chi)H(x - L_s)]\rho_f A_o \tag{13}$$

where χ pertains to the confined portion of the cantilevered pipe, $0 \le x \le L_s$, and $H(x - L_s)$ is the Heaviside step function.

As is shown in Figure 1a, both the external flow velocity U_o and the confinement degree vary along the length of the cantilevered pipe. For the sake of simplicity, it is assumed that U_o is zero along the unconfined part, $L_s < x \le L$, and immediately attains the value of U_o along the confined portion of the cantilevered pipe, $0 \le x < L_s$. The above assumptions can also be found in the literature [1]. Thus, the lateral inviscid hydrodynamic force F_A can be written as

$$F_A = \left(\frac{\partial}{\partial t} + U_o\frac{\partial}{\partial x} - U_oH(x - L_s)\frac{\partial}{\partial x}\right)\left\{[\chi + (1 - \chi)H(x - L_s)]\rho_f A_o\left(\frac{\partial w}{\partial t} + U_o\frac{\partial w}{\partial x} - U_oH(x - L_s)\frac{\partial w}{\partial x}\right)\right\}, \tag{14}$$

where the initial expression [16] and later [17] have been altered to explicate the spatial variation of both the external flow velocity and the added (hydrodynamic) mass Equation (13). After various simplifications, the formula given in Equation (14) was arrived at

$$F_A = (1 - \chi)\rho_f A_o H(x - L_s)\frac{\partial^2 w}{\partial t^2} + \rho_f A_o \chi\frac{\partial^2 w}{\partial t^2} - 2A_o U_o\rho_f\chi H(x - L_s)\frac{\partial^2 w}{\partial x\partial t} + 2A_o U_o\rho_f\chi\frac{\partial^2 w}{\partial x\partial t} - A_o U_o^2\rho_f\chi H(x - L_s)\frac{\partial^2 w}{\partial x^2} + A_o U_o^2\rho_f\chi\frac{\partial^2 w}{\partial x^2} \tag{15}$$

The fluid frictional viscous force F_L tangential to the cantilevered pipe centerline is

$$F_L = \frac{1}{2}C_f\rho_f D_o U_o^2[1 - H(x - L_s)], \tag{16}$$

where C_f is the frictional damping coefficient, with a value of 0.0125 to give an acceptable estimate of the fluid frictional viscous force F_L [8].

Equation (16) was modified from the literature [17] to take the spatial variation of the mean external flow velocity into account. Likewise, in terms of the literature [17], the fluid frictional viscous force F_N perpendicular to the cantilevered pipe centerline is

$$F_N = \frac{1}{2}C_f\rho_f D_o U_o[1 - H(x - L_s)]\left\{\frac{\partial w}{\partial t} + [1 - H(x - L_s)]U_o\frac{\partial w}{\partial x}\right\} + k\frac{\partial w}{\partial t}, \tag{17}$$

where the variation of the external flow velocity U_o has been taken into account; k is a viscous damping coefficient which only relates to lateral motions of the cantilever pipe, without average external flow U_o. Based on two-dimensional flow analysis, expressions used for k are derived in [13,18]. For a thin boundary layer, the coefficient k is calculated via the following expression

$$k = \frac{2\sqrt{2}}{\sqrt{S}}\frac{1 + \overline{\gamma}^3}{\left(1 - \overline{\gamma}^2\right)^2}\rho_f A_o \Re\{\Omega\} \tag{18}$$

In (10), $S = R\{\Omega\}D_o^2/(4\nu)$ represents the oscillatory Reynolds number (i.e., Stokes number), where ν and $R\{\Omega\}$ are the kinematic viscosity of the fluid and the circular frequency of oscillation, respectively; $\overline{\gamma} = D_o/D_{ch}$ is a measure of the radial confinement degree of the annular passage. Naturally, based on Figure 1a, the value of k varies along the cantilevered pipe. This variation of k is explained by letting

$$F_N = \tfrac{1}{2}C_f\rho_f D_o U_o\left\{[1 - H(x - L_s)]\tfrac{\partial w}{\partial t} + [1 - H(x - L_s)]U_o\tfrac{\partial w}{\partial x}\right\} + k_u\left[\tfrac{1+\overline{\gamma}^3}{(1-\overline{\gamma}^2)^2} + H(x - L_s)\left(1 - \tfrac{1+\overline{\gamma}^3}{(1-\overline{\gamma}^2)^2}\right)\right]\tfrac{\partial w}{\partial t}. \quad (19)$$

$$k_u = \frac{2\sqrt{2}}{\sqrt{S}}\rho_f A_o \Re\{\Omega\} \quad (20)$$

In (19) and (20), k_u represents the friction coefficient applicable to the cantilevered pipe without confined portion, $L_s < x \leq L$.

Finally, the forces (F_{px} and F_{pz}) caused by mean tensioning (induced by gravity) and pressurization are considered, respectively. F_{px} and F_{pz} are cleverly derived in [15,17] and they are expressed as

$$F_{px} = -\frac{\partial}{\partial x}(A_o p_o) + A_o\frac{\partial p_o}{\partial x} \quad (21)$$

and

$$F_{pz} = A_o\frac{\partial}{\partial x}\left(p_o\frac{\partial w}{\partial x}\right) \quad (22)$$

in the x- and z-directions respectively, where p_o represents the fluid pressure outside the cantilevered pipe.

For the purpose of analyzing the model shown in Figure 1a, the outlet effects of the flow exiting from the annular passage have been reduced to a tinily thin region at $x = L_s$. For $0 \leq x < L_s$, there is a pressure loss due to the friction of the flowing fluid inside the annulus, while there is a pressure increasing due to the gravity. For $L_s < x \leq L$, there is a purely hydrostatic pressure distribution along the unconfined portion of the cantilever pipe. For $0 \leq x < L_s$, an annular fluid element (i.e., the external flow of length δx) (Figure 5a) is considered, and a force balance per unit length is obtained as

$$- A_{ch}\frac{\partial p_o}{\partial x} - F_f + A_{ch}\rho_f g = 0, \quad (23)$$

where $A_{ch} = \pi(D_{ch}^2 - D_o^2)/4$ represents the cross-sectional flow area of the annular passage; F_f represents the total wall-friction force acting on the fluid element.

It is assumed that an equal wall shear stress acts on both annular passage surfaces, the total wall-friction force F_f can be expressed as

$$\frac{F_f}{S_{tot}} = \frac{F_L}{S_o} \quad (24)$$

in which $S_{tot} = \pi(D_{ch} + D_o)$ is the total wetted area of the annular flow per unit length, and S_o is the outside wetted pipe perimeter.

Combining Equations (23) and (24), multiplied by (A_o/A_{ch}), the resultant equation is rearranged as

$$A_o\frac{\partial p_o}{\partial x} = -F_L\left(\frac{D_o}{D_h}\right) + A_o\rho_f g, \quad (25)$$

where $D_h = 4A_{ch}/S_{tot} = (D_{ch} - D_o)$ is the hydraulic diameter of the external flow in the annular passage. Integration of Equation (25), where for the confined portion external fluid pressure at $x = 0$ is the reference pressure ($p_o|_{x=0} = 0$), with respect to x gives

$$p_o(x) = \left[-\frac{F_L}{A_o}\left(\frac{D_o}{D_h}\right) + \rho_f g\right]x, \quad (26)$$

for $0 \le x < L_s$.

For the unconfined part, i.e., $L_s < x \le L$, the external fluid pressure distribution is hydrostatic

$$\frac{\partial p_o}{\partial x} = \rho_f g, \tag{27}$$

which upon integration results in

$$p_o(x) = \rho_f g x + C_1, \tag{28}$$

in which C_1 is an integration constant determined in the following description.

It is assumed that $x_1 = L_s^-$ is the axial location just inside the annular passage, and $x_2 = L_s^+$ is the location just outside the annular passage. Based on Bernoulli's equation, an energy balance of the fluid is obtained from x_1 to x_2

$$p_o|_{x_2} = p_o|_{x_1} + \frac{1}{2}\rho_f U_o^2 - \rho_f g h_e, \; h_e = \frac{K_1 U_o^2}{2} \tag{29}$$

where the quantity h_e, with $K_1 = 1$, is the head-loss associated with sudden enlargement of the external flow in the annular passage into the surrounding fluid at $x = L_s^+$ [19]. Combination Equations (28) and (29) yields

$$C_1 = \left[-\frac{\|F_L\|}{A_o}\left(\frac{D_o}{D_h}\right)L_s \right] + \frac{1}{2}\rho_f U_o^2 - \rho_f g h_e, \tag{30}$$

in which $\|F_L\| = 0.5\rho_f C_f D_o U_o^2$.

Through the above analysis, the resulting fluid pressure gradient and fluid pressure distribution over the whole cantilevered pipe, for $0 \le x \le L$, are, respectively

$$\frac{\partial p_o}{\partial x} = -\frac{\|F_L\|}{A_v}\left(\frac{D_o}{D_h}\right)[1 - H(x - L_s)] + \rho_f g + \left(\frac{1}{2}\rho_f U_o^2 - \rho_f g h_e\right)\delta_D(x - L_s), \tag{31}$$

and

$$p_o = -\frac{\|F_L\|}{A_o}\left(\frac{D_o}{D_h}\right)x + \frac{\|F_L\|}{A_o}\left(\frac{D_o}{D_h}\right)(x - L_s)H(x - L_s) + \rho_f g x + \left(\frac{1}{2}\rho_f U_o^2 - \rho_f g h_e\right)H(x - L_s), \tag{32}$$

where $\delta_D(x - L_s)$ is the Dirac delta function.

Substituting Equations (16) and (21) in (10), and subsequently substituting (10) into (8), the force balance on the cantilevered pipe is obtained in the x-direction, namely

$$\frac{\partial}{\partial x}\left(F_T - A_f p_i + A_o p_o\right) + M_p g + \rho_f A_f g - A_o \frac{\partial p_o}{\partial x} + \frac{1}{2}C_f \rho_f D_o U_o^2[1 - H(x - L_s)] = 0 \tag{33}$$

Substituting Equations (15), (16), (19), and (22) in (11), and subsequently substituting the result in (1), the motion equation is obtained in the z-direction,

$$EI\frac{\partial^4 w}{\partial x^4} - \frac{\partial}{\partial x}\left[\left(F_T - A_f p_i + A_o p_o\right)\frac{\partial w}{\partial x}\right] + M_p\frac{\partial^2 w}{\partial t^2} + \left[M_f\left(\frac{\partial^2 w}{\partial t^2} - 2U_i\frac{\partial^2 w}{\partial x \partial t} + U_i^2\frac{\partial^2 w}{\partial x^2}\right)\right] + (1 - \chi)\rho_f A_o H(x - L_s)\frac{\partial^2 w}{\partial t^2} + \rho_f A_o \chi\frac{\partial^2 w}{\partial t^2}$$
$$- 2A_o U_o \rho_f \chi H(x - L_s)\frac{\partial^2 w}{\partial x \partial t} + 2A_o U_o \rho_f \chi \frac{\partial^2 w}{\partial x \partial t} - A_o U_o^2 \rho_f \chi H(x - L_s)\frac{\partial^2 w}{\partial x^2} + A_o U_o^2 \rho_f \chi \frac{\partial^2 w}{\partial x^2} + \frac{1}{2}C_f \rho_f D_o U_o[1 - H(x - L_s)]\frac{\partial w}{\partial t} \tag{34}$$
$$+ k_u\left[\frac{1 + \bar{\gamma}^3}{(1 - \bar{\gamma}^2)^2} + H(x - L_s)\left(1 - \frac{1 + \bar{\gamma}^3}{(1 - \bar{\gamma}^2)^2}\right)\right]\frac{\partial w}{\partial t} = 0.$$

In Equation (34), the tensioning and pressurization term $(F_T - p_i A_f + A_o p_o)$ is determined by integrating Equation (33) from x to L, which yields

$$\left(F_T - A_f p_i + A_o p_o\right) = \frac{1}{2}C_f \rho_f D_o U_o^2\left(\frac{D_o}{D_h} + 1\right)(L_s - x)[1 - H(x - L_s)] - A_o\left(\frac{1}{2}\rho_f U_o^2 - \rho_f g h_e\right)[1 - H(x - L_s)]$$
$$+ \left(M_p g + \rho_f A_f g - A_o \rho_f g\right)(L - x) + \left(F_T - A_f p_i + A_o p_o\right)|_L. \tag{35}$$

In Equation (35) $|_L$ represents an evaluation of the term in parentheses at $x = L$.

Hence, the final form of the motion equation is obtained in the z-direction by substituting Equation (35) in (34), as follows

$$
\begin{aligned}
&EI\frac{\partial^4 w}{\partial x^4} + \left\{ \left(M_p + \rho_f A_f - \rho_f A_o\right)g + \tfrac{1}{2}C_f\rho_f D_o U_o^2\left(\tfrac{D_o}{D_h} + 1\right)[1 - H(x - L_s)] - A_o\left(\tfrac{1}{2}\rho_f U_o^2 - \rho_f g h_e\right)\delta_D(x - L_s) \right\}\frac{\partial w}{\partial x} \\
&+ \left\{ \left(-M_p - \rho_f A_f + \rho_f A_o\right)g(L - x) - \tfrac{1}{2}C_f\rho_f D_o U_o^2\left(\tfrac{D_o}{D_h} + 1\right)(L_s - x)[1 - H(x - L_s)] + A_o\left(\tfrac{1}{2}\rho_f U_o^2 - \rho_f g h_e\right)[1 - H(x - L_s)] - \left(T - A_f p_i + A_o p_o\right)\Big|_L \right\}\frac{\partial^2 w}{\partial x^2} \\
&+ M_p\frac{\partial^2 w}{\partial t^2} + \rho_f A_f\frac{\partial^2 w}{\partial t^2} - 2U_i\rho_f A_f\frac{\partial^2 w}{\partial x \partial t} + \rho_f A_f U_i^2\frac{\partial^2 w}{\partial x^2} + \rho_f A_o \chi U_o^2[1 - H(x - L_s)]\frac{\partial^2 w}{\partial x^2} + 2A_o U_o\rho_f\chi[1 - H(x - L_s)]\frac{\partial^2 w}{\partial x \partial t} + (1 - \chi)\rho_f A_o H(x - L_s)\frac{\partial^2 w}{\partial t^2} \\
&+ \rho_f A_o\chi\frac{\partial^2 w}{\partial t^2} + \tfrac{1}{2}C_f\rho_f D_o U_o[1 - H(x - L_s)]\frac{\partial w}{\partial t} + k_u\left\{1 + [1 - H(x - L_s)]\left(\tfrac{1+\gamma^3}{(1-\gamma^2)^2} - 1\right)\right\}\frac{\partial w}{\partial t} = 0
\end{aligned}
\tag{36}
$$

According to the Bernoulli equation, the pressures $P_o|_L$ and $P_i|_L$ are related to the energy balance of the fluid at $x = L$

$$
p_i|_L = p_o|_L - \frac{1}{2}\rho_f U_i^2 - \rho_f g h_a, \qquad h_a = \frac{K_2 U_i^2}{2}.
\tag{37}
$$

In Equation (37), the quantity h_a is the headloss due to the stagnant fluid entering the cantilevered pipe at $x = L$ and acquiring internal flow velocity U_i, calculated with $0.8 \leq K_2 \leq 0.9$, independent of flow velocity [19]. Consequently, evaluation of Equation (32) at $x = L$ yields $P_o|_L$

$$
p_o|_L = \frac{1}{2A_o}C_f\rho_f D_o U_o^2\left(\frac{D_o}{D_h}\right)L' + \rho_f g L + \frac{1}{2}\rho_f U_o^2 - \rho_f g h_a.
\tag{38}
$$

Finally, the internal and external flow velocities are related through conservation of mass, i.e., $U_i A_f = U_o A_{ch}$, which after some transformation is expressed as

$$
U_o = U_i \frac{D_i^2}{D_{ch}^2 - D_o^2}.
\tag{39}
$$

2.2. Boundary Conditions

The motion Equation (36) is subjected to the boundary conditions

$$
w(0,t) = \left.\frac{\partial w}{\partial x}\right|_{x=0} = 0, \qquad \left.\frac{\partial^2 w}{\partial x^2}\right|_{x=L} = \left.\frac{\partial^3 w}{\partial x^3}\right|_{x=L} = 0.
\tag{40}
$$

2.3. Dimensionless Motion Equation and Boundary Conditions

The motion Equation (36) is non-dimensionalized using the following dimensionless quantities

$$
\begin{aligned}
&\zeta = \frac{x}{L}, \qquad \tau = \left[\frac{EI}{M_p + \rho_f A_f + \rho_f A_o}\right]^{\frac{1}{2}}\frac{t}{L^2}, \qquad \eta = \frac{w}{L}, \\
&u_i = \left(\frac{\rho_f A_f}{EI}\right)^{\frac{1}{2}}LU_i, \qquad u_o = \left(\frac{\rho_f A_o}{EI}\right)^{\frac{1}{2}}LU_o, \qquad \beta_o = \frac{\rho_f A_o}{M_p + \rho_f A_f + \rho_f A_o}, \\
&\beta_i = \frac{\rho_f A_f}{M_p + \rho_f A_f + \rho_f A_o}, \qquad \gamma = \frac{(M_p + \rho_f A_f - \rho_f A_o)gL^3}{EI}, \qquad \Gamma = \frac{T|_L L^2}{EI}, \\
&\Pi_{iL} = \frac{p_i|_L A_f L^2}{EI}, \qquad \Pi_{oL} = \frac{p_o|_L A_o L^2}{EI}, \qquad c_f = \frac{4C_f}{\pi}, \\
&\kappa_u = \frac{k_u L^2}{\left[EI\left(M_p + \rho_f A_f + \rho_f A_o\right)\right]^{\frac{1}{2}}}, \qquad \varepsilon = \frac{L}{D_o}, \qquad h = \frac{D_o}{D_h}, \\
&\alpha = \frac{D_i}{D_o}, \qquad \alpha_{ch} = \frac{D_{ch}}{D_o}, \qquad r_{ann} = \frac{L_s}{L}.
\end{aligned}
\tag{41}
$$

Thus, the resulting dimensionless form of Equation (36) is

$$
\begin{aligned}
&\frac{\partial^4 \eta}{\partial\zeta^4} + \left\{\gamma + \tfrac{1}{2}c_f\varepsilon u_o^2(h + 1)[1 - H(\zeta - r_{ann})] - \tfrac{1}{2}u_o^2(1 - K_1)\delta_D(\zeta - r_{ann})\right\}\frac{\partial\eta}{\partial\zeta} \\
&- \left\{\gamma(1 - \zeta) + \tfrac{1}{2}c_f\varepsilon u_o^2(h + 1)(r_{ann} - \zeta)[1 - H(\zeta - r_{ann})] - \tfrac{1}{2}u_o^2(1 - K_1)[1 - H(\zeta - r_{ann})] + (\Gamma - \Pi_{iL} + \Pi_{oL})\right\}\frac{\partial^2\eta}{\partial\zeta^2} \\
&+ \left\{u_i^2 + \chi u_o^2[1 - H(\zeta - r_{ann})]\right\}\frac{\partial^2\eta}{\partial\zeta^2} + \left\{1 + (\chi - 1)\beta_o[1 - H(\zeta - r_{ann})]\right\}\frac{\partial^2\eta}{\partial\tau^2} + 2\left(\chi u_o\beta_o^{\frac{1}{2}}[1 - H(\zeta - r_{ann})] - u_i\beta_i^{\frac{1}{2}}\right)\frac{\partial^2 w}{\partial\zeta\partial\tau} \\
&+ \tfrac{1}{2}c_f\varepsilon u_o\beta_o^{\frac{1}{2}}[1 - H(\zeta - r_{ann})]\frac{\partial\eta}{\partial\tau} + \kappa_u\left\{1 + [1 - H(\zeta - r_{ann})]\left(\frac{1 + \alpha_{ch}^{-3}}{(1 - \alpha_{ch}^{-2})^2} - 1\right)\right\}\frac{\partial\eta}{\partial\tau} = 0,
\end{aligned}
\tag{42}
$$

with corresponding dimensionless boundary conditions

$$\eta(0,\tau) = \left.\frac{\partial \eta}{\partial \xi}\right|_{\xi=0} = 0, \left.\frac{\partial^2 \eta}{\partial \xi^2}\right|_{\xi=1} = \left.\frac{\partial^3 \eta}{\partial \xi^3}\right|_{\xi=1} = 0. \tag{43}$$

Additionally, the dimensionless (complex) frequency of vibration, ω, is related to the dimensional one, Ω, by

$$\omega = \left[\frac{M_p + \rho_f A_f + \rho_f A_o}{EI}\right]^{\frac{1}{2}} L^2 \Omega. \tag{44}$$

It is noted that the external and internal fluid pressures in dimensionless form at the free end have the relationship

$$\Pi_{iL} = \alpha^2 \Pi_{oL} - \frac{1}{2}u_i^2 - A_f \rho_f g h_a \left(\frac{L^2}{EI}\right), \tag{45}$$

where

$$\Pi_{oL} = \frac{1}{2}c_f h r_{ann} \varepsilon u_o^2 + \frac{1}{2}u_o^2(1 - K_1) + \frac{A_o \rho_f g L^3}{EI}, \tag{46}$$

with h_a given in Equation (37), and K_1 as shown in Equation (29).

Finally, the dimensionless flow velocities are expressed as

$$u_o = \frac{\alpha}{\alpha_{ch}^2 - 1}u_i, \tag{47}$$

where α and α_{ch} are defined in (41).

3. Solution of Equations by a Galerkin Method

The governing Equation (42) of lateral vibration of the cantilever pipe contains the Heaviside step function, which is a discontinuous function, and a conventional Galerkin method is used here to discretize this system into an ordinary differential equation. The cantilever pipe is divided into N units. Consequently, solution of Equation (42) was effected as follows. Let an approximate solution be

$$\eta(\xi,\tau) \approx \tilde{\eta}(\xi,\tau) = \sum_{j=1}^{N} \Phi_j(\xi) q_j(\tau). \tag{48}$$

In (48), $q_j(\tau)$ is the generalized coordinate of the system and written as

$$q_j(\tau) = s_j e^{i\omega_j \tau} = \frac{s_j}{e^{\mathrm{Im}(\omega_j)\tau}} e^{i\mathrm{Re}(\omega_j)\tau}. \tag{49}$$

The $\Phi_j(\xi)$ are the appropriate comparison functions, in this case the normalized beam eigenfunctions for the fixed-free E–B beam, and satisfy the geometric and natural boundary conditions of the problem. The $q_j(\tau)$ are the generalized coordinates of the system. In Equation (49), ω_j are the eigenfrequencies; s_j is a complex amplitude; and $\mathrm{Im}(\omega_j)$ and $\mathrm{Re}(\omega_j)$ are the imaginary and real part of ω_j, respectively. It is assumed that the solution of Equation (42) is separable in terms of the dimensionless spatial variable ξ and the dimensionless time τ.

Equation (48) denotes a truncated series, in which N is finite positive integer and it represents the number of the appropriate comparison functions. For the sake of compactness of the notation, the following integral expressions are defined as

$$a_{ij_{(m,n)}} \equiv \int_m^n \Phi_i \Phi_j d\xi, \qquad b_{ij_{(m,n)}} \equiv \int_m^n \Phi_i \left(\frac{\partial \Phi_j}{\partial \xi}\right) d\xi,$$

$$c_{ij_{(m,n)}} \equiv \int_m^n \Phi_i \left(\frac{\partial^2 \Phi_j}{\partial \xi^2}\right) d\xi, \qquad d_{ij_{(m,n)}} \equiv \int_m^n \xi \Phi_i \left(\frac{\partial^2 \Phi_j}{\partial \xi^2}\right) d\xi, \tag{50}$$

where i, j are indices corresponding to the relevant quantity to be integrated; m and n of (m, n) are the lower/upper bounds of integration.

Substituting Equations (48) and (50) into (42), simultaneous equations of the form are obtained by pre-multiplying by Φ_i and integrating from $\xi = 0$ to $\xi = 1$

$$\mathbf{M}\frac{d^2}{d\tau^2}\left(\left\{\begin{array}{c} q_1 \\ q_2 \\ \vdots \\ q_3 \end{array}\right\}\right) + \mathbf{C}\frac{d}{d\tau}\left(\left\{\begin{array}{c} q_1 \\ q_2 \\ \vdots \\ q_3 \end{array}\right\}\right) + \mathbf{K}\left(\left\{\begin{array}{c} q_1 \\ q_2 \\ \vdots \\ q_3 \end{array}\right\}\right) = 0. \tag{51}$$

In Equation (51), $\mathbf{M}$, $\mathbf{C}$, and $\mathbf{K}$ matrices are mass matrix, damping matrix and stiffness matrix, respectively. The components of the $\mathbf{M}$, $\mathbf{C}$, and $\mathbf{K}$ matrices are

$$\mathbf{M}_{ij} = a_{ij_{(0,1)}} - \beta_o(1-\chi)\,a_{ij_{(0,rann)}}, \tag{52}$$

$$\mathbf{C}_{ij} = -2u_i\beta_i^{\frac{1}{2}}b_{ij_{(0,1)}} + 2\chi u_o\beta_o^{\frac{1}{2}}b_{ij_{(0,rann)}} + \tfrac{1}{2}c_f\varepsilon u_o\beta_o^{\frac{1}{2}}a_{ij_{(0,rann)}} + \kappa_u a_{ij_{(0,1)}} + \kappa_u\left(\frac{1+\alpha_{ch}^{-3}}{(1-\alpha_{ch}^{-2})^2} - 1\right)a_{ij_{(0,rann)}}, \tag{53}$$

$$\begin{aligned}\mathbf{K}_{ij} = {} & \lambda_j^4 a_{ij_{(0,1)}} + \gamma b_{ij_{(0,1)}} - \tfrac{1}{2}c_f\varepsilon u_o^2(h-1)b_{ij_{(0,rann)}} - \tfrac{1}{2}u_o^2(1-K_1)\left(\Phi_i|_{\zeta=r_{ann}}\left.\frac{\partial\Phi_j}{\partial\zeta}\right|_{\zeta=r_{ann}}\right) - (\Gamma - \Pi_{iL} + \Pi_{oL})c_{ij_{(0,1)}} - \gamma\left(c_{ij_{(0,1)}} - d_{ij_{(0,1)}}\right) \\ & + \tfrac{1}{2}c_f\varepsilon u_o^2(h-1)\left(r_{ann}c_{ij_{(0,rann)}} - d_{ij_{(0,rann)}}\right) + \tfrac{1}{2}u_o^2(1-K_1)c_{ij_{(0,1)}} + u_i^2 c_{ij_{(0,1)}} + \chi u_o^2 c_{ij_{(0,rann)}}; \end{aligned} \tag{54}$$

where λ_j is the jth dimensionless eigenvalue of the fixed-free E–B beam.

Solutions of Equation (51) constitute approximate solutions of Equation (42) with boundary conditions in Equation (43). In order to obtain a non-trivial solution of Equation (51), it is required that the determinant of coefficient matrix is equal to zero. This corresponds to a generalized eigenvalue problem, and the stability of the system can be determined by the generalized eigenvalue (i.e., eigenfrequency) ω of matrix $\mathbf{E}$ which consists of $\mathbf{M}$, $\mathbf{C}$, and $\mathbf{K}$, as

$$\mathbf{E} = \begin{bmatrix} -\mathbf{M}^{-1}\mathbf{C} & -\mathbf{M}^{-1}\mathbf{K} \\ \mathbf{I} & 0 \end{bmatrix}. \tag{55}$$

For a given internal dimensionless flow velocity u_i, the stability of the cantilever pipe is determined by $\mathrm{Im}(\omega)$. When $\mathrm{Im}(\omega) < 0$, the $q_j(\tau)$ (shown in Equation (49)) grows exponentially in time, so $\mathrm{Im}(\omega) < 0$ indicates instability. Consequently, for $\mathrm{Im}(\omega) < 0$ the cantilevered pipe becomes unstable, and for $\mathrm{Im}(\omega) = 0$, divergence or buckling of the cantilever pipe happens. Divergence represents a static loss of stability, vulgarly known as buckling and, in the nonlinear dynamics milieu, as a static pitchfork bifurcation. In the following passages, any discussion of the system stability involves only the stability of the internal cantilevered pipe.

4. Theoretical Analysis for Slender, Leaching-Tubing-Like Systems

The geometry shown in Figure 1 has been applied in the salt cavern underground energy storage industry, called leaching-tubing systems. Salt caverns serving as underground energy storage are generally located at a depth between 500 and 2000 m [3], and storage volumes of salt caverns range from 5×10^4 m^3 to 2×10^5 m^3 [20,21]. The typical depths of a salt cavern height and the salt cavern ceiling are about 400 m and 600 m, respectively, learned from the industry survey described in [2]. Hence common lengths of the leaching tubing are on the order of one kilometer.

Based on information from industrial applications, theoretical analyses for slender leaching-tubing-like systems are carried out in this section. Sample dimensional and associated dimensionless parameters for the leaching tubing are given in Table 1. In the salt cavern storage industry, the geometric parameters α_{ch}, r_{ann} and L shown in Figure 1 can be varied by operators, as required by the engineering practice. Therefore, it is physically meaningful to discuss the effects of the geometric parameters α_{ch}, r_{ann}, and L on the pipe system. The effects of varying α_{ch}, r_{ann} and L are discussed from Sections 4.1 to 4.3.

Table 1. Sample properties of real-life leaching tubings and associated dimensionless parameters [1].

Dimensional parameters	D_i (m)	D_o (m)	D_{ch} (m)	L (m)	L_s (m)	EI (N·m^2)	M_p (kg/m)	
	0.159	0.1778	0.298	1283	1085	3.47×10^6	38.7	
Dimensionless parameters	α	α_{ch}	ε	β_i	β_o	h	γ	r_{ann}
	0.897	1.676	7216	0.239	0.297	1.479	2.019×10^5	0.85

4.1. Effect of the Radial Confinement α_{ch}

Here, it is interesting to see the effect of different radial confinement values α_{ch} on the leaching-tubing-like system behavior. According to Equation (47), decreasing α_{ch} corresponds to an increase in radial confinement and to a related increase in the dimensionless annular flow velocity u_o.

The leaching-tubing-like systems behavior with the variation of α_{ch} from 1.10 to 20 is summarized in Table 2, based on $r_{ann} = 0.50$, and $\varepsilon = 1124.9$ ($L = 200$ m) and the remaining parameters shown in Table 1. Theoretical results of both u_{cr} and ω_{cr} are shown in Figure 6. u_{cr} is a dimensionless critical instability flow velocity and indicates the onset of instability of the internal pipe. For example, Figure 7 shows the vibration and stability of the system with $\alpha_{ch} = 1.20$, for the first three modes. From Figure 7, we can see that when $u_{cr} = 4.34$ the first predicted instability of the system occurs, and the associated frequency $\omega_{cr} = 0$, which indicates the onset of divergence. Naturally, these results presented here are related to this particular parameters selection, and only indicate what the leaching-tubing-like systems behavior could be.

Table 2. Summary of the leaching-tubing-like systems behavior with the vibration of αch from 1.10 to 20.

Behavior [a]	D1	F2	F3	F1	F2
α_{ch}	1.10 1.20 1.27 1.32 1.35	$\cdots$	4.46	$\cdots$	20 $\cdots$

F and D represent two kinds of unstable modes, respectively, i.e., flutter instability and divergence. As an example, D1 denotes first-mode divergence and F1 denotes first-mode flutter. [a] [$r_{ann} = 0.50$, $\varepsilon = 1124.9$ ($L = 200$ m)].

As shown in Table 2, when $\alpha_{ch} \approx 20$, i.e., the leaching-tubing-like system is effectively unconfined, flutter instability occurs in the second mode. For $1.35 \leq \alpha_{ch} < 20$, the decrease in α_{ch} results in flutter instability in which the modes are sequentially lowered. However, for $1.32 \leq \alpha_{ch} < 1.35$, flutter instability occurs in the third mode. A further decrease of α_{ch} below $\alpha_{ch} < 1.32$ can still result in flutter instability in which the modes are sequentially lowered, until first-mode divergence (i.e., a static instability) arises for $1.10 \leq \alpha_{ch} < 1.27$. The vanishing of Re($\omega$) indicates the beginning of divergence, as clearly seen in Figure 6.

With regard to the dimensionless critical flow velocity u_{cr}, and with the very confined case ($\alpha_{ch} = 1.10$) as a reference, the increase of α_{ch} leads initially to the rapid increase of u_{cr}. This is followed by a steep and significant decrease of u_{cr} as α_{ch} is increased continuously. Moreover, when $\alpha_{ch} > 1.48$, u_{cr} is less than 1.00. However, with a further increase of α_{ch} above $\alpha_{ch} > 2.00$ approximately, u_{cr} remains almost unaffected. As mentioned above, when $1.10 \leq \alpha_{ch} < 1.27$, a first-mode divergence arises, so ω_{cr} is equal to 0. An increase of α_{ch} to $1.27 \leq \alpha_{ch} < 1.35$ causes a significant increase of ω_{cr}. However, when $\alpha_{ch} > 1.35$, decreased confinement causes a very steep reduction in ω_{cr}. When $1.35 \leq \alpha_{ch} < 4.46$, ω_{cr} remains almost unaffected with the increase of α_{ch}. When $4.46 \leq \alpha_{ch} < 20$, the system instability state changes from first-order flutter to second-order flutter, ω_{cr} steeply increases and remains almost unaffected with the increase of α_{ch}.

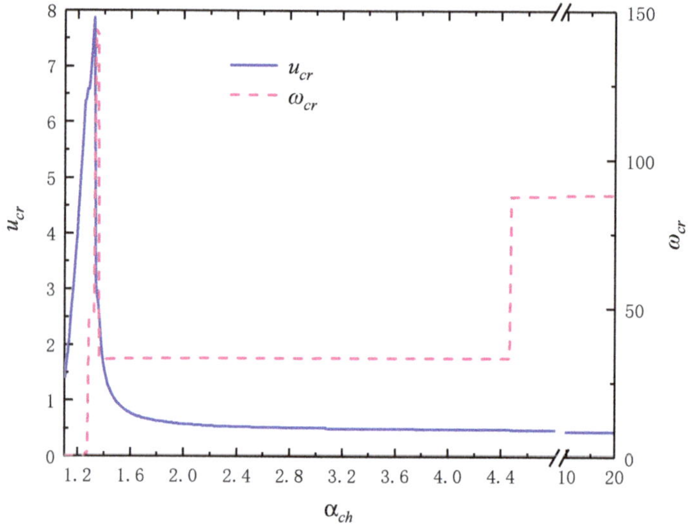

Figure 6. Leaching-tubing-like systems with $r_{ann} = 0.50$ and $\varepsilon = 1124.9$ ($L = 200$ m), theoretical variation of u_{cr} and ω_{cr} with the radial confinement α_{ch}. The results are based on the instability of the first prediction. A zero frequency (i.e., $\omega_{cr} = 0$) indicates divergence.

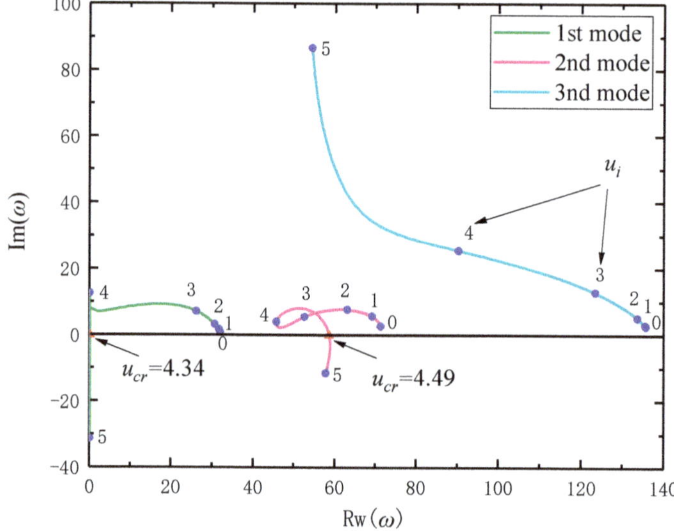

Figure 7. Argand diagram of the dimensionless complex eigenfrequencies of the leaching-tubing-like system, as a function of the nondimensional internal flow velocity u_i for $\alpha_{ch} = 1.20$. Blue solid circles represent the nondimensional internal flow velocity u_i. Red solid triangles represent the critical flow velocity when flutter instability occurs in the corresponding mode.

4.2. Effect of the Confinement Length r_{ann}

The effect of the confinement length r_{ann} (i.e., confined-flow length) on the slender leaching-tubing-like system is studied next. A larger value of the confinement length r_{ann} corresponds to a larger part of the cantilevered pipe being subjected to annular flow. In the schematic view of the theoretical model shown in Figure 1, the inflexible external pipes cannot be longer than the internal cantilevered pipe. Consequently, the studied values of r_{ann} are less than 1 and the studied maximum value of r_{ann} is 0.9.

The leaching-tubing-like systems behavior with the variation of r_{ann} from 0 to 0.90 is summarized in Table 3, respectively based on $\alpha_{ch} = 1.676$, $\varepsilon = 562.5$ ($L = 100$ m) and $\alpha_{ch} = 1.676$, $\varepsilon = 1124.9$ ($L = 200$ m) and the remaining parameters shown in Table 1. Theoretical results of both u_{cr} and ω_{cr} are shown in Figure 8. The value of $\alpha_{ch} = 1.676$ is used, due to the typical degree of confinement in the real application. Here, two different leaching-tubing lengths (i.e., $L = 100$ m and $L = 200$ m) are used to show the qualitative difference in system behavior.

Table 3. Summary of the leaching-tubing-like systems behavior with the vibration of r_{ann} from 0 to 0.90.

Behavior [a]	F1					F2				
r_{ann}	0	0.15	0.30	0.45	0.60	0.72	0.75	0.80	0.85	0.90
Behavior [b]	F2	F1				F2				
r_{ann}	0	0.15	0.19	0.45	0.60	0.69	0.75	0.80	0.85	0.90

F represents a kind of unstable mode, i.e., flutter instability. As an example, F1 denotes first-mode flutter.
[a] $[\alpha_{ch} = 1.676, \varepsilon = 562.5\ (L = 100\ \text{m})]$. [b] $[\alpha_{ch} = 1.676, \varepsilon = 1124.9\ (L = 200\ \text{m})]$.

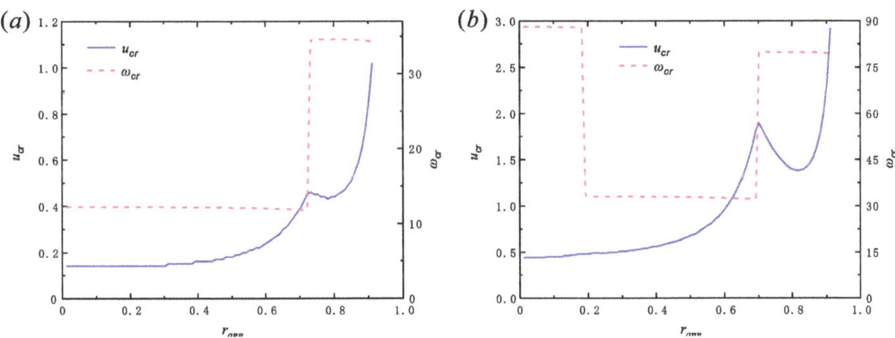

Figure 8. Leaching-tubing-like systems, theoretical variation of u_{cr} and ω_{cr} with the confinement length r_{ann}. The results are based on the instability of the first prediction. (**a**) Leaching-tubing with $\alpha_{ch} = 1.676$ and $\varepsilon = 562.5$ ($L = 100$ m). (**b**) Leaching-tubing with $\alpha_{ch} = 1.676$ and $\varepsilon = 1124.9$ ($L = 200$ m).

The slender leaching-tubing-like system behavior is discussed by considering r_{ann} values increased from $r_{ann} = 0$. For the shorter, ($L = 100$ m) leaching-tubing-like system initially loses stability by the first-mode flutter. A further increase of r_{ann} results in the second-mode flutter. For the longer, ($L = 200$ m) leaching-tubing-like system, an increasing r_{ann} leads initially to loss of stability by the second-mode flutter. A further increase of r_{ann} leads to first-mode flutter and second-mode flutter.

For the shorter leaching-tubing length considered, in terms of the dimensionless critical flow velocity u_{cr}, an increase of r_{ann} from 0 to 0.50 results in a slight stabilization. u_{cr} increases nonlinearly with a further increase of r_{ann} from 0.50 to 0.72. When r_{ann} increases from 0.72 to 0.90, u_{cr} begins to decrease and reaches a minimum at $r_{ann} = 0.78$, after which u_{cr} continues to increase. As shown in Figure 8a, ω_{cr} is almost unaffected by r_{ann} during the first-mode flutter phase and the second-mode flutter phase. However, when the unstable state changes from the first-mode flutter phase to the second-mode flutter phase, ω_{cr} increases steeply. For the longer leaching-tubing length considered, the dimensionless critical flow velocity u_{cr} remains almost unaffected with the increase of r_{ann} from

0 to 0.30. Then, u_{cr} increases nonlinearly with a further increase of r_{ann} from 0.30 to 0.69. When r_{ann} increases from 0.69 to 0.90, u_{cr} begins to decrease and reaches a minimum at $r_{ann} = 0.81$, after which u_{cr} continues to increase. As shown in Figure 8b, ω_{cr} is almost unaffected by r_{ann} during the first-mode flutter phase and the second-mode flutter phase. However, when the unstable state changes from the second-mode flutter phase to the first-mode flutter phase, ω_{cr} steeply decreases. When the unstable state returns from the first-mode flutter back to the second-mode flutter, ω_{cr} steeply increases. Hence, for both leaching-tubing lengths considered, the curve of u_{cr} has a U shape, and the curve of ω_{cr} is stepped.

4.3. Effect of the Cantilevered Pipe Length L

Dimensional length L of the cantilevered pipe has been used in the nondimensionalization of both the flow velocity u_i and the frequency ω. Consequently, the dimensionless parameters given in Equation (41) are unsuitable for studying the dependence of the dynamics on the system length. Another set of dimensionless parameters have been defined in [9] to study the dynamic characteristics of slender hanging pipes conveying fluid and in [22] to study the dynamic characteristics of cylinders in axial flow. In the present paper, no new dimensionless quantities have been defined, but rather the dimensional critical flow velocity U_{cr} and associated frequency Ω_{cr} have been used to carried out a brief analysis. Notwithstanding the loss of generality, this approach facilitates illustrating the variation of the U_{cr} and Ω_{cr} with increasing the pipe system length L. Based on diameters, mass, and stiffness typical of a leaching-tubing, as shown in Table 1, numerical results are obtained to predict dynamical behaviors of the leaching-tubing-like systems with increasing lengths L for various values of r_{ann}.

As shown in Table 4, the leaching-tubing-like systems behavior with the variation of L from 5 m to 500 m is summarized, based on $r_{ann} = 0.125, 0.25, 0.50$, and 0.85. When $5\text{ m} \leq L \leq 500\text{ m}$, with increasing of the length, the leaching-tubing loses stability by first-mode flutter, for $r_{ann} = 0.125, 0.25$, and 0.50, while the leaching-tubing loses stability by second-mode flutter, for $r_{ann} = 0.85$. Theoretical results of both U_{cr} and Ω_{cr} are shown in Figures 9 and 10. When the length L is below 40 m, an increasing of L leads initially to a nearly negligible stabilization. After that, U_{cr} increases significantly with a further increase of L from 40 m to 500 m. U_{cr} increases approximately linearly, for $r_{ann} = 0.125, 0.25$, and 0.50, while U_{cr} increases approximately nonlinearly, for $r_{ann} = 0.85$. For all studied values of r_{ann}, with increasing the pipe system length L, the critical frequency Ω_{cr} decrease. Especially when $L < 50$ m, Ω_{cr} decreases sharply. Moreover, for a long enough system, the frequency Ω_{cr} almost vanishes, which indicates a divergence is about to begin.

Table 4. Summary of the leaching-tubing-like systems behavior with increasing L from 5 to 500 m.

Behavior [a, b, c]	F1				
L (m)	5	150	300	450	500
Behavior [d]	F2				
L (m)	5	150	300	450	500

F represents a kind of unstable mode, i.e., flutter instability. As an example, F1 denotes first-mode flutter.
[a] $[\alpha_{ch} = 1.676, r_{ann} = 0.125]$, [b] $[\alpha_{ch} = 1.676, r_{ann} = 0.25]$. [c] $[\alpha_{ch} = 1.676, r_{ann} = 0.50]$, [d] $[\alpha_{ch} = 1.676, r_{ann} = 0.85]$.

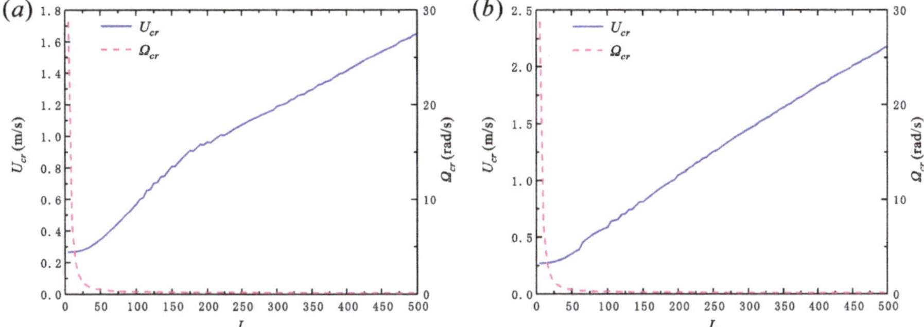

Figure 9. Leaching-tubing-like systems, theoretical variation of U_{cr} and Ω_{cr} with the increase of the pipe system length L. The results are based on the instability of the first prediction. (**a**) Leaching-tubing with $\alpha_{ch} = 1.676$ and $r_{ann} = 0.125$. (**b**) Leaching-tubing with $\alpha_{ch} = 1.676$ and $r_{ann} = 0.25$.

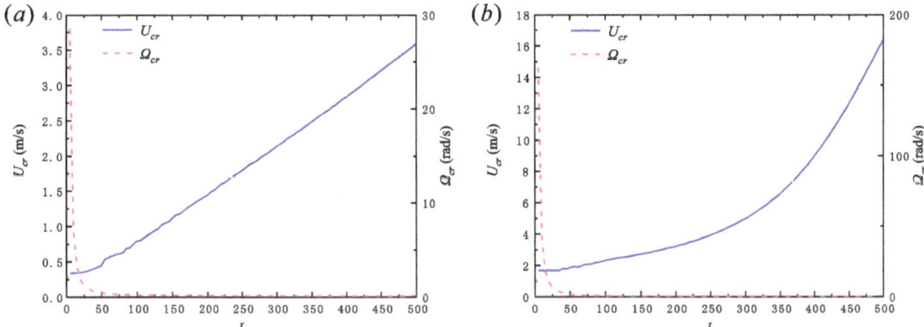

Figure 10. Leaching-tubing-like systems, theoretical variation of U_{cr} and Ω_{cr} with the increase of the pipe system length L. The results are based on the instability of the first prediction. (**a**) Leaching-tubing with $\alpha_{ch} = 1.676$ and $r_{ann} = 0.50$. (**b**) Leaching-tubing with $\alpha_{ch} = 1.676$ and $r_{ann} = 0.85$.

5. Conclusions

A linear theoretical model has been formulated for a hanging vertical cantilevered pipe which is subjected concurrently to two dependent axial flows. It should be pointed out that after the external flow exits from the outside annular passage, fluid is conveyed upwards as the internal flow in the cantilever pipe. The motion equation of the system is solved by means of a Galerkin method, and eigenfunctions of Euler–Bernoulli beam are used as comparison functions. Theoretical predictions were obtained for a long leaching-tubing-like system with parameters related to the salt cavern energy storage.

A theoretical study into the effect of radial confinement (i.e., the degree of radial confinement of the annular passage) has shown that, the system loses stability in flutter and even in divergence regardless of length, as the radial confinement degree and consequently the annular flow velocity is increased. When α_{ch} is greater than a certain value, u_{cr} decreases as α_{ch} increases, even to a small value. Considering that the flow velocity in industrial applications is relatively large, the value of α_{ch} should not be too large. Additionally, only when values of α_{ch} exceed a certain parameter-dependent threshold does the radial confinement has no appreciable effect on the pipe system.

The confinement length r_{ann} (i.e., confined-flow length), was shown to have a destabilizing or stabilizing effect on the pipe system, depending on value ranges of r_{ann}. The effect is a clear and significant stabilization, especially when the annular flow extends over most portion of the cantilevered pipe.

Furthermore, investigations have been conducted in dimensional terms on the effect of the pipe system length L, for the leaching-tubing-like system. The pipe system length L was shown to have a stabilizing effect on the pipe system. Increasing the pipe system length L results in an increase of dimensional critical flow velocity U_{cr}, while it results in a decrease of the associated dimensional frequency Ω_{cr}. For a long enough system, the frequency Ω_{cr} almost vanishes, which indicates a divergence is about to begin.

The next step is to conduct experiments to check the validity of the linear theoretical model, based on the device described in [23], and further research results are deferred to another paper.

Author Contributions: X.G. conceived the study, derived formulas, interpreted the results and wrote the paper under the guidance of Y.L. and C.Y. X.C. and N.Z. helped in the numerical calculation. X.S. and H.M. reviewed the study plan, edited the manuscript, and corrected the grammar mistakes. H.Y. participated in data analysis.

Funding: This research was funded by National Natural Science Foundation of China grant number 51874273, 51874274, 51774266 and by National Key Research and Development Program of China grant number 2018YFC0808401. And The APC was funded by 51874274.

Acknowledgments: The authors would like to be sincerely grateful to Jaak J.K. Daemen from University of Nevada, USA, for his English help and thoughtful review of this paper. The authors would like to gratefully acknowledge the financial support from National Natural Science Foundation of China (nos. 51874273, 51874274, 51774266), and the National Key Research and Development Program of China (grant no. 2018YFC0808401).

Conflicts of Interest: The authors declare no conflict of interest.

References

1. Moditis, K.; Païdoussis, M.P.; Ratigan, J. Dynamics of a partially confined, discharging, cantilever pipe with reverse external flow. *J. Fluids Struct.* **2016**, *63*, 120–139. [CrossRef]
2. Ratigan, J.L. Brine string integrity and model evaluation. In Proceedings of the Processes of SMRI Fall Meeting, Galveston, TX, USA, 13–14 October 2008; pp. 273–293.
3. Li, Y.; Yang, C.; Qu, D.; Yang, C.; Shi, X. Preliminary study of dynamic characteristics of tubing string for solution mining of oil/gas storage salt caverns. *Rock Soil Mech.* **2012**, *33*, 681–686.
4. Cesari, F.; Curioni, S. Buckling instability in tubes subject to internal and external axial fluid flow. In Proceedings of the 4th Conference on Dimensioning, Hungarian Academy of Science, Budapest, Hungary, October 1971; pp. 301–311.
5. Hannoyer, M.; Païdoussis, M.P. Instabilities of tubular beams simultaneously subjected to internal and external axial flows. *J. Mech. Des.* **1978**, *100*, 328–336. [CrossRef]
6. Paidoussis, M.P.; Besancon, P. Dynamics of arrays of cylinders with internal and external axial flow. *J. Sound Vib.* **1981**, *76*, 361–379. [CrossRef]
7. Wang, X.; Bloom, F. Dynamics of a submerged and inclined concentric pipe system with internal and external flows. *J. Fluids Struct.* **1999**, *13*, 443–460. [CrossRef]
8. Païdoussis, M.P.; Luu, T.; Prabhakar, S. Dynamics of a long tubular cantilever conveying fluid downwards, which then flows upwards around the cantilever as a confined annular flow. *J. Fluids Struct.* **2008**, *24*, 111–128. [CrossRef]
9. Doaré, O.; De Langre, E. The flow-induced instability of long hanging pipes. *Eur. J. Mech. A. Solids* **2002**, *21*, 857–867. [CrossRef]
10. Lemaitre, C.; Hémon, P.; De Langre, E. Instability of a long ribbon hanging in axial air flow. *J. Fluids Struct.* **2005**, *20*, 913–925. [CrossRef]
11. Kuiper, G.; Metrikine, A. Dynamic stability of a submerged, free-hanging riser conveying fluid. *J. Sound Vib.* **2005**, *280*, 1051–1065. [CrossRef]
12. Kuiper, G.; Metrikine, A. Experimental investigation of dynamic stability of a cantilever pipe aspirating fluid. *J. Fluids Struct.* **2008**, *24*, 541–558. [CrossRef]
13. Chen, S.; Wambsganss, M.T.; Jendrzejczyk, J. Added mass and damping of a vibrating rod in confined viscous fluids. *J. Appl. Mech.* **1976**, *43*, 325–329. [CrossRef]
14. Païdoussis, M.P. *Fluid-Structure Interactions: Slender Structures and Axial Flow*, 2nd ed.; Academic Press: Oxford, UK, 2014; Volume 1.

15. Païdoussis, M.P. *Fluid-Structure Interactions: Slender Structures and Axial Flow*, 2nd ed.; Academic Press: Oxford, UK, 2016; Volume 2.
16. Lighthill, M. Note on the swimming of slender fish. *J. Fluid Mech.* **1960**, *9*, 305–317. [CrossRef]
17. Paidoussis, M.P. Dynamics of cylindrical structures subjected to axial flow. *J. Sound Vib.* **1973**, *29*, 365–385. [CrossRef]
18. Sinyavskii, V.; Fedotovskii, V.; Kukhtin, A. Oscillation of a cylinder in a viscous liquid. *Sov. Appl. Mech.* **1980**, *16*, 46–50. [CrossRef]
19. Brater, E.F.; King, H.W.; Lindell, J.E.; Wei, C.Y. *Handbook of Hydraulics for the Solution of Hydraulic Engineering Problems*; McGraw-Hill: Boston, MA, USA, 1996.
20. Liu, W.; Muhammad, N.; Chen, J.; Spiers, C.; Peach, C.; Deyi, J.; Li, Y. Investigation on the permeability characteristics of bedded salt rocks and the tightness of natural gas caverns in such formations. *J. Nat. Gas Sci. Eng.* **2016**, *35*, 468–482. [CrossRef]
21. Liu, W.; Chen, J.; Jiang, D.; Shi, X.; Li, Y.; Daemen, J.K.; Yang, C. Tightness and suitability evaluation of abandoned salt caverns served as hydrocarbon energies storage under adverse geological conditions (AGC). *Appl. Energy* **2016**, *178*, 703–720.
22. De Langre, E.; Paidoussis, M.; Doaré, O.; Modarres-Sadeghi, Y. Flutter of long flexible cylinders in axial flow. *J. Fluid Mech.* **2007**, *571*, 371–389. [CrossRef]
23. Ge, X.; Li, Y.; Shi, X.; Chen, X.; Ma, H.; Yang, C.; Shu, C.; Liu, Y. Experimental device for the study of liquid–solid coupled flutter instability of salt cavern leaching tubing. *J. Nat. Gas Sci. Eng.* **2019**, in press. [CrossRef]

MDPI

St. Alban-Anlage 66

4052 Basel

Switzerland

Tel. +41 61 683 77 34

Fax +41 61 302 89 18

www.mdpi.com

Applied Sciences Editorial Office

E-mail: applsci@mdpi.com

www.mdpi.com/journal/applsci